Plant Ecophysiology

Volume 11

Series Editors
Luit J. De Kok, *University of Groningen, The Netherlands*
Malcolm J. Hawkesford, *Rothamsted Research, United Kingdom*

Aims & Scope:

The Springer Series in *Plant Ecophysiology* comprises a series of volumes that deals with the impact of biotic and abiotic factors on plant functioning and physiological adaptation to the environment. The aim of the *Plant Ecophysiology* series is to review and integrate the present knowledge on the impact of the environment on plant functioning and adaptation at various levels: from the molecular, biochemical and physiological to a whole plant level. This series is of interest to scientists who like to be informed of new developments and insights in plant ecophysiology, and can be used as advanced textbooks for biology students.

More information about this series at http://www.springer.com/series/6193

Elizabeth A.H. Pilon-Smits • Lenny H.E. Winkel
Zhi-Qing Lin
Editors

Selenium in plants

Molecular, Physiological, Ecological and Evolutionary Aspects

Springer

Editors
Elizabeth A.H. Pilon-Smits
Department of Biology
Colorado State University
Fort Collins, CO, USA

Zhi-Qing Lin
Environmental Sciences Program and Department of Biological Sciences
Southern Illinois University Edwardsville
Edwardsville, IL, USA

Lenny H.E. Winkel
Eawag, Swiss Federal Institute of Aquatic Science and Technology
Duebendorf, Switzerland

Institute of Biogeochemistry and Pollutant Dynamics
ETH Zurich
Zurich, Switzerland

ISSN 1572-5561 ISSN 2405-4321 (electronic)
Plant Ecophysiology
ISBN 978-3-319-56248-3 ISBN 978-3-319-56249-0 (eBook)
DOI 10.1007/978-3-319-56249-0

Library of Congress Control Number: 2017939733

Printed on acid-free paper

This Springer imprint is published by Springer Nature
The registered company is Springer International Publishing AG
The registered company address is: Gewerbestrasse 11, 6330 Cham, Switzerland

Preface

The element selenium (Se) is fascinating and important for several reasons. It is an essential micronutrient for many life forms including bacteria, archaea, some algae, and most animals. Essential Se metabolism appears to have been lost, however, in plants and fungi. Organisms that require Se for their survival utilize Se in the form of selenocysteine (SeCys) in the active site of the so-called selenoproteins (humans have 25), which have redox functions and are involved in immune function, thyroid metabolism, and spermatogenesis. Interestingly, SeCys in proteins is encoded by the opal stop codon, if it is flanked by a specific SeCys insertion sequence, which is different in prokaryotes and eukaryotes. At high tissue levels, Se becomes toxic to organisms, due to oxidative stress and to the replacement of S amino acids in proteins by their Se analogs. Relative to other micronutrients, Se has a very narrow window between adequacy and toxicity, which is around one order of magnitude. As a consequence, both Se deficiency and toxicity are prevalent worldwide and coincide with soils that are naturally low and high in Se, respectively. More than a billion people have been estimated to be affected, particularly by Se deficiency, which compromises their immune system (higher chance of infections including HIV and cancer), thyroid activity and male fertility, as well as mental function.

While not essential for plants, Se is a beneficial nutrient that enhances plant growth and antioxidant activity. Plants readily take up Se due to its similarity to sulfur (S) and assimilate it into a variety of organic Se compounds, analogous to S. Different plant species vary in the degree to which they take up and metabolize Se and in their capacity to tolerate Se. Some plants native to seleniferous soils even have evolved the capacity to hyperaccumulate Se to levels between 0.1 and 1.5% of their dry weight. This variation in plant Se accumulation, metabolism, and tolerance is not only of intrinsic interest but important because of its relevance for human, animal, and environmental health. Different forms of Se have different nutritional values. Crop Se content and Se speciation (chemical forms) are highly relevant for consumers, since the majority of the world's population directly depends on plants for its dietary Se, and all people obtain their Se ultimately from plants, even if indirectly via animals that feed on plants. Crops with optimal Se concentration and speciation may be selected or bred via classical or genetically enhanced breeding strategies. The levels and forms of Se in plants are not only a function of the genetic properties of the plants but also of their growth substrate. Soils vary in the concentration and forms of Se and Se bioavailability, which is influenced by abiotic and biotic factors. Selenium may be added in chemical or green fertilizer to crops growing on low-Se soils, a practice called biofortification. On the other end of the spectrum, if soils are particularly high in Se, this may cause toxicity, especially if Se-rich soil is used for irrigated agriculture or if Se-rich fossil fuels are used for energy. Plants may be used to extract excess Se from soil or water, a technology termed phytoremediation. In the absence of other contaminants, the resulting Se-rich plant material may be used to supply dietary Se in low-Se areas.

Apart from these applications, there are also many intrinsically interesting aspects of plant Se accumulation. Why and how do different plant species differ so much in the degree to which they accumulate and metabolize Se? Which processes underlie Se hyperaccumulation? Which benefits and disadvantages may Se accumulation confer to the plant? In other words, which selection pressures may favor or constrain Se accumulation? How does plant Se affect ecological interactions with herbivores, pollinators, microbes, and other plants, and how does this drive the evolution of both plant and ecological partners? How may plant Se accumulation and metabolism affect Se movement through the food chain, ecosystem, and ultimately global Se cycling?

In this book about Se, the plant has center stage, but the plant is placed in the context of its environment and its evolutionary history. The book starts with an overview of global Se distribution and the geological and biological processes that affect global Se movement in water, air, or soil particles. In subsequent chapters, the reader can follow Se from the point when it becomes bioavailable; is taken up by algae and plants and then metabolized, translocated, and sequestered; and is ultimately volatilized, consumed, and moved up in the food chain or redeposited in litter to the soil and recycled after decomposition. Relevant for Se bioavailability and for Se movement in the food chain, Se metabolism is also reviewed in prokaryotes and in mammalian consumers, and the nutritional effects of the consumed plant Se for consumers are discussed. The reader will learn about the profound ecological effects of plant Se on interactions with herbivores, pollinators, microbes, and other plants and the likely selection pressures that drive the evolution of Se hyperaccumulation. Furthermore, the latest knowledge is presented about the molecular processes involved in Se uptake, metabolism, tolerance, and (hyper)accumulation, as well as successful approaches to optimize Se accumulation and speciation via classical crop breeding and genetic engineering. The book concludes by highlighting global Se deficiency and toxicity issues in the world and successful applications of plant Se accumulation for biofortification and phytoremediation.

We are happy to be able to present this book in this special year, 200 years after the discovery of Se by Jons Jacob Berzelius in Sweden. We thank everyone who has contributed to this book and helped highlight the many fascinating facets of Se in plants from their respective vantage point and area of expertise. We hope the book will be of use for the Se research community and anyone who is interested in learning about this interesting and important element.

Fort Collins, CO, USA
Zurich, Switzerland
Edwardsville, IL, USA
January 24, 2017

Elizabeth A.H. Pilon-Smits
Lenny H.E. Winkel
Zhi-Qing Lin

Contents

Contributors

Elias S.J. Arnér Division of Biochemistry, Department of Medical Biochemistry and Biophysics, Karolinska Institutet, Stockholm, Sweden

Gary S. Bañuelos USDA, Agricultural Research Service, San Joaquin Valley Agricultural Sciences Center, Parlier, CA, USA

Martin Broadley University of Nottingham, School of Biosciences, Sutton Borrington Campus, Sutton Borrington, Leicestershire, UK

Ozgur Çakir Department of Molecular Biology and Genetics, Faculty of Science, Istanbul University, Istanbul, Turkey

Jennifer J. Cappa Biology Department, Colorado State University, Fort Collins, CO, USA

Alyssa T. Cochran Department of Biology, Colorado State University, Fort Collins, CO, USA

Karaj S. Dhillon Department of Soil Science, Punjab Agricultural University, Ludhiana, Punjab, India

André Rodrigues dos Reis UNESP – São Paulo State University, Tupã, SP, Brazil

Ali F. El Mehdawi Biology Department, Colorado State University, Fort Collins, CO, USA

Hassan El-Ramady Faculty of Agriculture, Kafrelsheikh University, Kafr el-Sheikh, Egypt

Priscila Lupino Gratão UNESP – São Paulo State University, Jaboticabal, SP, Brazil

Zackary Guignardi Biology Department, Colorado State University, Fort Collins, CO, USA

Malcolm J. Hawkesford Rothamsted Research, Harpenden, UK

Ying Jiang Department of Biology, Colorado State University, Fort Collins, CO, USA

College of Agronomy and Biotechnology, China Agricultural University, Beijing, China

Gerrad D. Jones Eawag, Swiss Federal Institute of Aquatic Science and Technology, Duebendorf, Switzerland

Current: Department of Biological & Ecological Engineering, Oregon State University, Corvallis, OR, USA

Leonardo Warzea Lima Department of Biology, Colorado State University, Fort Collins, CO, USA

Zhi-Qing Lin Environmental Sciences Program and Department of Biological Sciences, Southern Illinois University Edwardsville, Edwardsville, IL, USA

Akiko Maruyama-Nakashita Kyushu University, Fukuoka, Japan

Hisaaki Mihara Laboratory of Applied Molecular Microbiology, Department of Biotechnology, College of Life Sciences, Ritsumeikan University, Kusatsu, Shiga, Japan

Ronald S. Oremland US Geological Survey, Menlo Park, CA, USA

Elizabeth A.H. Pilon-Smits Biology Department, Colorado State University, Fort Collins, CO, USA

Colin F. Quinn Biology Department, Colorado State University, Fort Collins, CO, USA

R. Jason B. Reynolds Biology Department, Colorado State University, Fort Collins, CO, USA

Elcio Ferreira Santos Center of Energy in Agriculture, University of São Paulo, Piracicaba, SP, Brazil

Michela Schiavon Department of Biology, Colorado State University, Fort Collins, CO, USA

Lutz Schomburg Institute for Experimental Endocrinology, Charité – University Medical School, Berlin, Germany

Lucian C. Staicu Faculty of Applied Chemistry and Materials Science, University Politehnica of Bucharest, Bucharest, Romania

Masanori Tamaoki National Institute for Environmental Studies, Tsukuba, Ibaraki, Japan

Ryuta Tobe Laboratory of Applied Molecular Microbiology, Department of Biotechnology, College of Life Sciences, Ritsumeikan University, Kusatsu, Shiga, Japan

Doug Van Hoewyk Department of Biology, Coastal Carolina University, Conway, SC, USA

Francesca Dalla Vecchia Department of Biology, University of Padova, Padova, Italy

Philip J. White Ecological Science Group, The James Hutton Institute, Invergowrie, Dundee, UK

Lenny H.E. Winkel Eawag, Swiss Federal Institute of Aquatic Science and Technology, Duebendorf, Switzerland

Institute of Biogeochemistry and Pollutant Dynamics, ETH Zurich, Zurich, Switzerland

About the Editors

Dr. Elizabeth A.H. Pilon-Smits is professor in the Biology Department at Colorado State University, Fort Collins, CO, USA. She obtained her Ph.D. in biology at Utrecht University in the Netherlands in 1992, followed by a 2-year postdoctoral research position at the same university. From 1994–1998, Dr. Pilon-Smits did postdoctoral work at the UC Berkeley with Prof. Norman Terry in the Department of Plant and Microbial Biology, where she genetically engineered Indian mustard plants for enhanced selenium tolerance and accumulation. As of 1998, she holds a faculty position at Colorado State University, where she teaches courses in phytoremediation and in plant physiology and continues to study plant selenium physiology, molecular biology, ecology, and evolution. Her research has been funded by the National Science Foundation, the US Department of Agriculture, and the Environmental Protection Agency. As of January 2017, Dr. Pilon-Smits has authored 136 publications, including 100 on the topic of selenium.

Dr. Lenny H.E. Winkel is assistant professor at ETH Zurich (Swiss Federal Institute of Technology in Zurich) and group leader at Eawag (Swiss Federal Institute of Aquatic Science and Technology), Duebendorf, Switzerland. She obtained her Ph.D. in geochemistry at ETH Zurich, in a joint appointment with the Paul Scherrer Institute, Villigen, Switzerland, in 2006, followed by postdoctoral research positions at Eawag and the University of Grenoble, France; the University of Aberdeen, UK; and the Technical University of Crete, Greece, in the frame of an EU-funded Marie Curie project. Her postdoctoral research focused on broadscale predictions of arsenic in groundwater and pathways of methylated selenium species in the natural environment. In 2011, she was awarded a Swiss National Science Foundation (SNFS) professorship for her project on the global biogeochemical cycle of selenium. In her current position, she combines experimental work, field measurements, and modeling studies so as to (1) obtain an improved understanding of the processes controlling the biogeochemical cycling and environmental distribution of trace elements and (2) predict the effects of climate and environmental changes on these processes. Her research has been published in top-ranked disciplinary scientific journals (e.g., *Environmental Science and Technology* and *Nature Geoscience*) and interdisciplinary journals (e.g., *Nature Communications* and *PNAS*).

Dr. Zhi-Qing Lin is professor in the Department of Biological Sciences and director of the environmental sciences program at Southern Illinois University Edwardsville in Illinois. Dr. Lin received his B.S. degree in environmental biology from Liaoning University in China, M.S. degree in pollution ecology from the Chinese Academy of Sciences, and Ph.D. degree in renewable resources from McGill University in Canada. He was a postgraduate researcher on biogeochemistry of selenium at the University of California, Berkeley. His research is related to the biogeochemistry of environmentally important trace elements, with a special focus on biological accumulation, transformation, and volatilization processes of selenium in the air-water-soil-plant system. He is the author or coauthor of over 85 peer-reviewed articles published in scientific journals, proceedings, and monographs.

Part I
Selenium Distribution, Bioavailability and Metabolism in Plants

Chapter 1
Multi-scale Factors and Processes Controlling Selenium Distributions in Soils

Gerrad D. Jones and Lenny H.E. Winkel

Abstract Selenium (Se) is an essential trace element for maintaining homeostasis in humans and is characterized by a narrow range of recommended dietary intake levels. The main dietary sources of Se are food crops and therefore human intake levels largely depend on total concentrations and forms of Se in those food products. Important factors controlling Se uptake by plants are concentrations and speciation of Se in soils. Generally, Se concentrations in soils are driven by gradients in chemical and physical variables, which are in turn controlled by multiple biotic and abiotic processes that simultaneously span multiple spatial and temporal scales. This chapter discusses the factors and processes that control soil Se distributions on different spatial scales (i.e. from molecular to global scales) and how these gradients can be affected over time. In addition, it discusses how increased environmental scales lead to increased interactions among multi-scale factors and processes as well as to non-linear patterns between soil Se concentrations and environmental variables. Finally, it will be discussed how these patterns can be analyzed using sophisticated statistical techniques and how multi-scale variables and their interactions can be used to make predictions of soil Se concentrations in areas where this information is not available.

Keywords Selenium • Soils • Speciation • Multi-scale • Broad-scale predictions

G.D. Jones
Eawag, Swiss Federal Institute of Aquatic Science and Technology,
Ueberlandstrasse 133, P.O. Box 611, CH-8600 Duebendorf, Switzerland

Current: Department of Biological & Ecological Engineering, Oregon State University,
116 Gilmore Hall, Corvallis, OR 97331, USA
e-mail: jonesger@oregonstate.edu

L.H.E. Winkel (✉)
Eawag, Swiss Federal Institute of Aquatic Science and Technology,
Ueberlandstrasse 133, P.O. Box 611, CH-8600 Duebendorf, Switzerland

Institute of Biogeochemistry and Pollutant Dynamics,
ETH Zurich, Universitaetsstrasse 16, CH-8092 Zurich, Switzerland
e-mail: lenny.winkel@eawag.ch; lwinkel@ethz.ch

E.A.H. Pilon-Smits et al. (eds.), *Selenium in plants*, Plant Ecophysiology 11,
DOI 10.1007/978-3-319-56249-0_1

1.1 Introduction

Selenium (Se) is an essential trace element for maintaining homeostasis in humans (Rayman 2000; Combs 2001) and is characterized by a narrow range of recommended dietary intake levels. Current estimates indicate that a Se intake of >900 µg/day is potentially harmful for humans while an intake of <30 µg/day is inadequate (Fairweather-Tait et al. 2011). While Se toxicity occurs locally in areas with high soil Se concentrations, low Se intake is more common and is thought to affect up to one billion people globally (Combs 2001). Low serum Se levels in humans have been associated with negative effects on the immune system (Arthur et al. 2003; Hoffmann and Berry 2008) and in extreme cases, diseases related to Se-deficiency (e.g. in central China (Stone 2009; Fairweather-Tait et al. 2011). In populations with low dietary Se intake, Se supplementation has been shown to substantially reduce these health problems (Steinnes 2009).

The main dietary sources of Se are food crops, and human intake levels largely depend on total concentrations and forms of Se in these food products (Fairweather-Tait et al. 2011). Plant Se levels are determined by Se uptake as well as translocation of Se from roots to the edible part of crops. Furthermore, plant Se uptake is largely determined by total soil Se concentrations, Se speciation, competing ions present in soils and other soil physicochemical factors (Abrams et al. 1990; Terry et al. 2000; Fernández-Martínez and Charlet 2009; Li et al. 2010; Weng et al. 2011; Le Hécho et al. 2012; Kikkert and Berkelaar 2013; Nakamaru and Altansuvd 2014). While uptake is controlled by several factors, all things being equal, increases in bioavailable Se concentrations in soils have been shown to increase Se uptake in plants (Hart et al. 2011; Nothstein et al. 2016). Therefore, understanding the processes governing Se concentrations and speciation in soils is essential for understanding Se uptake in plants and thus Se supply levels to humans.

Factors, or gradients, controlling soil Se concentrations have been thoroughly evaluated in the literature, primarily through small-scale experimentation. This body of research has been summarized previously (Masscheleyn and Patrick 1993; Nakamaru and Altansuvd 2014; Winkel et al. 2015), but it is worthwhile to briefly review the dominant factors that were found to control soil Se concentrations in small-scale experiments. In general, Se speciation and the physicochemical composition of soils govern Se retention to soil particles. Sorption is heavily dependent on redox state as well as pH (Frost and Griffin 1977; Bar-Yosef and Meek 1987; Rovira et al. 2008). In soils, Se can be present in both inorganic and organic forms as well as in different oxidation states (i.e. −II, −I, 0, IV, VI). In well drained oxic soils at typical environmental pH values (i.e., ~4–9), Se is primarily present as the oxyanions selenite (Se[IV], $HSeO_3^-$ or SeO_3^{2-}, pKa = 2.4, 7.3) or selenate (Se[VI], SeO_4^{2-}, pKa = −3, 1.9) (Seby et al. 2001; Winkel et al. 2015). In general, sorption of selenite to soil particles is greater than selenate (Alemi et al. 1991; Sharmasarkar and Vance 2002; Hyun et al. 2006; Singh et al. 1981). For both oxyanions, retention in soils increases with decreasing pH (Frost and Griffin 1977; Bar-Yosef and Meek 1987; Rovira et al. 2008). Elemental Se (i.e. Se[0]) is thermodynamically favored in many natural environments (Oremland et al. 2004; Stolz et al. 2006; Lenz et al. 2008, 2009) although its formation and presence in soils is still largely unknown. The most reduced form of Se (i.e. Se (−II) is present as gaseous hydrogen selenide (H_2Se), as well as numerous metallic selenides (e.g. iron selenide), which are the most insoluble forms of Se (Elrashidi et al. 1989; Masscheleyn et al. 1990). Thus, Se mobility is strongly decreased in reducing environments.

In addition to pH and redox potential (Eh) dependent retention in soils, Se preferentially partitions to specific binding sites, e.g. on metal oxide/hydroxide surfaces of soil particles (Frost and Griffin 1977; Rovira et al. 2008), and therefore, the mineralogical structure of the soil plays an important role in the retention of Se. In addition, Se complexes with organic matter and forms organo-mineral colloids (Fernández-Martínez and Charlet 2009; Weng et al. 2011; Le Hécho et al. 2012). Nevertheless, there is still a lot of missing information on the different forms of Se in soils, which is clear given the large fraction of unidentified Se species in soils, which have been hypothesized to be largely organic forms (Li et al. 2010; Tolu et al. 2011, 2014; Stroud et al. 2012).

In the broadest sense, soil Se concentrations are driven by gradients of "sources" and "sinks" in the environment. Although the existing body of literature has focused on quantifying these gradients in small scales, concentrations of Se in soil are driven by multiple processes that simultaneously span multiple spatial scales. For example, climate plays an important role in governing soil Se distributions on broad scales (Jones et al. 2017), but as the investigative scale decreases, climate becomes increasingly homogeneous and is uniform at small scales (e.g. 1 m^2). At some point, gradients in soil Se concentrations can no longer be explained by climate and are instead controlled by some other process. This concept of scale has been virtually ignored within the literature. In order to understand the dominant processes driving Se retention and distributions, the influence of scale must be examined. The goal of this chapter is to explore the scale dependent processes (e.g. respiration, precipitation) that drive the gradients (e.g. pH, soil organic carbon content) controlling Se sources and sinks in the environment.

1.2 Se Sources and Sinks

Generally, variations in soil Se concentrations are governed by the uneven distribution of sources and sinks in the environment. Before discussing the influence of scale, it is important to clarify the terminology used to describe Se sources and sinks. Unlike anthropogenic contaminants (e.g. polychlorinated biphenyls and other organic pollutants), which have a clear point of origin into the environment, Se and other trace elements cycle continuously through the environment. As a result, there is no readily identifiable point of entry, and at best, sources should be considered intermediate sources. For example, biotic and abiotic processes can change the redox state of soils (McLatchey and Reddy 1998; Olivie-Lauquet et al. 2001), and thereby change the concentrations of Se in environmental compartments. For example, biomethylation can drastically increase the mobility of Se as (micro) organisms (e.g. the green alga *Chlamydomonas reinhardtii*; (Vriens et al. 2016)) can intracellularly produce volatile methylated Se species that are subsequently excreted from the cell. The volatile species are subsequently prone to atmospheric transformations and transport. In such instances, a particular process acts as a source for a particular Se species (in the example above, methylated Se species such as dimethyl selenide (DMSe)). Similarly, Se sinks are only temporary given long enough time scales. For example, while Se is known to partition to clay and organic carbon, Se can desorb following environmental change (e.g. changes in pH or redox state) suggesting that these could also be considered sources depending on the time scale in question. Finally, some processes are termed sources but actually only serve to facilitate redistribution of Se in the environment. For example, volcanoes are often reported as Se sources (Wen and Carignan 2007) but have no intrinsic relationship to Se. Especially considering time scales and the variability in environmental gradients over varying time scales, source and sink may be inappropriate terms for conceptually describing processes governing the Se cycle. Instead, we suggest using the following terminology to more explicitly describe the processes affecting soil Se:

Initial Sources add Se to the global system and could include asteroids (meteors, meteorites) and nuclear processes. For most environmental Se studies, the initial Se sources have not been relevant.

Intermediate Sources are processes that distinctly change the speciation or structure of Se and drastically alter its environmental behavior (e.g. atmospheric Se oxidation, biomethylation). Intermediate sources should have a clearly defined point of entry into the environment (e.g. volatilization from organisms). For most environmental Se studies, the intermediate Se sources are most relevant.

Transport Conduits are physical processes that have no unique role in modifying Se but facilitate its transport (e.g., volcanoes, colloid facilitated transport, deposition via rainfall (wet deposition) and dust (dry deposition, leaching). Some transport conduits are considered to be sources in the literature (e.g., volcanoes).

Reservoirs temporarily store Se, which can be released over relatively short time scales (e.g. sorption and desorption from particles or organic carbon) and over longer time scales (e.g. bedrock weathering). In literature, these reservoirs are often considered to be sinks (e.g. clay) or sources (e.g. geology).

Terminal Sinks permanently remove Se atoms from the global system (nuclear decay, planetary discharge). In most environmental studies, terminal sinks of Se have been irrelevant.

1.3 Retention of Se in Soils

While many environmental variables have been identified to affect Se retention in soils, the distribution and concentration of Se in the environment is likely governed by the gradients of these variables, which can vary across 13 orders of magnitude. For example, Se partitioning on clay particles largely depends on the mineralogical structure (angstrom scale), while continental/global climate patterns have been suggested to affect its broad-scale distributions (100 km scale). Therefore, depending on environmental heterogeneity of these gradients, the importance of some mechanisms are likely to give way to others as the environmental scale changes. Moreover, spatial distributions are likely to vary temporally. For example, Se could be affected by regular (i.e. daily or annual) temperature-dependent cycles or irregular cycles (e.g. storm events or inter-annual droughts). In the following sections, we examine various mechanisms that have been proposed to control soil Se concentrations at different spatial and temporal scales.

1.3.1 Molecular/Micro Scale (10^{-7}–10^{-3} m) Processes

On micro scales, Se concentrations have been shown to vary over ~4 orders of magnitude on individual particles (Eiche 2015). Several mechanisms have been identified to be important for Se retention on such small scales. For example, Se readily partitions to metal oxides/hydroxides and clay minerals (e.g. aluminum (Al) and iron (Fe) within the clay matrix) (Bar-Yosef and Meek 1987; Balistrieri and Chao 1990; Saeki and Matsumoto 1994; Ippolito et al. 2009; Nakamaru and Altansuvd 2014). The distribution of these high affinity binding sites on clay particles varies depending on the degree of weathering that occurs during clay formation. Clays are formed from the long-term chemical weathering of silicate rocks (Moronta 2004). Chemically, clays are hydrous aluminum silicate sheets, wherein the silicon (Si^{4+})/aluminum (Al^{3+}) atoms can be permanently replaced by other lower valent metals including Mg^{2+}, Fe^{2+}, Fe^{3+}, and Al^{3+} (Otterstedt and Brandreth 1998), creating favorable sorption sites for Se. Although weathering rates vary depending on a variety of factors, the chemical composition of particle surfaces are governed by weathering patterns that can be measured on longer term time scales (e.g. >1000 years (Velde and Meunier 2008). Compared to human lifespans, gradients in preferential sorption sites on particle surfaces caused by chemical weathering are likely to be virtually constant and are thus unlikely to drive changes in soil Se distribution/concentrations on shorter time scales.

While chemical weathering is relatively slow, variations in pH and redox conditions can greatly affect partitioning to metal oxides/hydroxides on shorter time scales. For example, Se mobility decreases under increasingly reduced conditions, especially at low pH (Frost and Griffin 1977; Bar-Yosef and Meek 1987; Rovira et al. 2008). Microscale gradients of pH and Eh can form as a result of both biotic

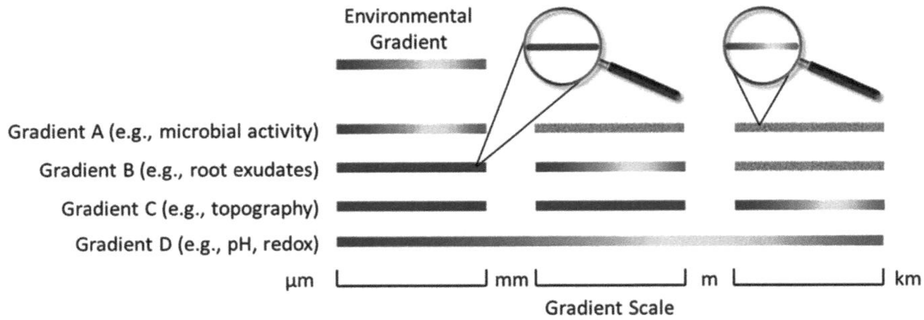

Fig. 1.1 Environmental gradients that govern soil Se concentrations are scale dependent. At larger scales, smaller-scale gradients (e.g. *gradient A*) appear homogeneous. Conversely, at smaller scales, larger-scale gradients (e.g. *gradients B* and *C*) appear homogeneous. Some gradients (e.g. *gradient D*) are measurable at multiple scales because multiple processes operating at different scales can control a single *gradient*

and abiotic processes (e.g. microbial respiration; Sexstone et al. 1985; McLatchey and Reddy 1998; Olivie-Lauquet et al. 2001). In soils, anaerobic microsites have been linked to oxygen consumption by microbes (Sexstone et al. 1985). These microsites are maintained by microbial activity and oxygen diffusion limitations. In these microbially mediated anaerobic zones within soil aggregates, Se immobilization can occur via increased sorption to metal surfaces or (co-)precipitation (Stolz and Oremland 1999; Kausch et al. 2012; Kausch and Pallud 2013; Eiche 2015).

As microbial processes are relatively dynamic, they likely contribute to the microscale changes in soil Se distributions on shorter time scales (e.g. daily-seasonally). Microbially mediated redox gradients exist on a microscale, but as scale increases, these microscale gradients are likely to appear more homogenous as the bulk redox state of the soil becomes increasingly fixed by broad scale factors such as climate (Fig. 1.1).

1.3.2 Local Scale (10^{-3}–10^{1} m) Processes

Local heterogeneity in biotic (e.g. plant distributions) and abiotic (e.g. topography, land use) factors could greatly affect the local distribution of Se in soils. Algae and microbes are known to decrease soil Se concentrations through volatilization (Karlson and Frankenberger 1989; Frankenberger and Arshad 2001; Vriens et al. 2014; Sun et al. 2016; Vriens et al. 2016). Plants release exudates from roots in order to manipulate pH and thus increase the bioavailability of nutrients (Hinsinger et al. 2003). While Se is not known to be required by plants, the effects of plants on soils can greatly affect Se behavior in the soil. Selenium in the form of selenite and selenate is inadvertently taken up through various ion transporters (e.g. sulfate, phosphate) and is thus removed from soils (Gissel-Nielsen 1971; Barrow and

Whelan 1989; Johnsson 1991; Zayed et al. 1998; Terry et al. 2000; Ellis and Salt 2003; Chilimba et al. 2011; De Temmerman et al. 2014). Furthermore, plants can create favorable microclimates for nutrient retention. For example, canopies created by "nurse plants" in arid regions change soils due to increased shade, soil moisture, and soil organic carbon content (Callaway 1995; Padilla and Pugnaire 2006), which could increase soil Se concentrations locally. In addition, plants reduce soil wind/water erosion, which could result in reduced losses of Se from top soils (Ravi et al. 2010). Depending on the distribution of plants, the distribution of Se could thus be spatially heterogeneous on a local scale or even landscape scale, in addition to being temporally heterogeneous as plants grow and die. In addition, nutrient leaching decreases with increasing plant transpiration by limiting the volume of water that is allowed to percolate through the soil (Boulding and Ginn 2003). As leaching is likely to be reduced in warmer conditions, it could result in seasonal heterogeneity of distributions of Se and other compounds.

Land use heterogeneity can also cause gradients in soil physicochemical properties over short distances. For example, strong gradients can exist in organic carbon between adjacent agricultural and non-agricultural (e.g. forest and grassland) areas (Knights et al. 2000; Jones et al. in press). Changes in land use from agricultural land to forest and grassland resulted in a gain of soil organic carbon, which in turn was a dominant contributor to increases in soil Se (Jones et al. 2017). In addition, the application of Se fertilizers to agricultural soils can increase local soil Se concentrations immediately following application. Assuming soil concentrations are always approaching equilibrium following system changes, added Se is likely to be removed via leaching or plant uptake unless soil reservoirs are simultaneously manipulated to increase Se retention.

In addition to biotic processes, abiotic processes can affect local distributions of Se. Fluctuations in water level can greatly affect the redox conditions of soils as well as pH, both of which can result in alternating periods of release and retention of elements in soils (Gambrell 1994; Dwire et al. 2006). For example, changes in the water table from 10 to 30 cm below the soil surface have been shown to change top soil conditions from reducing to oxidizing conditions in riparian soils following rainfall events (Hefting et al. 2004). Although soil Se concentrations have not been measured following such rapid changes in soil moisture, laboratory evidence indicates that the reduction and oxidation of Se can occur rapidly (Tokunaga et al. 1996), implying that the mobility of soil Se could change in concordance to rising and falling water levels following storm events.

1.3.3 Field Scale (10^1–10^3 m) Processes

Many gradients that determine soil Se concentrations occur on both local and field scales; however, compared to local gradients, the field-scale gradients are more diffuse and occur over longer distances. For example, soil moisture can affect redox gradients, but these gradients can be short (wet riparian soils adjacent to dry upland soils) or elongated (dry desert soils transitioning to wet montane soils). Many field-scale gradients are determined by the physical features of the landscape. For example, topography plays a large role in the biological, chemical, and physical processes that govern pedogenesis (Baumann et al. 2009, 2014). Topography can create soil textural gradients and is also responsible for the creation of field scale microclimates, which are largely determined by aspect (e.g. north vs south facing slopes) and its relation to incoming solar radiation (Moore et al. 1993; Harvey 2002). Aspect in turn affects many biotic and abiotic factors including plant community composition, soil moisture, redox conditions, and organic carbon content (Franzmeier et al. 1969; Macyk et al. 1978; Chen et al. 1997; Tsui et al. 2004; Yimer et al. 2006), which in turn can affect distributions of Se and other soil chemical constituents.

As scales increase, environmental heterogeneity is more likely to result in interactions among processes. For example, topography can affect soil physical properties (e.g. sand/silt/clay content) which structures plant communities, which in turn affect soil physicochemical properties (e.g. pH, organic carbon, etc.; see previous discussion) that govern soil Se concentrations (Jones et al. in press). As the interdependence of the processes governing soil Se distribution increases, it becomes increasingly difficult to tease these factors and processes apart. For example, elevation, precipitation, temperature, and soil organic carbon are correlated (Amundson et al. 1989; Jobbagy and Jackson 2000) and therefore it is difficult to isolate the individual effects of each variable. As scale increases, the relationships between these variables become highly complex and nonlinear, necessitating the use of statistical models that can deal with variable interactions (Jones et al. 2017).

1.3.4 Regional/Broad Scale (10^3–10^6 m) Processes

As previously mentioned, much of the literature has examined how small scale gradients affect soil Se concentrations. However, at these smaller scales, broad-scale gradients appear homogeneous and can thus not be recognized (Fig. 1.1). Although some broad-scale observational studies have been carried out which identified pH, clay, soil organic carbon, and precipitation as important drivers of soil Se concentrations (Låg and Steinnes 1974, 1978; Wu and Lag 1988; Shand et al. 2012; Blazina et al. 2014), little is still known about how broad-scale gradients drive soil Se patterns. Furthermore, given the strong interdependence of these drivers as well as their dependence on climate, their individual contributions have been difficult to characterize. Recently, we used data on soil Se concentrations from different continents and machine learning tools to analyze the factors controlling soil Se concentrations on a broad scale (Jones et al. 2017). These analyses indicate that on this scale, soil Se was dominated by climate variables, namely aridity and precipitation, and soil properties. In contrast to climate and soil parameters, geology, which has been traditionally viewed as a major factor driving the broad scale distributions of Se in soils, was not found to play a large role on this scale and its influence may be more localized (Jones et al. 2017). It is worth noting that compared to continuous variables, categorical variables including geology present computational challenges making it more difficult to quantify their effects. Because geology is known to influence soil Se concentrations at local scales (e.g. Dhillon and Dhillon 2014), some of the poor performance of geology in Jones et al. (2017) may be attributable to these computational challenges.

Jones et al. (2017) also found strong interactions between climate and soil variables at broad scales, which were concluded to be major drivers of soil Se concentrations. For example, although pH was the most important soil variable (decreases in pH resulted in increases in soil Se), the effects of pH were strongly suppressed in arid environments. Conversely, Se concentrations were enhanced in low pH/high clay environments. In addition, global precipitation gradients were shown to be very complex by both indirectly and directly affecting soil Se concentrations. Precipitation had a strong positive effect on soil organic carbon and strong negative effects on pH and clay content, all of which subsequently affected soil Se. Independent of other variables, precipitation was directly responsible for decreases in soil Se. Although precipitation and aridity are inversely related, both variables are hypothesized to affect leaching, where precipitation affects the transport of selenium in the subsurface and aridity drives soil redox conditions which subsequently control mobility.

Clearly, soil-climate interactions and their effects on soil Se concentrations are complex and therefore still merit further investigation. This complexity is likely a result of increased environmental heterogeneity at larger scales compared to smaller scales (Winkel et al. 2015). However, the influence of scale on the relative importance of soil Se retention mechanisms is virtually unknown. On most scales, pH and redox conditions play an important role, but several processes govern these gradients. Environmental gradients that occur over regional landscapes (e.g. bulk clay and soil organic carbon content) are governed by processes (e.g. pedogenesis) that occur over large time scales (e.g. centennially). Such large scale processes likely determine the environmental gradients that set the upper and lower bounds of the bulk soil Se concentration at larger scales. Conversely, at small spatial scales, the microscale processes that affect Se distributions are likely to be more dynamic as they appear to operate at smaller time scales (daily-seasonally-annually; e.g. plant/microbial respiration, soil water content) and may be responsible for driving deviations from the average broad-scale concentration. This multi-scale aspect has been virtually ignored in the literature but is nevertheless essential for better understanding the processes and factors that govern spatial and temporal soil Se distributions.

1.4 Predictions of Soil Se in Areas Where Se Distributions Are Unknown

Traditionally, relationships between soil Se concentrations and predictor variables are evaluated using correlations and linear regression (Shand et al. 2012). However, linear regression and other similar techniques assume that gradients controlling soil Se concentrations are independent. If the influence of scale is ignored in an analysis and if variables are not allowed to interact in models, our understanding of the processes driving soil Se distributions will be skewed. As previously discussed, many of the gradients responsible for controlling soil Se concentrations are highly correlated (e.g. pH, soil organic carbon, and precipitation). At smaller scales, broader scale gradients appear homogeneous and variable interdependencies seem to play less of a role in controlling soil Se concentrations. Therefore, in order to understand how this interdependency affects soil Se, appropriate mathematical tools must be used.

To illustrate interactive and non-interactive effects, two random variables were created (A [range = 0–2] and B [range 0–1]), and hypothetical soil Se concentrations were determined by multiplying or adding A and B together. In a multiplicative scenario (i.e. [Se] = A*B), variable A or B can suppress the effect of the other variable (i.e. [Se] = 0, if A or B = 0), whereas in an additive scenario (i.e. [Se] = A + B), changes in one variable does not influence the effects of the other (i.e. [Se] = A, if B = 0 and vice versa). When variables A or B are plotted against the hypothetical soil Se, the multiplicative and additive models resulted in very different patterns. The multiplicative model, which assumes variable interactions, resulted in a wedged shaped pattern with heteroscedastic variance, whereas the additive model, which assumes variable independence, resulted in a pattern with uniform change with homoscedastic variance (Fig. 1.2). These homoscedastic and heteroscedastic patterns have been observed with field and broad scale datasets and could be indicative of interactions between predictor variables (Fig. 1.2, see also Shand et al. 2012). Therefore, suitable models must be selected in order to accurately describe the factors controlling soil Se concentrations.

Depending on the strength of direct or indirect effects of predictor variables on soil Se, confounding variables can lead to erroneous conclusions about the influence of other variables. For example, on a global scale, Jones et al. (2017) found an overall positive relationship between precipitation and soil Se, but this was driven by the effect of precipitation on soil organic carbon, clay, and pH which subsequently affected soil Se. When controlling for these confounding variables, it was found that precipitation had a negative effect on soil Se. Tools such as machine learning models and structural equation models are useful for teasing apart complex interactions between variables, which cannot be evaluated using linear regression (Jones et al. 2017).

Recently, several standardized, nation-wide soil geochemical analyses have been conducted (e.g. within China, Europe, the US, the UK (Zheng 1994; Rawlins et al. 2012; Reimann et al. 2014; Smith et al. 2014; Martin et al. 2016)). These databases can be used to test hypotheses regarding processes that may control distributions of

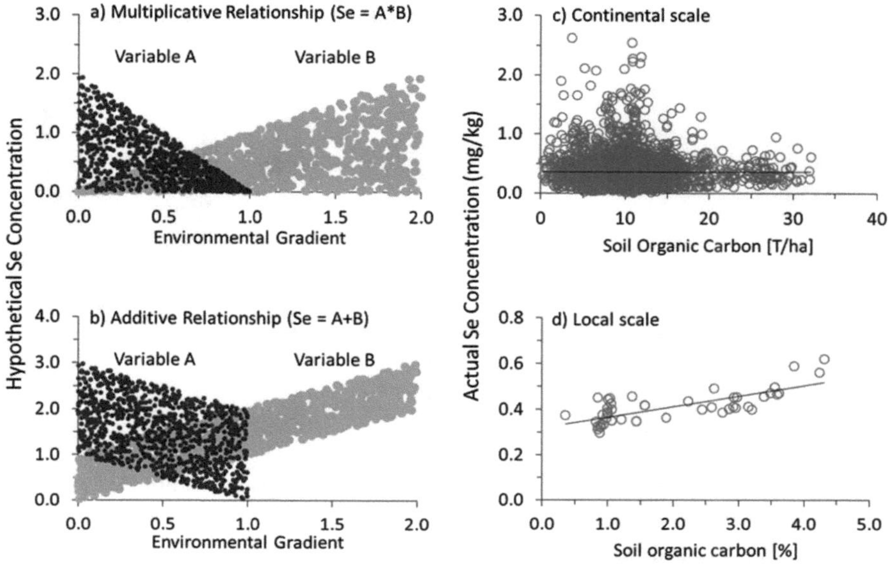

Fig. 1.2 Soil Se concentrations as a function of hypothetical environmental gradients (**a**, **b**) and a measured gradient (i.e. soil organic carbon content) on continental (**c**) and field (**d**) scales. When predictor variables interact (e.g. under multiplicative scenarios), scatter plots of individual variables appear wedge shaped (**a**). Conversely, when predictor variables are independent (e.g., in additive scenarios), scatter plots of change uniformly (**b**). These patterns appear in continental (**c**, United States (Smith et al. 2014), Europe (Rawlins et al. 2012; Reimann et al. 2014), and China (Zheng 1994); SOC data from (Gottschalk et al. 2012) and field (**d** Rothamsted, England) scale field experiments (Jones et al. 2017), which suggests that these patterns could be indicative of variable interactions

trace elements in soils and to make predictions of trace elements in soils. Sophisticated statistical techniques for analyzing complex heterogeneous data collected across multiple spatial and temporal scales are already available in other fields, such as community and landscape ecology. In such fields, analogous questions are investigated (i.e. what are the dominant gradients/processes governing the distribution of selected species and/or communities). These fields could thus serve as a valuable template for better understanding and predicting distributions of Se and other trace elements in soils. Furthermore, broad-scale spatial predictions have previously been established for trace elements such as arsenic (As) in groundwaters. Arsenic contaminations pose a major health threat to hundreds of millions of people worldwide (Winkel et al. 2008, 2011b) but in many areas around the world, possible occurrences of As contaminated groundwaters are still unrecognized/unknown. Therefore, maps have been created that identify untested areas that are likely vulnerable to As contamination of groundwater using broad-scale data of geology and soil variables (Amini et al. 2008; Lado et al. 2008; Winkel et al. 2008, 2011b; Yang et al. 2014). The same approach (Fig. 1.3) has also been used to predict soil Se distributions on a global scale (Jones et al. 2017).

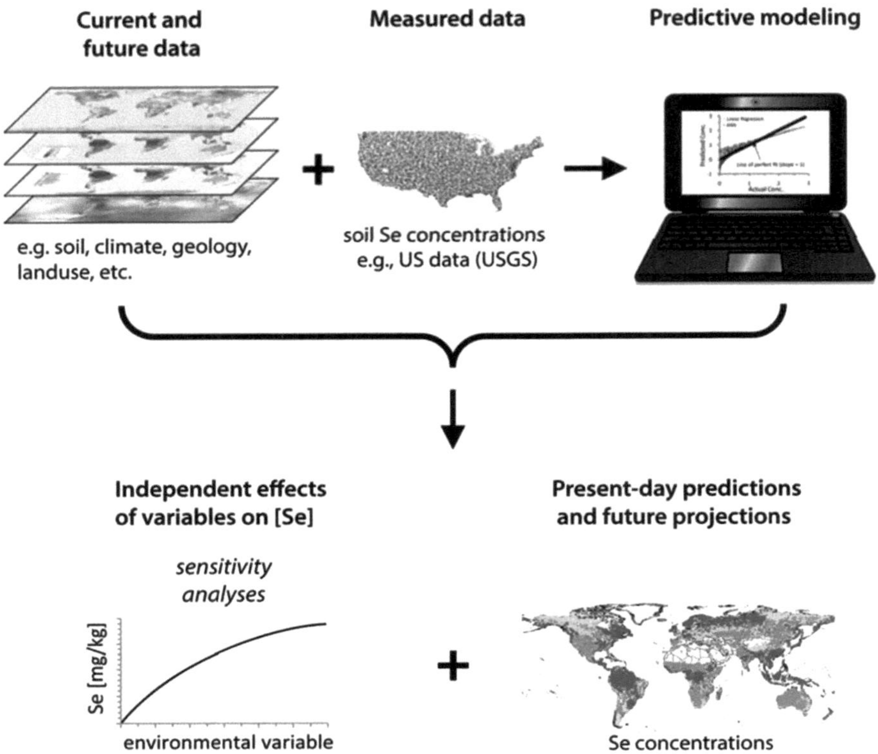

Fig. 1.3 Schematic diagram of the different steps in analyzing interactions between soil Se concentrations and environmental variables and in developing broad-scale predictive models for soil Se concentrations. The example of soil Se data is from Smith et al. 2014

1.5 Predictions of Future Changes in Soil Se Distributions

As indicated above, climate parameters were found to play a major role in controlling the global distribution of Se in soils, in addition to soil properties (Jones et al. 2017). Given future climate change, the magnitude of these variables is expected to change in the (near) future. For example, global soil organic carbon stocks are expected to increase (Gottschalk et al. 2012) suggesting an increased potential for Se accumulation in soils. Nevertheless, changes in precipitation across the land surface are complex, with some areas getting drier and others wetter (Greve et al. 2014). These analyses suggest that soil Se concentrations will change in response to changing climate conditions, which could have an important effect on the Se status of crops. In order to project soil Se concentrations in the future, predictive models for soil Se can be established based on climate data (precipitation, evapotranspiration, potential evapotranspiration) and soil organic carbon data assuming that the processes and mechanisms governing Se distributions today will remain unchanged in the future. Recently (Jones et al. 2017), a future model (an average of three machine learning models) for global Se distribution in soils for 2080–2099 was established based on moderate climate change scenarios (RCP 6.0 for precipitation, aridity index, and evapotranspiration (Taylor et al. 2012) and ECHAM5-A1B for SOC (Gottschalk et al. 2012). This model indicated that average soil Se concentrations are expected to drop from 0.331 ± 0.003 mg Se/kg in 1980–1999 to 0.316 ± 0.002 mg Se/kg in 2080–2099. Areas with notable losses include croplands of Europe and India, pastures of China, southern South America, Southern Africa, and the southwestern United States (Jones et al. 2017). Many of the gradients established by climate (e.g. precipitation, aridity, soil organic carbon) are likely to change more slowly (e.g. decennial time scale) compared to other time scales previously discussed. However, given the rapid changes in soil properties (e.g. soil organic carbon) following land use change, it is likely that Se levels in soils will change on time scales relevant to human health.

The ability to establish future models of changes in Se and other trace element concentrations in soils largely depends on the availability of future data that is projected for the same or comparable climate scenarios. Currently, future projections are not available on a global scale for pH, clay content, or for potential changes in Se source contributions (e.g. spatial changes in emission sources). Therefore, future work should focus on establishing future projections of these variables. Furthermore, the effect of atmospheric Se deposition on soil Se distributions is largely unknown and atmospheric Se transport models are currently not available. Therefore, additional research is needed to better understand the effect of atmospheric processes on Se distributions in soils.

1.6 Outlook

Selenium deficiency still remains a public health concern (FAO 2001, EFSA 2014). Therefore it is critical to understand the biogeochemical pathways that govern Se cycling in the environment. Improved knowledge of the relative importance of changes in intermediate sources and reservoirs along spatial scales will help identify the most important factors governing soil Se concentrations. Present-day broad-scale predictions of soil Se are essential to identify areas that may have elevated risk of over/under exposure to Se and to prioritize regions of geochemical testing. Such information will be extremely beneficial for poor rural communities that depend on local produce and are thus particularly vulnerable (Winkel et al. 2011a). Future global projections of soil organic carbon stocks in soils indicate that these stocks are expected to change under influence of climate change in many areas. As many studies report significant relationships between Se and soil organic carbon concentrations in soils, it is expected that Se levels (in addition to other trace elements) are likely to change as well.

Future projections of changes in soil Se concentrations based on projected changes in soil organic carbon and other variables will be an important tool to forecast such changes and pave the way for early measures to prevent potential health problems. Initial future predictions indicate potential losses of soil Se, which suggests that the nutritional quality of food will decrease, thereby increasing the worldwide risk of Se deficiency. Given the importance of climate-soil interactions in governing soil Se distributions, it is likely that other trace elements with similar control mechanisms will also experience losses (Jones et al. 2017). Although there is a clear link between soil Se concentrations and Se status of crops (FAO 2001), Se concentrations within edible plant material are not only controlled by total soil Se concentrations but are also a function of the bioavailability of soil Se and the efficiency of uptake and distribution in plant tissues. Therefore, human risks cannot be assessed based on soil Se concentrations alone. However, when more information becomes available on the quantitative relationship between total soil Se, speciation and Se levels in food crops, a next generation of future predictions can be developed that indicate the human health risks related to Se availability in soils and crops.

References

Abrams MM, Shennan C, Zasoski RJ et al (1990) Selenomethionine uptake by wheat seedlings. Agron J 82:1127–1130

Alemi MH, Goldhamer DA, Nielsen DR (1991) Modeling selenium transport in steady-state, unsaturated soil columns. J Environ Qual 20:89–95

Amini M, Abbaspour KC, Berg M et al (2008) Statistical modeling of global geogenic arsenic contamination in groundwater. Environ Sci Technol 42:3669–3675

Amundson R, Chadwick O, Sowers J (1989) A comparison of soil climate and biological activity along an elevation gradient in the eastern Mojave Desert. Oecologia 80:395–400

Arthur JR, Mckenzie RC, Beckett GJ (2003) Selenium in the immune system. J Nutr 133:1457S–1459S

Balistrieri LS, Chao TT (1990) Adsorption of selenium by amorphous iron oxyhydroxide and manganese-dioxide. Geochim Cosmochim Acta 54:739–751

Barrow NJ, Whelan BR (1989) Testing a mechanistic model. VIII. The effects of time and temperature of incubation on the sorption and subsequent desorption of selenite and selenate by a soil. J Soil Sci 40:29–37

Bar-Yosef B, Meek D (1987) Selenium sorption by kaolinite and montmorillonite. Soil Sci 144:11–19

Baumann F, He JS, Schmidt K et al (2009) Pedogenesis, permafrost, and soil moisture as controlling factors for soil nitrogen and carbon contents across the Tibetan Plateau. Glob Chang Biol 15:3001–3017

Baumann F, Schmidt K, Dorfer C et al (2014) Pedogenesis, permafrost, substrate and topography: plot and landscape scale interrelations of weathering processes on the central-eastern Tibetan Plateau. Geoderma 226:300–316

Blazina T, Sun Y, Voegelin A et al (2014) Terrestrial selenium distribution in China is potentially linked to monsoonal climate. Nat Commun 5:1–7

Boulding JR, Ginn JS (2003) Practical handbook of soil, vadose zone, and ground-water contamination: assessment, prevention, and remediation. CRC Press, London

Callaway RM (1995) Positive interactions among plants. Bot Rev 61:306–349

Chen ZS, Hsieh CF, Jiang FY et al (1997) Relations of soil properties to topography and vegetation in a subtropical rain forest in southern Taiwan. Plant Ecol 132:229–241

Chilimba ADC, Young SD, Black CR et al (2011) Maize grain and soil surveys reveal suboptimal dietary selenium intake is widespread in Malawi. Sci Rep 1:1–9
Combs GF (2001) Selenium in global food systems. Br J Nutr 85:517–547
De Temmerman L, Waegeneers N, Thiry C et al (2014) Selenium content of Belgian cultivated soils and its uptake by field crops and vegetables. Sci Total Environ 468:77–82
Dhillon KS, Dhillon SK (2014) Development and mapping of seleniferous soils in northwestern India. Chemosphere 99:56–63
Dwire KA, Kauffman JB, Baham JE (2006) Plant species distribution in relation to water-table depth and soil redox potential in montane riparian meadows. Wetlands 26:131–146
EFSA (2014) Scientific opinion on dietary reference values for selenium. EFSA J 12:3846
Eiche E (2015) Microscale distribution and elemental associations of Se in seleniferous soils in Punjab, India. Environ Sci Pollut R 22:5425–5436
Ellis DR, Salt DE (2003) Plants, selenium and human health. Curr Opin Plant Biol 6:273–279
Elrashidi M, Adriano D, Lindsay W (1989) Solubility, speciation, and transformations of selenium in soils. In: Jacobs LW (ed) Selenium in agriculture and the environment. SSSA, Madison, pp 51–63
Fairweather-Tait SJ, Bao YP, Broadley MR et al (2011) Selenium in human health and disease. Antioxid Redox Signal 14:1337–1383
FAO (2001) Human vitamin and mineral requirements: report of a joint FAO/WHO expert consultation. Food and Nutrition Division, Rome
Fernández-Martínez A, Charlet L (2009) Selenium environmental cycling and bioavailability: a structural chemist point of view. Rev Environ Sci Biotechnol 8:81–110
Frankenberger WT, Arshad M (2001) Bioremediation of selenium-contaminated sediments and water. Biofactors 14:241–254
Franzmeier D, Pedersen E, Longwell T et al (1969) Properties of some soils in the Cumberland plateau as related to slope aspect and position. Soil Sci Soc Am J 33:755–761
Frost RR, Griffin RA (1977) Effect of pH on adsorption of arsenic and selenium from landfill leachate by clay minerals. Soil Sci Soc Am J 41:53–57
Gambrell RP (1994) Trace and toxic metals in wetlands – a review. J Environ Qual 23:883–891
Gissel-Nielsen G (1971) Influence of pH and texture of the soil on plant uptake of added selenium. J Agric Food Chem 19:1165–1167
Gottschalk P, Smith JU, Wattenbach M et al (2012) How will organic carbon stocks in mineral soils evolve under future climate? Global projections using RothC for a range of climate change scenarios. Biogeosciences 9:3151–3171
Greve P, Orlowsky B, Mueller B et al (2014) Global assessment of trends in wetting and drying over land. Nat Geosci 7:716–721
Hart D, Fairweather-Tait S, Broadley M et al (2011) Selenium concentration and speciation in biofortified flour and bread: retention of selenium during grain biofortification, processing and production of Se-enriched food. Food Chem 126:1771–1778
Harvey A (2002) Factors influencing the geomorphology of dry-region alluvial fans: a review. Aportaciones a la Geomorfologıa de Espana en el inicio del tercer milenio. Publicaciones del Instituto Geológico y Minero de Espana, Serie: Geologıa 1: 59–75
Hefting M, Clement JC, Dowrick D et al (2004) Water table elevation controls on soil nitrogen cycling in riparian wetlands along a European climatic gradient. Biogeochemistry 67:113–134
Hinsinger P, Plassard C, Tang CX et al (2003) Origins of root-mediated pH changes in the rhizosphere and their responses to environmental constraints: a review. Plant Soil 248:43–59
Hoffmann PR, Berry MJ (2008) The influence of selenium on immune responses. Mol Nutr Food Res 52:1273–1280
Hyun S, Burns PE, Murarka I et al (2006) Selenium(IV) and (VI) sorption by soils surrounding fly ash management facilities. Vadose Zone J 5:1110–1118
Ippolito JA, Scheckel KG, Barbarick KA (2009) Selenium adsorption to aluminum-based water treatment residuals. J Colloid Interface Sci 338:48–55

Jobbagy EG, Jackson RB (2000) The vertical distribution of soil organic carbon and its relation to climate and vegetation. Ecol Appl 10:423–436
Johnsson L (1991) Selenium uptake by plants as a function of soil type, organic matter content and pH. Plant Soil 133:57–64
Jones GD, Droz B, Greve P et al. (2017) Selenium deficiency risk predicted to increase under future climate change. P Natl Acad Sci USA 114:2848–2853. DOI: 10.1073/pnas.1611576114
Karlson U, Frankenberger WT (1989) Accelerated rates of selenium volatilization from California soils. Soil Sci Soc Am J 53:749–753
Kausch MF, Pallud CE (2013) Modeling the impact of soil aggregate size on selenium immobilization. Biogeosciences 10:1323–1336
Kausch M, Ng P, Ha J et al (2012) Soil-aggregate-scale heterogeneity in microbial selenium reduction. Vadose Zone J 11
Kikkert J, Berkelaar E (2013) Plant uptake and translocation of inorganic and organic forms of selenium. Arch Environ Contam Toxicol 65:458–465
Knights JS, Zhao FJ, Spiro B et al (2000) Long-term effects of land use and fertilizer treatments on sulfur cycling. J Environ Qual 29:1867–1874
Lado LR, Polya D, Winkel L et al (2008) Modelling arsenic hazard in Cambodia: a geostatistical approach using ancillary data. Appl Geochem 23:3010–3018
Låg J, Steinnes E (1974) Soil selenium in relation to precipitation. Ambio:237–238
Låg J, Steinnes E (1978) Regional distribution of selenium and arsenic in humus layers of Norwegian forest soils. Geoderma 20:3–14
Le Hécho I, Tolu J, Thiry Y et al (2012) Influence of selenium speciation and fractionation on its mobility in soils. In: Selim HM (ed) Competitive sorption and transport of heavy metals in soils and geological media. CRC Press, Boca Raton, pp 215–230
Lenz M, Smit M, Binder P et al (2008) Biological alkylation and colloid formation of selenium in methanogenic UASB reactors. J Environ Qual 37:1691–1700
Lenz M, Enright AM, O'flaherty V et al (2009) Bioaugmentation of UASB reactors with immobilized Sulfurospirillum barnesii for simultaneous selenate and nitrate removal. Appl Microbiol Biotechnol 83:377–388
Li HF, Lombi E, Stroud JL et al (2010) Selenium speciation in soil and rice: influence of water management and Se fertilization. J Agric Food Chem 58:11837–11843
Macyk TM, Pawluk S, Lindsay JD (1978) Relief and microclimate as related to soil properties. Can J Soil Sci 58:421–438
Martin A, Turnbull R, Rattenbury M et al (2016) The regional geochemical baseline soil survey of southern New Zealand: design and initial interpretation. J Geochem Explor 167:70–82
Masscheleyn PH, Patrick WH (1993) Biogeochemical processes affecting selenium cycling in wetlands. Environ Toxicol Chem 12:2235–2243
Masscheleyn PH, Delaune RD, Patrick WH (1990) Transformations of selenium as affected by sediment oxidation reduction potential and pH. Environ Sci Technol 24:91–96
Mclatchey GP, Reddy KR (1998) Regulation of organic matter decomposition and nutrient release in a wetland soil. J Environ Qual 27:1268–1274
Moore ID, Gessler PE, Nielsen GA et al (1993) Soil attribute prediction using terrain analysis. Soil Sci Soc Am J 57:443–452
Moronta A (2004) Catalytic and adsorption properties of modified clay surfaces. Interface Sci Technol 1:321–344
Nakamaru YM, Altansuvd J (2014) Speciation and bioavailability of selenium and antimony in non-flooded and wetland soils: a review. Chemosphere 111:366–371
Nothstein AK, Eiche E, Riemann M et al (2016) Tracking Se assimilation and speciation through the rice plant – nutrient competition, toxicity and distribution. PLoS One 11:e0152081
Olivie-Lauquet G, Gruau G, Dia A et al (2001) Release of trace elements in wetlands: role of seasonal variability. Water Res 35:943–952
Oremland RS, Herbel MJ, Blum JS et al (2004) Structural and spectral features of selenium nanospheres produced by Se-respiring bacteria. Appl Environ Microbiol 70:52–60

Otterstedt J-E, Brandreth DA (1998) Clays and colloidal silicas. In: Small particles technology. Springer, New York, pp 155–183
Padilla FM, Pugnaire FI (2006) The role of nurse plants in the restoration of degraded environments. Front Ecol Environ 4:196–202
Ravi S, Breshears DD, Huxman TE, D'Odorico P (2010) Land degradation in drylands: interactions among hydrologic–aeolian erosion and vegetation dynamics. Geomorphology 116:236–245
Rawlins B, Mcgrath S, Scheib A et al (2012) The advanced soil geochemical atlas of England and Wales. British Geological Survey, Nottingham
Rayman MP (2000) The importance of selenium to human health. Lancet 356:233–241
Reimann C, Demetriades A, Birke M et al (2014) Distribution of elements/parameters in agricultural and grazing land soil in Europe. In: Reimann C, Birke M, Demetriades A, Filzmoser P, O'Connor P (eds) Chemistry of Europe's agricultural soils. Part a: methodology and interpretation of the GEMAS data set. Bundesanstalt für Geowissenschaften und Rohstoffe, Hannover, pp 103–474
Rovira M, Gimenez J, Martinez M et al (2008) Sorption of selenium(IV) and selenium(VI) onto natural iron oxides: goethite and hematite. J Hazard Mater 150:279–284
Saeki K, Matsumoto S (1994) Selenite adsorption by a variety of oxides. Commun Soil Sci Plant Anal 25:2147–2158
Seby F, Potin-Gautier M, Giffaut E et al (2001) A critical review of thermodynamic data for selenium species at 25 degrees C. Chem Geol 171:173–194
Sexstone AJ, Revsbech NP, Parkin TB et al (1985) Direct measurement of oxygen profiles and denitrification rates in soil aggregates. Soil Sci Soc Am J 49:645–651
Shand CA, Eriksson J, Dahlin AS et al (2012) Selenium concentrations in national inventory soils from Scotland and Sweden and their relationship with geochemical factors. J Geochem Explor 121:4–14
Sharmasarkar S, Vance GF (2002) Selenite-selenate sorption in surface coal mine environment. Adv Environ Res 7:87–95
Singh M, Singh N, Relan PS (1981) Adsorption and desorption of selenite and selenate selenium on different soils. Soil Sci 132:134–141
Smith DB, Cannon WF, Woodruff LG et al. (2014) Geochemical and mineralogical maps for soils of the conterminous United States. US Geological Survey No. 2014–1082.
Steinnes E (2009) Soils and geomedicine. Environ Geochem Health 31:523–535
Stolz JF, Oremland RS (1999) Bacterial respiration of arsenic and selenium. FEMS Microbiol Rev 23:615–627
Stolz JE, Basu P, Santini JM et al (2006) Arsenic and selenium in microbial metabolism. Annu Rev Microbiol 60:107–130
Stone R (2009) A medical mystery in middle China. Science 324:1378–1381
Stroud JL, Mcgrath SP, Zhao FJ (2012) Selenium speciation in soil extracts using LC-ICP-MS. Int J Environ Anal Chem 92:222–236
Sun G-X, Meharg AA, Li G et al (2016) Distribution of soil selenium in China is potentially controlled bydeposition and volatilization? Sci Rep 6:1–9
Taylor KE, Stouffer RJ, Meehl GA (2012) An overview of CMIP-5 and the experiment design. B Am Meteorol Soc 93:485–498
Terry N, Zayed AM, De Souza MP et al (2000) Selenium in higher plants. Annu Rev Plant Physiol Plant Mol Biol 51:401–432
Tokunaga TK, Pickering IJ, Brown GE (1996) Selenium transformations in ponded sediments. Soil Sci Soc Am J 60:781–790
Tolu J, Le Hecho I, Bueno M et al (2011) Selenium speciation analysis at trace level in soils. Anal Chim Acta 684:126–133
Tolu J, Thiry Y, Bueno M et al (2014) Distribution and speciation of ambient selenium in contrasted soils, from mineral to organic rich. Sci Total Environ 479:93–101

Tsui CC, Chen ZS, Hsieh CF (2004) Relationships between soil properties and slope position in a lowland rain forest of Southern Taiwan. Geoderma 123:131–142
Velde BB, Meunier A (2008) The origin of clay minerals in soils and weathered rocks. Springer, Heidelberg
Vriens B, Lenz M, Charlet L et al (2014) Natural wetland emissions of methylated trace elements. Nat Commun 5:1–8
Vriens B, Behra R, Voegelin A et al (2016) Selenium uptake and methylation by the microalga *Chlamydomonas reinhardtii*. Environ Sci Technol 50:711–720
Wen H, Carignan J (2007) Reviews on atmospheric selenium: emissions, speciation and fate. Atmos Environ 41:7151–7165
Weng LP, Vega FA, Supriatin S et al (2011) Speciation of Se and DOC in soil solution and their relation to se bioavailability. Environ Sci Technol 45:262–267
Winkel L, Berg M, Amini M et al (2008) Predicting groundwater arsenic contamination in Southeast Asia from surface parameters. Nat Geosci 1:536–542
Winkel LHE, Johnson CA, Lenz M et al (2011a) Environmental selenium research: from microscopic processes to global understanding. Environ Sci Technol 46:571–579
Winkel LHE, Pham TKT, Vi ML et al (2011b) Arsenic pollution of groundwater in Vietnam exacerbated by deep aquifer exploitation for more than a century. Proc Natl Acad Sci U S A 108:1246–1251
Winkel LHE, Vriens B, Jones GD et al (2015) Selenium cycling across soil-plant-atmosphere interfaces: a critical review. Forum Nutr 7:4199–4239
Wu XF, Lag J (1988) Selenium in Norwegian farmland soils. Acta Agric Scand 38:271–276
Yang NF, Winkel LHE, Johannesson KH (2014) Predicting geogenic arsenic contamination in shallow groundwater of South Louisiana, United States. Environ Sci Technol 48:5660–5666
Yimer F, Ledin S, Abdelkadir A (2006) Soil organic carbon and total nitrogen stocks as affected by topographic aspect and vegetation in the Bale Mountains, Ethiopia. Geoderma 135:335–344
Zayed A, Lytle CM, Terry N (1998) Accumulation and volatilization of different chemical species of selenium by plants. Planta 206:284–292
Zheng C (1994) Atlas of soil environmental background value in the People's Republic of China. China Environmental Science Press, Beijing

Chapter 2
Biochemistry of Plant Selenium Uptake and Metabolism

Zackary Guignardi and Michela Schiavon

Abstract Selenium (Se) is a trace element indispensable for humans, animals and some microorganisms. For plants its essentiality has not yet been established, despite its responsibility for a number of beneficial effects in several plant species. Plants take up Se mainly as selenate and selenite, using root high-affinity membrane transporters that normally mediate the influx of sulfate and phosphate ions, respectively. Once inside cells, Se can access the sulfur (S) assimilation pathway and be incorporated into the Se-amino acids Se-cysteine (SeCys) and Se-methionine (SeMet). Studies with transgenics showed that some enzymes working in this pathway are rate limiting for Se uptake, tolerance and accumulation in plants. Selenium at high concentration is toxic for plants, both due to oxidative stress and because Se-amino acids are non-specifically incorporated into proteins, which lose their folding and function as a result. Therefore, plants have evolved different strategies to cope with Se toxicity. They usually involve the conversion of Se-amino acids into less harmful volatile compounds. Specifically, plants that do not accumulate Se at high levels produce dimethylselenide (DMSe) using SeMet as a precursor, while Se-hyperaccumulators, i.e. plants able to tolerate and accumulate significant amounts of Se in their tissues, generate dimethyldiselenide (DMDSe) starting from the amino acid SeCys. Selenium hyperaccumulators have additional mechanisms to prevent SeCys misincorporation into protein, like methylation of SeCys to methylselenocysteine (MeSeCys) via SeCys methyltransferase (SMT), and breaking down of SeCys into elemental Se and alanine. In this chapter we review the main mechanisms implied in Se acquisition, assimilation and detoxification in plants.

Keywords Selenium • Assimilation • ATP sulfurylase • Selenoamino acids • Volatilization

Z. Guignardi (✉) • M. Schiavon
Biology Department, Colorado State University,
1878 Campus Delivery, Fort Collins, CO, USA
e-mail: zguignardi@gmail.com; michela.schiavon@colostate.edu

E.A.H. Pilon-Smits et al. (eds.), *Selenium in plants*, Plant Ecophysiology 11,
DOI 10.1007/978-3-319-56249-0_2

2.1 Uptake and Translocation of Selenium

Plants take up Se primarily in two forms, either as selenate (SeO_4^{2-}) or selenite (SeO_3^{2}), but they have the capacity to take up organic Se compounds as well. However, plants are unable to take up elemental Se or metal selenide compounds (White and Broadley 2009). Selenate is the most common form of Se taken up by plants and is the predominant bioavailable form in alkaline and well-oxidized soils, while selenite is the main identifiable bioavailable form in anaerobic soils and wetlands (Mikkelsen et al. 1989; White et al. 2007; Fordyce 2012). Due to its chemical similarities to sulfur (S), Se in the form of selenate is transported throughout the plant via the sulfate transport system. Sulfate transporters were first characterized in *Arabidopsis thaliana* selenate-resistant mutants (Shibagaki et al. 2002) and can be clustered into 4 main groups. Group 1 includes high affinity sulfate transporters, SULTR1;1 and SULTR1;2, which are the best-characterized and primarily found in the roots (Buchner 2004). Group 2 transporters have a low affinity for sulfate, are found throughout the plant, and have a role in sulfate loading into the vascular systems, and thus in translocation. Two isoforms have been identified in *A. thaliana*, SULTR2:1 and SULTR2:2, both expressed in leaves and roots. AtSULTR2;1 localizes to the xylem parenchyma, as well as the phloem cells in leaves and pericycle cells in roots, while AtSULTR2;2 is found in the phloem cells in roots and the bundle sheath cells in leaves (Takahashi et al. 2000; Buchner 2004). Group 3 sulfate transporters are only found in leaves, and do not show responsiveness to the sulfur status of the plant (Buchner 2004). AtSULTR3;1 localizes to the chloroplasts, and loss of this transporter greatly reduced the sulfate uptake capacity of these organelles (Cao et al. 2012). Group 4 includes sulfate transporters localized in tonoplasts. In *A. thaliana*, AtSULTR4;1 and AtSULTR4;2 have been characterized as low affinity sulfate transporters playing a role in sulfate vacuolar efflux, which may make sulfate more available for export via the vasculature; thus, AtSULTR4;1 and AtSULTR4;2 have been implicated to contribute to root-shoot translocation and the delivery of sulfate to developing seeds (Zuber et al. 2010).

Selenate enters the roots through the high affinity sulfate transporters SULTR1;1 and SULTR1;2, which are proton-sulfate symporters; for every molecule of selenate that enters the roots, 3 protons are also taken up (Lass and Ullrich-Eberius 1984; Hawkesford et al. 1993). The expression of SULTR1;1 and SULTR1;2 is controlled by the sulfur status of the plant. SULTR1;1 expression is lower and upregulated under S-deficient conditions, while SULTR1;2 is highly expressed under both S-sufficient and S-deficient conditions (White et al. 2007; El Kassis et al. 2007). Both SULTR1 transporters have the capacity to mediate selenate transport from the soil into the root cells, but there is unequal functional redundancy between these two transporters (Barberon et al. 2008). *Arabidopsis thaliana sultr1;2* mutants displayed a higher tolerance to selenate compared to *sultr1;1* mutants and wild-type plants, while *sultr1;1-sultr1;2* double mutants exhibited the greatest tolerance to selenate (Barberon et al. 2008). This suggests that SULTR1;2 is the main portal for selenate entry into the plant, compared to SULTR1;1. SULTR1;2 shares 70% amino acid homology with other high-affinity plant sulfate transporters, and is localized in the root hairs as well as the root epidermis and cortex (Yoshimoto 2002). AtSULTR1;2 was found to complement the function of two yeast sulfate transporters located in the plasma membrane (Yoshimoto 2002).

Recent research suggests that SULTR1 homologs found in Se hyperaccumulator species may have a preference for selenate transport over sulfate, which may explain the high Se/S ratio and Se hyperaccumulator status of these plants (White 2015). SULTR1 sequences isolated from several hyperaccumulator species in the genus *Astragalus* (Fabaceae) contain an alanine residue instead of the glycine found in SULTR1 isoforms of non-accumulating angiosperms, which may play a role in the preferential uptake of selenate over sulfate reported in these species (White 2015; Cabannes et al. 2011).

While the high-affinity sulfate transporters are responsible for the transport of selenate into the plant, selenite is taken up through a separate pathway. It is believed that selenite uptake is mediated by root phosphate transporters. Studies in perennial ryegrass (*Lolium perenne* L. cv. Evening Shade) and strawberry clover (*Trifolium fragiferrum* L. cv. O'Conner) showed that selenite uptake was reduced by up to 50% in response to a 10-fold increase in phosphate treatment (Hopper and Parker 1999). Another study has shown that the K_m of selenite influx increased in the presence of phosphate in wheat (*Triticum aestivum*) (Li et al. 2008). These results indicate the existence of competition for uptake between selenite and phosphate, suggesting the two molecules share a common transporter, as has been reported for yeast (Lazard et al. 2010).

Plants also have the capacity to take up organic forms of Se via amino acid permeases, which are plasma membrane-localized transporters mediating the uptake of amino acids in the cell. Two common forms of organic Se are selenocysteine (SeCys) and selenomethionine (SeMet). Normally, these products are formed from inorganic pools of Se through the S assimilation pathway, but there is evidence that plants can take up organic selenocompounds directly. Studies in durum wheat (*Triticum turgidum*) and spring canola (*Brassica napus*) showed that organic forms of Se, specifically selenomethionine and selenocystine, were taken up at rates over 20-fold higher than selenate or selenite (Zayed et al. 1998; Kikkert and Berkelaar 2013). A broad specificity amino acid permease isolated from *A. thaliana* complemented proline uptake in yeast mutant strains, with the strongest competitors for proline uptake being cysteine and methionine (Frommer et al. 1993). It is conceivable that selenocysteine and selenomethionine are taken up by this amino acid transporter as well.

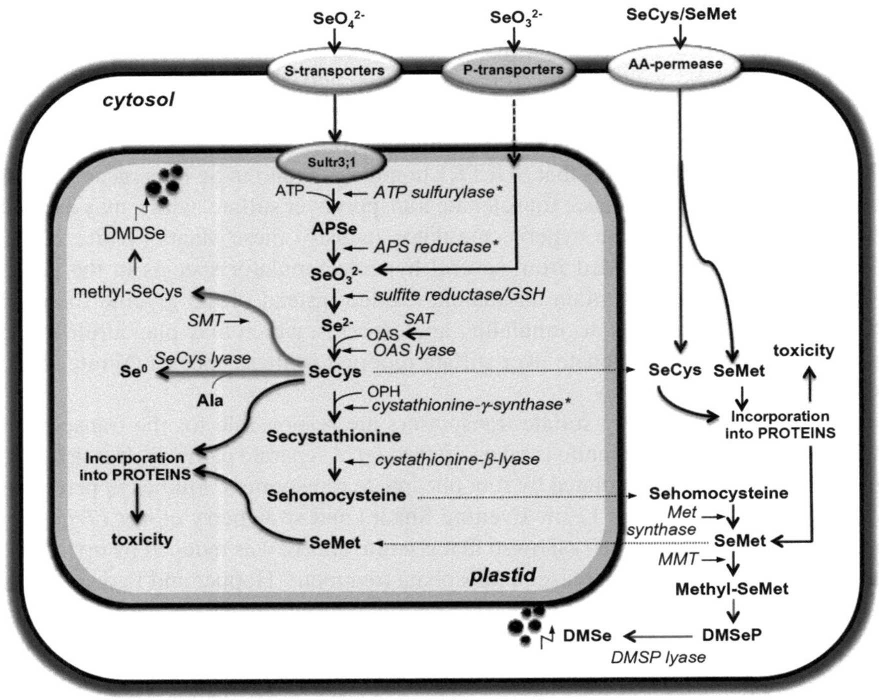

Fig. 2.1 Schematic model of Se assimilation and metabolism in plant mesophyll cells. *Red text* and *arrows* indicate Se hyperaccumulator processes. *Asterisks* indicate enzymes overexpressed via genetic engineering. *Sultr* sulfate/selenate cotransporters, *APSe* adenosine phosphoselenate, *GSH* glutathione, *SAT* serine acetyltransferase, *OAS* O-acetylserine, *(Se)Cys* (seleno)cysteine, *OPH* O-phosphohomoserine, *(Se)Met* (seleno)methionine, *MMT* methylmethionine methyltransferase, *DMSeP* dimethylselenoproprionate, *DM(D)Se* dimethyl(di)selenide (volatile), *SMT* selenocysteine methyltransferase

2.2 Conversion of Inorganic Selenium into Organic Forms: The First Steps of Selenium Assimilation

After uptake into the roots, selenate needs to be converted into a biologically active form for assimilation into the plant (Fig. 2.1). This is carried out by the enzyme ATP sulfurylase, which couples selenate (or sulfate) to ATP, forming adenosine 5′-phosphoselenate (APSe) or adenosine 5′-phosphosulfate (APS) (Leustek 1994; Pilon-Smits and Le Duc 2009; Schiavon et al. 2015). This step, which was found to be rate limiting in Se assimilation (Pilon-Smits and Le Duc 2009) occurs in both the cytosol and plastids (White et al. 2007; Pilon-Smits and Le Duc 2009; Pilon-Smits 2012). First characterized during studies of S assimilation, ATP sulfurylase was found to be derepressed by a selenate concentration 1/10th that of sulfate, indicating it is responsible for the assimilation of both molecules (Reuveny 1977). There have been 4 isoforms of ATP sulfurylase identified in *A. thaliana* (APS1–4), all localizing to the plastids of cells (Anjum et al. 2015), but *A. thaliana* APS2 was found to have dual localization to both the plastids and cytosol (Bohrer et al. 2015). ATP sulfurylase has been a target for genetic engineering of plants with higher Se uptake capacity, with the aim of developing plants for use in phytoremediation. Transgenic Indian mustard (*Brassica juncea*) overexpressing APS1 from *A. thaliana* showed increased selenate reduction, with roots and shoots containing mostly organic Se compounds compared to wild-type plants which mostly accumulated selenate (Pilon-Smits et al. 1999). Greenhouse experiments conducted with *B. juncea* APS transgenics grown on naturally seleniferous soils demonstrated that these plants accumulated Se up to 3-fold higher than wild type plants (Van Huysen et al. 2004). Field experiments in California on Se-contaminated soil confirmed these findings, with APS transgenics accumulating 4-fold more Se than wild type plants (Bañuelos et al. 2005).

APSe is converted to selenite by the activity of APS reductase (APR). This reaction happens exclusively in the plastids. APR is an essential enzyme and is reported to be another rate-limiting step in selenate assimilation (Setya et al. 1996; Suter et al. 2000; Sors et al. 2005a, b). The reaction equilibrium of ATP sulfurylase favors the reverse direction, and so the products of this reaction need to be converted rapidly in order for assimilation to proceed (Sors et al. 2005a, b; Saito 2004). While native expression of APR in several *Astragalus* species was not found to correlate with Se hyperaccumulation, transgenic experiments have shown that overexpression of APR enhances selenate reduction into organic forms, thus suggesting a role for this enzyme in selenate assimilation (Sors et al. 2005a, b). APR's role in the Se assimilation pathway is also supported by the fact that increased activity of this enzyme contributed to increased Se flux through the plant (Sors et al. 2005a, b). *Apr2–1 Arabidopsis* mutants showed enhanced levels of selenate, but decreased levels of selenite, implicating APR2 in converting APSe into selenite (Grant et al. 2011). These mutants also had decreased selenate tolerance due to decreased levels of glutathione, which helps to prevent the formation of damaging superoxides in the cell (Grant et al. 2011).

The next step in the Se assimilation pathway is the reduction of selenite to selenide, for incorporation into organic molecules such as amino acids. The conversion of selenite into selenide may occur either enzymatically or non-enzymatically. Sulfite reductase (SiR) is responsible for the conversion of sulfite to sulfide during reductive sulfate assimilation (Yarmolinsky et al. 2012), so it is not out of the question for the same enzyme to catalyze the reduction of selenite (Pilon-Smits 2012; White 2015). There is a single copy of the gene coding for SiR in *A. thaliana* (Khan et al. 2010), and it has been found to localize to plastids (Bork et al. 1998). The conversion to selenide may also occur non-enzymatically through an interaction between selenite and reduced glutathione (GSH) (Anderson and Mcmahon 2001; Terry et al. 2007; Pilon-Smits 2012). This conversion takes place in multiple steps, with selenite first converted to the organic molecule GSSeSG non-enzymatically, which is then converted to GSSeH and finally to selenide through the action of glutathione reductase (GR) using NADPH as a reductant (Hsieh and Ganther 1975). In support of a GR role in Se assimilation, yeast glutathione was shown to reduce selenite to selenide (Hsieh and Ganther 1975). Thus, while the reduction of selenite may be non-enzymatic, the regeneration of reduced glutathione is mediated by the enzyme GR. It belongs to the oxidoreductase family of proteins, which require NADP+ or NAD+ to transfer electrons from one molecule to another (Price and Stevens 1999). Glutathione reductase is responsible for converting glutathione from its oxidized state back to its reduced form, which is essential in numerous cellular processes such as combating oxidative stress, promoting enzyme stability, and the regulation of cell metabolism (Jocelyn 1972; Williams 1976). In plants, this enzyme is active in chloroplasts and cytosol (Foyer and Halliwell 1976). The reduction of oxidized glutathione by GR in chloroplasts has been reported to be coupled to photosynthetic electron transport (Jablonski and Anderson 1978; Schaedle and Bassham 1977) and may suggest that the reduction of selenite to selenide occurs in the chloroplasts as part of a light-dependent reaction (Ng and Anderson 1979).

Selenium toxicity in plants can be attributed to many factors, including oxidative stress, but the main cause is considered to be the misincorporation of selenoamino acids into proteins (Pilon-Smits 2012). Selenium can replace sulfur in the amino acids cysteine (Cys) and methionine (Met) to produce selenocysteine (SeCys) and selenomethionine (SeMet). The prevention of incorporating these selenoamino acids into proteins is a key feature of Se hyperaccumulator species, and is instrumental for their high Se tolerance (Brown and Shrift 1982).

2.3 Formation and Processing of Seleno-Amino Acids: Mechanisms of Preventing Selenium Toxicity

The first step in the formation of selenoamino acids is carried out by the enzyme complex Cysteine synthase (CS), which catalyzes the formation of SeCys from O-acetylserine (OAS) and selenide (White 2015; Pilon-Smits 2012). This process occurs in the chloroplasts of cells, but also in the cytosol and mitochondria (Ng and Anderson 1979; Wirtz et al. 2001). During + assimilation, Cys is formed by the reaction between OAS and hydrogen sulfide (Giovanelli 1990). Selenocysteine formation is identical to this reaction, with the substitution of hydrogen selenide as a reactant. Cysteine synthase is a complex formed by the association of two enzymes, serine acetyltransferase (SAT) and OAS thiol-lyase (OAS-TL) (Bogdanova and Hell 1997). SeCys can be incorporated into proteins nonspecifically, which can lead to disruption of protein function and thus Se toxicity (Stadtman 1990; Neuhierl and Bock 1996; Van Huysen et al. 2003). The prevention of non-specific incorporation of SeCys into proteins is crucial in preventing Se toxicity. The methylation of SeCys to form methyl-SeCys (MeSeCys) is a key mechanism used by hyperaccumulator species to reduce the amount of SeCys available for incorporation into proteins (Pilon-Smits and Le Duc 2009). The enzyme SeCys methyltransferase (SMT) is responsible for this conversion (Neuhierl and Bock 1996). SMT is homologous to other enzymes with similar functions, such as YagD in *Escherichia coli*, a homocysteine methyltransferase (HMT) able to methylate both SeCys and homocysteine, and belongs to a class of methyltransferases involved in the metabolism of S-methylmethionine (Neuhierl et al. 1999; Sors et al. 2005a, b). SMT was also found to be highly homologous to HMTs isolated from *A. thaliana* and *Oryza sativa* (Sors et al. 2005a, b), and is localized in the chloroplasts (Sors et al. 2009). SMT also shows a preference for the methylation of SeCys over Cys by at least 3 orders of magnitude (Neuhierl and Bock 1996), further solidifying its role in conferring Se tolerance to plants (Neuhierl et al. 1999). SMT has been identified in multiple non-accumulator and Se hyperaccumulator species of *Astragalus* but only the isoform from the hyperaccumulators had the ability to produce MeSeCys, indicating its essential role in the ability to tolerate and accumulate high levels of Se (Sors et al. 2009). In fact, the main form of Se found in the hyperaccumulators *A. bisulcatus* and *Stanleya pinnata* is MeSeCys, due to the high activity of the SMT enzyme (Neuhierl et al. 1999; Birringer et al. 2002; Pickering 2003; Sors et al. 2005a, b; Freeman 2006, 2010; Lindblom et al. 2013; Alford et al. 2014; White 2015), while selenate was the major Se compound found in related non-accumulator species (de Souza et al. 1998; Freeman 2006; Pilon-Smits 2012). Although SMT is found to be highly expressed specifically in hyperaccumulators (Sors et al. 2009), some Se accumulator species, such as *Brassica oleracea (*Broccoli) also have an SMT enzyme, but it is expressed only in the presence of Se (Lyi et al. 2007; Pilon-Smits 2012). SMT has been used in transgenic studies to confer increased Se accumulation and tolerance in non-accumulating species. SMT isolated from *A. bisulcatus* induced the accumulation of MeSeCys and γ-glutamyl-MeSeCys in *A. thaliana*, and increased Se accumulation and volatilization in *B. juncea* (Leduc et al. 2006; Ellis et al. 2004).

While the production of MeSeCys is critical to Se tolerance in plants, further processing of this molecule into volatile compounds serves as another mechanism by which plants tolerate high levels of Se. The volatile compound dimethyldiselenide (DMDSe) is formed by oxidation and methylation of MeSeCys (Meija et al. 2002; Sors et al. 2005a, b). First, MeSeCys is converted to methylselenocysteineselenideoxide (MeSeCysSeO), whose sulfur analog methylcysteinesulfoxide (MeCysSO) is responsible for many *Brassica* varieties' characteristic flavors (Chin and Lindsay 1994). This compound is then converted to another key intermediate methaneselenol (CH_3SeH) via the action of the enzyme Cys sulfoxide lyase (Chin and Lindsay 1994; Griffiths et al. 2002; Ellis and Salt 2003). DMDSe production occurs in the leaves, and has been detected in the Se hyperaccumulator *Astragalus racemosus* (Evans et al. 1968). Volatile Se compounds have been hypothesized to aid in defense against herbivory. This is supported not only by the fact that the production of these volatiles occurs in the leaves, but that it also occurs primarily after tissue injury (Ellis and Salt 2003).

The formation of SeMet occurs through the enzymatic conversion of SeCys. There are multiple steps involved in the synthesis of SeMet, which include potential targets for transgenic phytoremediation efforts. First, SeCys is converted to Se-cystathionine by the enzyme cystathionine-γ-synthase (CGS) (Pilon-Smits 2012). CGS catalyzes the formation of Se-cystathionine via the condensation of

O-phosphohomoserine (OPH) and SeCys (Van Huysen et al. 2013; Sors et al. 2005a, b). CGS was shown to be a rate-limiting enzyme in the conversion of SeCys to volatile DMSe (Van Huysen et al. 2003). Transgenic *B. juncea* overexpressing CGS had 2–3 fold higher Se volatilization rates and concurrent 20–40% lower shoot and 50–70% lower root Se levels compared to wild type plants, highlighting the value of this approach for applications in Se phytoremediation (Van Huysen et al. 2003, 2004). Se-cystathionine is converted to Se-homocysteine via a reaction between Se-cystathionine and water, mediated by the enzyme cystathionine beta-lyase. This enzyme is shared in both the Se and S assimilation pathways, evidenced by the fact that cystathionine beta-lyase isolated from both Se hyperaccumulator and non-accumulator plant species had the capacity to cleave both Se-cystathionine and cystathionine into Se-homocysteine and homocysteine, respectively (Sors et al. 2005a, b; McCluskey et al. 1986). Finally, the conversion of Se-homocysteine to SeMet is catalyzed by the enzyme Met synthase. Met synthase has been isolated from plants from various angiosperm taxa, including *A. thaliana, Catharanthus roseus*, and *Coleus blumei* (Eichel et al. 1995; Petersen et al. 1995; Ravanel et al. 1998). Using methyl-tetrahydrofolate as a carbon donor, Met synthase catalyzes the conversion of Se-homocysteine to SeMet (Cossins and Chen 1997).

Like SeCys, SeMet is subject to further processing steps that reduce its incorporation into proteins. The volatile Se compound DMSe is synthesized via the S volatilization pathway starting from SeMet (Tagmount 2002). Enzymes involved in the S volatilization pathway and formation of dimethyl sulfide (DMS) have also been discovered to be involved in the production of DMSe (Terry and Zayed 1994; Tagmount 2002). The production of DMSe in plants is important not only as a defense against herbivores, but it also diverts large pools of potentially toxic SeMet to the significantly less toxic DMSe. DMSe was found to be almost 600 times less toxic than inorganic Se compounds (McConnell and Portman 1952; Wilber 1980). DMSe is the main volatile Se compound isolated from non-accumulator plant species, while DMDSe is primarily produced in hyperaccumulators (Pilon-Smits and Le Duc 2009). The first step in the synthesis of DMSe is the methylation of SeMet to form Se-methyl Se-Met (SeMM) by the enzyme S-adenosyl-L-Met:Met-S-methyltransferase (MMT) (Tagmount 2002). SeMM can be converted to DMSe by one of two pathways. SeMM may first be converted to the intermediate molecule 3-dimethylselenoniopropionate (DMSeP) (Kocsis 1998). The sulfur analog DMSP is a biologically important molecule, playing important roles in osmoprotection of plants and bacteria (Mason and Blunden 1989; Hanson et al. 1994; Kocsis 1998). The synthesis of DMSP has been detected in members of the family Poaceae, such as *Spartina alterniflora* (Kocsis 1998), as well as members of the Asteraceae including *Melanthera biflora* (syn. = *Wollastonia biflora*) (Hanson et al. 1994; James et al. 1995) and *Ratibida pinnata* (Paquet et al. 1995). The synthesis of DMSe may also proceed directly from SeMM via the enzyme methylmethionine hydrolase (Mudd and Datko 1990; Meija et al. 2002; Ellis and Salt 2003).

Aside from volatilization, plants have another mechanism to help prevent Se toxicity. Selenocysteine lyase (SL) is an enzyme that breaks down SeCys into elemental Se and alanine, reducing the amount of free SeCys available for misincorporation into proteins (Van Hoewyk et al. 2005). Selenocysteine lyases are analogous to NifS-like Cys desulfurase proteins characterized in Arabidopsis (Ye et al. 2005), whose main role is to generate free S from Cys for the formation of FeS clusters (Pilon-Smits et al. 2002). There are two isoforms of SeCys lyase found in plants, with different subcellular localization patterns; one isoform localizes to the cytosol (Kushnir et al. 2001), and the other to mitochondria and plastids (Pilon-Smits et al. 2002). Overexpression of a chloroplast-localizing NifS protein from Arabidopsis (AtCpNifS) was found to increase Se tolerance by 1.9-fold and increased Se accumulation by 2.2-fold (Van Hoewyk 2013). Similarly, expression of a mouse SL caused a 2-fold reduction in Se incorporation into proteins and a 1.5-fold increase in shoot Se concentration in Arabidopsis (Pilon et al. 2003), as well as a 2-fold increase in Se accumulation in Indian mustard in both lab (Garifullina et al. 2003) and field (Bañuelos et al. 2007) studies. Selenocysteine lyases not only help to reduce Se toxicity in plants, but also appear to be promising enzymes to exploit for phytoremediation purposes.

The mechanisms by which plants accumulate, assimilate, and tolerate Se mirror aspects of the S assimilation pathway, but the roles these two elements play in the plant are very different. By better understanding the pathways of Se assimilation, new approaches to developing plants for phytoremediation and biofortification can be exploited, and mechanisms that hyperaccumulator species exploit in their uptake and assimilation of Se can be further elucidated.

References

Alford ER, Lindblom SD, Pittarello M, Freeman JL, Fakra SC, Marcus MA, Broeckling C, Pilon-Smits EAH, Paschke MW (2014) Roles of rhizobial symbionts in selenium hyperaccumulation in *Astragalus* (Fabaceae). Amer J Bot 101:1895–1905

Anderson JW, Mcmahon PJ (2001) The role of glutathione in the uptake and metabolism of sulfur and selenium. In: Grill D, Tausz MM, de Kok LJ (eds) Significance of glutathione to plant adaptation to the environment, Plant ecophysiology, vol 2. Springer, Dordrecht, pp 57–99

Anjum NA, Gill R, Kaushik M, Hasanuzzaman M, Pereira E, Ahmad I, Tuteja N, Gill SS (2015) ATP-sulfurylase, sulfur-compounds, and plant stress tolerance. Front Plant Sci 6:210

Bañuelos G, Terry N, Leduc D, Pilon-Smits EAH, Mackey B (2005) Field trial of transgenic indian mustard plants shows enhanced phytoremediation of selenium-contaminated sediment. Environ Sci Technol 39:1771–1777

Bañuelos G, Leduc D, Pilon-Smits EAH, Terry N (2007) Transgenic indian mustard overexpressing selenocysteine lyase or selenocysteine methyltransferase exhibit enhanced potential for selenium phytoremediation under field conditions. Environ Sci Technol 41:599–605

Barberon M, Berthomieu P, Clairotte M, Shibagaki N, Davidian J, Gosti F (2008) Unequal functional redundancy between the two *Arabidopsis thaliana* high-affinity sulphate transporters SULTR1;1 and SULTR1;2. New Phytol 180:608–619

Birringer M, Pilawa S, Flohé L (2002) Trends in selenium biochemistry. Nat Prod Rep 19:693–718

Bogdanova N, Hell R (1997) Cysteine synthesis in plants: protein-protein interactions of serine acetyltransferase from Arabidopsis thaliana. Plant J 11:251–262

Bohrer A-S, Kopriva S, Takahashi H (2015) Plastid-cytosol partitioning and integration of metabolic pathways for APS/PAPS biosynthesis in Arabidopsis thaliana. Front Plant Sci 5:1–4

Bork C, Schwenn JD, Hell R (1998) Isolation and characterization of a gene for assimilatory sulfite reductase from *Arabidopsis thaliana*. Gene 212:147–153

Brown TA, Shrift A (1982) Selenium: toxicity and tolerance in higher plants. Biol Rev 57:59–84

Buchner P (2004) Regulation of sulfate uptake and expression of sulfate transporter genes in *Brassica oleracea* as affected by atmospheric H2S and pedospheric sulfate nutrition. Plant Physiol 136:3396–3408

Cabannes E, Buchner P, Broadley MR, Hawkesford MJ (2011) A comparison of sulfate and selenium accumulation in relation to the expression of sulfate transporter genes in *Astragalus* species. Plant Physiol 157:2227–2239

Cao MJ, Wang Z, Wirtz M, Hell R, Oliver DJ, Xiang CB (2012) SULTR3;1 is a chloroplast-localized sulfate transporter in *Arabidopsis thaliana*. Plant J 73:607–616

Chin HW, Lindsay RC (1994) Mechanisms of formation of volatile sulfur compounds following the action of cysteine sulfoxide lyases. J Agric Food Chem 42:1529–1536

Cossins EA, Chen L (1997) Folates and one-carbon metabolism in plants and fungi. Phytochemistry 45:437–452

de Souza MP, Pilon-Smits EA, Lytle CM, Hwang S, Tai J, Honma TS, Yeh L, Terry N (1998) Rate-limiting steps in selenium assimilation and volatilization by Indian mustard. Plant Physiol 117:1487–1494

Eichel J, Gonzalez JC, Hotze M, Matthews RG, Schroder J (1995) Vitamin-B12-independent methionine synthase from a higher plant (*Catharanthus roseus*). Molecular characterization, regulation, heterologous expression, and enzyme properties. Eur J Biochem 230:1053–1058

Ellis DR, Salt DE (2003) Plants, selenium and human health. Curr Opin Plant Biol 6:273–279

Ellis DR, Sors TG, Brunk DG, Albrecht C, Orser C, Lahner B, Wood KV, Harris HH, Pickering IJ, Salt DE (2004) Production of Se-methylselenocysteine in transgenic plants expressing selenocysteine methyltransferase. BMC Plant Biol 4:1471–2229

Evans CS, Asher CJ, Johnson CM (1968) Isolation of dimethyl diselenide and other volatile selenium compounds from *Astragalus racemosus* (Pursh.) Aust J Biol Sci 21:13–20

Fordyce FM (2012) Selenium deficiency and toxicity in the environment. In: Selinus O (ed) Essentials of medical geology, 3rd edn. Springer, Dordrecht, pp 375–416

Foyer CH, Halliwell B (1976) The presence of glutathione and glutathione reductase in chloroplasts: a proposed role in ascorbic acid metabolism. Planta 133:21–25

Freeman JL (2006) Spatial imaging, speciation, and quantification of selenium in the hyperaccumulator plants *Astragalus bisulcatus* and *Stanleya pinnata*. Plant Physiol 142:124–134

Freeman JL, Tamaoki M, Stushnoff C, Quinn CF, Cappa JJ, Devonshire J, Fakra SC, Marcus MA, Mcgrath SP, Van Hoewyk D, Pilon-Smits EAH (2010) Molecular mechanisms of selenium tolerance and hyperaccumulation in *Stanleya pinnata*. Plant Physiol 153:1630–1652

Frommer WB, Hummel S, Riesmeier JW (1993) Expression cloning in yeast of a cDNA encoding a broad specificity amino acid permease from *Arabidopsis thaliana*. Proc Natl Acad Sci U S A 90:5944–5948

Garifullina GF, Owen JD, Lindblom SD, Tufan H, Pilon M, Pilon-Smits EAH (2003) Expression of a mouse selenocysteine lyase in *Brassica juncea* chloroplasts affects selenium tolerance and accumulation. Physiol Plant 118:538–544

Giovanelli J (1990) Regulatory aspects of cysteine and methionine synthesis. In: Rennenberg H et al (eds) Sulfur nutrition and sulfur assimilation in higher plants: fundamental environmental and agricultural aspects. SPB Academic Pub, The Hague

Grant K, Carey NM, Mendoza M, Schulze J, Pilon M, Pilon-Smits EAH, Van Hoewyk D (2011) Adenosine 5'-phosphosulfate reductase (APR2) mutation in Arabidopsis implicates glutathione deficiency in selenate toxicity. Biochem J 438:325–335

Griffiths G, Trueman L, Crowther T, Thomas B, Smith B (2002) Onions–a global benefit to health. Phytother Res 16:603–615

Hanson A, Rivoal J, Paquet L, Gage DA (1994) Biosynthesis of 3-dimethylsulfoniopropionate in *Wollastonia biflora* (L.) DC. Evidence that S-methylmethionine is an intermediate. Plant Physiol 105:103–110

Hawkesford M, Davidian JC, Grignon C (1993) Sulphate/proton cotransport in plasma-membrane vesicles isolated from roots of *Brassica napus* L.: increased transport in membranes isolated from Sulphur-starved plants. Planta 190:297–304

Hopper JL, Parker DR (1999) Plant availability of selenite and selenate as influenced by the competing ions phosphate and sulfate. Plant Soil 210:199–207

Hsieh SH, Ganther HE (1975) Acid-volatile selenium formation catalyzed by glutathione reductase. Biochemist 14:1632–1636

Huysen T, Abdel-Ghany SE, Hale KL, Leduc D, Terry N, Pilon-Smits EAH (2003) Overexpression of cystathionine-γ-synthase enhances selenium volatilization in *Brassica juncea*. Planta 218:71–78

Huysen T, Terry N, Pilon-Smits EAH (2004) Exploring the selenium phytoremediation potential of transgenic indian mustard overexpressing ATP sulfurylase or cystathionine-γ-synthase. Int J Theor Phys 6:111–118

Jablonski PP, Anderson JW (1978) Light-dependent reduction of oxidized glutathione by ruptured chloroplasts. Plant Physiol 61:221–225

James F, Paquet L, Sparace SA, Gage DA, Hanson AD (1995) Evidence implicating dimethylsulfoniopropionaldehyde as an intermediate in dimethylsulfoniopropionate biosynthesis. Plant Physiol 108:1439–1448

Jocelyn PC (1972) Biochemistry of the SH group; the occurrence, chemical properties, metabolism and biological function of thiols and disulphides. Academic Press, London/New York

Kassis E, Cathala RNH, Fourcroy P, Berthomieu P, Terry N, Davidian JC (2007) Characterization of a selenate-resistant Arabidopsis mutant. Root growth as a potential target for selenate toxicity. Plant Physiol 143:1231–1241

Khan MS, Haas FH, Samami AA, Gholami AM, Bauer A, Fellenberg K, Reichelt M, Hansch R, Mendel RR, Meyer AJ, Wirtz M, Hell R (2010) Sulfite reductase defines a newly discovered bottleneck for assimilatory sulfate reduction and is essential for growth and development in Arabidopsis thaliana. Plant Cell 22:1216–1231

Kikkert J, Berkelaar E (2013) Plant uptake and translocation of inorganic and organic forms of selenium. Arch Environ Contam Toxicol 65:458–465

Kocsis MG (1998) Dimethylsulfoniopropionate biosynthesis in *Spartina alterniflora* (L.) Evidence that S-methylmethionine and dimethylsulfoniopropylamine are intermediates. Plant Physiol 117:273–281

Kushnir S, Babiychuk E, Storozhenko S, Davey MW, Papenbrock J, De Rycke R, Engler G, Stephan UW, Lange H, Kispal G, Lill R, Van Montagu M (2001) A mutation of the mitochondrial ABC transporter Sta1 leads to dwarfism and chlorosis in the *Arabidopsis* mutant starik. Plant Cell 13:89–100

Lass B, Ullrich-Eberius CI (1984) Evidence for proton/sulfate cotransport and its kinetics in *Lemna gibba* G1. Planta 161:53–60

Lazard M, Blanquet S, Fisicaro P, Labarraque G, Plateau P (2010) Uptake of selenite by *Saccharomyces cerevisiae* involves the high and low affinity orthophosphate transporters. J Biol Chem 285:32029–32037

Leduc DL, Abdelsamie M, Móntes-Bayon M, Wu CP, Reisinger SJ, Terry N (2006) Overexpressing both ATP sulfurylase and selenocysteine methyltransferase enhances selenium phytoremediation traits in indian mustard. Environ Pollut 144:70–76

Leustek T (1994) Cloning of a cDNA encoding ATP sulfurylase from *Arabidopsis thaliana* by functional expression in *Saccharomyces cerevisiae*. Plant Physiol 105:897–902

Li HF, Mcgrath SP, Zhao FJ (2008) Selenium uptake, translocation and speciation in wheat supplied with selenate or selenite. New Phytol 178:92–102

Lindblom SD, Valdez-Barillas JR, Fakra SC, Marcus MA, Wangeline AL, Pilon-Smits EAH (2013) Influence of microbial associations on selenium localization and speciation in roots of *Astragalus* and *Stanleya hyperaccumulators*. Environ Exp Bot 88:33–42

Lyi SM, Zhou X, Kochian LV, Li L (2007) Biochemical and molecular characterization of the homocysteine S-methyltransferase from broccoli (*Brassica oleracea var. italica*). Phytochemistry 68:1112–1119

Mason TG, Blunden G (1989) Quaternary ammonium and tertiary sulphonium compounds of algal origin as alleviators of osmotic stress. Bot Mar 32:313–316

McCluskey TJ, Scarf AR, Anderson JW (1986) Enzyme catalysed α,β-elimination of selenocystathionine and selenocystine and their Sulphur isologues by plant extracts. Phytochemistry 25:2063–2068

McConnell KP, Portman OW (1952) Toxicity of dimethyl selenide in the rat and mouse. Exp Biol Med 79:230–231

Meija J, Montes-Bayón M, Le Duc D, Terry N, Caruso JA (2002) Simultaneous monitoring of volatile selenium and sulfur species from Se accumulating plants (wild type and genetically modified) by GC/MS and GC/ICPMS using solid-phase microextraction for sample introduction. Anal Chem 74:5837–5844

Mikkelsen RL, Page AL, Bingham FT (1989) Factors affecting selenium accumulation by agricultural crops. In: Jacobs L (ed) Selenium in agriculture and the environment. Soil Science Society of America & American Society of Agronomy, Madison, pp 65–94

Mudd SH, Datko AH (1990) The S-methylmethionine cycle in *Lemna paucicostata*. Plant Physiol 93:623–630

Neuhierl B, Bock A (1996) On the mechanism of selenium tolerance in selenium-accumulating plants. Purification and characterization of a specific selenocysteine methyltransferase from cultured cells of *Astragalus bisculatus*. Eur J Biochem 239:235–238

Neuhierl B, Thanbichler M, Lottspeich F, Bock A (1999) A family of S-methylmethionine-dependent thiol/selenol methyltransferases: role in selenium tolerance and evolutionary relation. J Biol Chem 274:5407–5414

Ng B, Anderson JW (1979) Light-dependent incorporation of selenite and sulphite into selenocysteine and cysteine by isolated pea chloroplasts. Phytochemistry 18:573–580

Paquet L, Lafontaine PJ, Saini HS, James F, Hanson AD (1995) Évidence en faveur de la présence du 3-diméthylsulfoniopropionate chez une large gamme d'angiospermes. Can J Bot 73:1889–1896

Petersen M, Van Der Straeten D, Bauw G (1995) Full-length cDNA clone from *Coleus blumei* (accession no. Z49150) with high similarity to cobalamin-independent methionine synthase. Plant Physiol 109:338

Pickering IJ (2003) Chemical form and distribution of selenium and sulfur in the selenium hyperaccumulator *Astragalus bisulcatus*. Plant Physiol 131:1460–1467

Pilon M, Owen JD, Garifullina GF, Kurihara T, Mihara H, Esaki N, Pilon-Smits EAH (2003) Enhanced selenium tolerance and accumulation in transgenic *Arabidopsis* expressing a mouse selenocysteine lyase. Plant Physiol 131:1250–1257

Pilon-Smits EAH (2012) Plant selenium metabolism. In: Wong M (ed) Environmental contamination: health risks and ecological restoration. CRC Press, Boca Raton, pp 295–312

Pilon-Smits EAH, Le Duc D (2009) Phytoremediation of selenium using transgenic plants. Curr Opin Biotechnol 20:207–212

Pilon-Smits EAH, De Souza MP, Hong G, Amini A, Bravo RC, Payabyab SB, Terry N (1999) Selenium volatilization and accumulation by twenty aquatic plant species. J Environ Qual 28:1011–1018

Pilon-Smits EAH, Garifullina GF, Abdel-Ghany SE, Kato SI, Mihara H, Hale KL, Burkhead JL, Esaki N, Kurihara T, Pilon M (2002) Characterization of a NifS-like chloroplast protein from *Arabidopsis*. Implications for its role in sulfur and selenium metabolism. Plant Physiol 130:1309–1318

Price NC, Stevens L (1999) Fundamentals of enzymology: the cell and molecular biology of catalytic proteins. Oxford Press, Oxford
Ravanel S, Gakiere B, Job D, Douce R (1998) The specific features of methionine biosynthesis and metabolism in plants. Proc Natl Acad Sci U S A 95:7805–7812
Reuveny Z (1977) Derepression of ATP sulfurylase by the sulfate analogs molybdate and selenate in cultured tobacco cells. Proc Natl Acad Sci U S A 74:619–622
Saito K (2004) Sulfur assimilatory metabolism. The long and smelling road. Plant Physiol 136:2443–2450
Schaedle M, Bassham JA (1977) Chloroplast glutathione reductase. Plant Physiol 59:1011–1012
Schiavon M, Pilon M, Malagoli M, Pilon-Smits EAH (2015) Exploring the importance of sulfate transporters and ATP sulphurylases for selenium hyperaccumulation: a comparison of *Stanleya pinnata* and *Brassica juncea* (Brassicaceae). Front Plant Sci 6:1–13
Setya A, Murillo M, Leustek T (1996) Sulfate reduction in higher plants: molecular evidence for a novel 5′-adenylylsulfate reductase. Proc Natl Acad Sci U S A 93:13383–13388
Shibagaki N, Rose A, Mcdermott JP, Fujiwara T, Hayashi H, Yoneyama T, Davies JP (2002) Selenate-resistant mutants of *Arabidopsis thaliana* identify sultr1;2, a sulfate transporter required for efficient transport of sulfate into roots. Plant J 29:475–486
Sors TG, Ellis DR, Nam Na G, Lahner B, Lee S, Leustek T, Pickering IJ, Salt DE (2005a) Analysis of sulfur and selenium assimilation in *Astragalus* plants with varying capacities to accumulate selenium. Plant J 42:785–797
Sors TG, Ellis DR, Salt DE (2005b) Selenium uptake, translocation, assimilation and metabolic fate in plants. Photosynth Res 86:373–389
Sors TG, Martin CP, Salt DE (2009) Characterization of selenocysteine methyltransferases from *Astragalus* species with contrasting selenium accumulation capacity. Plant J 59:110–122
Stadtman T (1990) Selenium biochemistry. Annu Rev Biochem 59:111–127
Suter M, Von Ballmoos P, Kopriva S, Den Camp RO, Schaller J, Kuhlemeier C, Schurmann P, Brunold C (2000) Adenosine 5′-phosphosulfate sulfotransferase and adenosine 5′-phosphosulfate reductase are identical enzymes. J Biol Chem 275:930–936
Tagmount A (2002) An essential role of S-adenosyl-L-methionine:L-methionine S-methyltransferase in selenium volatilization by plants. Methylation of selenomethionine to selenium-methyl-L-selenium-methionine, the precursor of volatile selenium. Plant Physiol 130:847–856
Takahashi H, Watanabe-Takahashi A, Smith FW, Blake-Kalff M, Hawkesford MJ, Kazuki S (2000) The roles of three functional sulphate transporters involved in uptake and translocation of sulphate in *Arabidopsis thaliana*. Plant J 23:171–182
Terry N, Zayed MA, De Souza MP, Tarun AS (2007) Selenium in higher plants. Annu Rev Plant Physiol Plant Mol Biol 51:401–432
Van Hoewyk D (2013) A tale of two toxicities: malformed selenoproteins and oxidative stress both contribute to selenium stress in plants. Ann Bot 112:965–972
Van Hoewyk D, Garifullina GF, Ackley AR, Abdel-Ghany SE, Marcus MA, Fakra S, Ishiyama K, Inoue E, Pilon M, Takahashi H, Pilon-Smits EAH (2005) Overexpression of AtCpNifS enhances selenium tolerance and accumulation in *Arabidopsis*. Plant Physiol 139:1518–1528
White PJ (2015) Selenium accumulation by plants. Ann Bot 117(2):217–235
White PJ, Broadley MR (2009) Biofortification of crops with seven mineral elements often lacking in human diets – iron, zinc, copper, calcium, magnesium, selenium and iodine. New Phytol 182:49–84
White PJ, Broadley MR, Bowen HC, Johnson SE (2007) Selenium and its relationship with sulfur. In: Hawkesford MK, Kok D, Luit J (eds) Plant ecophysiology, Sulfur in plants: an ecological perspective, vol 6. Springer, Dordrecht, pp 225–252
Wilber CG (1980) Toxicology of selenium: a review. Clin Toxicol 17:171–230
Williams CH (1976) 3 Flavin-containing dehydrogenases. In: Boyer P (ed) The enzymes, vol 13. Elsevier, Dordrecht, pp 89–173

Wirtz M, Berkowitz O, Droux M, Hell R (2001) The cysteine synthase complex from plants. Eur J Biochem 268:686–693

Yarmolinsky D, Brychkova G, Fluhr R, Sagi M (2012) Sulfite reductase protects plants against sulfite toxicity. Plant Physiol 161:725–743

Ye H, Garifullina GF, Abdel-Ghany SE, Lihong Z, Pilon-Smits EAH, Pilon M (2005) The chloroplast NifS-like protein of Arabidopsis thaliana is required for iron – sulfur cluster formation in ferredoxin. Planta 220:602–608

Yoshimoto N, Takahashi H, Smith FW, Yamaya T, Saito K (2002) Two distinct high-affinity sulfate transporters with different inducibilities mediate uptake of sulfate in Arabidopsis roots. Plant J 29:465–473

Zayed AM, Terry N (1994) Selenium volatilization by plants. Plant Physiol 143:8–14

Zayed AM, Mel Lytle C, Terry N (1998) Accumulation and volatilization of different species of selenium by plants. Planta 206:284–292

Zuber H, Davidian JC, Wirtz M, Hell R, Belghazi M, Thompson R, Gallardo K (2010) Sultr4;1 mutant seeds of *Arabidopsis* have an enhanced sulphate content and modified proteome suggesting metabolic adaptations to altered sulphate compartmentalization. BMC Plant Biol 10:78

Chapter 3
Molecular Mechanisms of Selenium Responses and Resistance in Plants

Masanori Tamaoki and Akiko Maruyama-Nakashita

Abstract Selenium (Se) is an essential nutrient for many organisms but is toxic at high levels. A better understanding of plant responses to Se is important to optimize the use of plants in alleviating dietary Se deficiency or for the cleanup of Se-polluted areas. Genetic analysis among accessions of *Arabidopsis thaliana* showed that several genes involved in sulfur (S) assimilation may be responsible for the differences in Se resistance and accumulation, and resistance to selenite and selenate may be regulated by different genes. Molecular and biochemical studies of non-accumulator plants revealed that defense responses mediated by phytohormones (such as ethylene, jasmonic acid, and salicylic acid) play an important role in acquiring Se resistance and accumulation. Production of these phytohormones is enhanced via signal pathways of reactive oxygen species (ROS), and the signal pathways of phytohormones act in a cooperative or antagonistic manner to induce stress and S-uptake and S-metabolic genes. In this chapter, the contribution of ROS and phytohormone signaling in the acquisition of Se resistance and accumulation in Se hyper-accumulator plants was discussed, and the application of Se-responsive genes to generate transgenic plants that can detect Se in the environment was also introduced.

Keywords *Arabidopsis thaliana* • Ethylene • Jasmonic acid • Reactive oxygen species • Selenium

---

M. Tamaoki (✉)
National Institute for Environmental Studies,
Onogawa 16-2, Tsukuba, Ibaraki 305-8506, Japan
e-mail: mtamaoki@nies.go.jp

A. Maruyama-Nakashita
Kyushu University, 6-10-1 Hakozaki, Higashi-ku, Fukuoka 812-8581, Japan
e-mail: amaru@agr.kyushu-u.ac.jp

E.A.H. Pilon-Smits et al. (eds.), *Selenium in plants*, Plant Ecophysiology 11,
DOI 10.1007/978-3-319-56249-0_3

3.1 Introduction

Selenium (Se) is a naturally occurring element commonly found in sedimentary rocks formed during the Carboniferous to Quaternary periods. Selenate, the most common soluble form of Se in soil, leaches into shallow groundwater when Se-rich soils are irrigated (Wilber 1980; McNeal et al. 1989). In addition, selenite is a common contaminant in oil-refinery wastewater (Hansen et al. 1998). The accumulation of Se at a higher level in surface water or soil can be toxic for livestock, humans, and wild organisms, including higher plants (Terry et al. 2000; Hamilton 2004; Hira et al. 2004). Selenium is chemically similar to sulfur (S), and hence, its metabolic pathways could be similar to those of S. Thus, plants can take up Se as selenate non-specifically via sulfate transporters, and assimilate it to the cysteine (Cys) and methionine (Met) analogs seleno-Cys (SeCys) and seleno-Met, respectively (Läuchli 1993; Terry et al. 2000; Sors et al. 2005). Non-specific replacement of the two essential S amino acids Cys and Met in proteins by the Se analogs is toxic, as they inhibit the formation of S bridges and/or affect protein synthesis (Eustice et al. 1981; Stadtman 1990; Gromer and Gross 2002). For these reasons, much of the research on plant Se resistance has investigated Se interactions in S metabolism. However, recent studies have shown that Se-induced defense responses, which resemble the hypersensitive response observed in plants treated with biotic or abiotic stresses (Tamaoki 2008; Denance et al. 2013), play important roles in Se resistance not only in non-accumulator but also in hyper-accumulator plants (Tamaoki et al. 2008a, b; Freeman et al. 2010).

In this chapter, recent advances in the acquisition of Se resistance in plants are presented and discussed from a perspective of defense responses such as biosynthesis and signaling of phytohormones, the role of phytohormones in plant responses to Se, and multiple interactions between the phytohormones in Se-treated plants. In addition, applications are discussed of the knowledge of Se-responsive genes to develop an *in vivo* system for monitoring Se in the environment.

3.2 Genetic Analysis of Se Resistance in Non-accumulator Plants

Recently, Se toxicity and resistance mechanisms in non-accumulator plants were investigated by comparisons of genetic, physiological, and molecular responses to Se in relatively Se-resistant *Arabidopsis thaliana* accessions with those in non-resistant accessions. Three quantitative trait loci (QTL), which co-segregated with higher selenate resistance, were identified using a population of recombinant inbred lines prepared from *A. thaliana* accession Columbia (Col) (selenate resistant) compared with accession Landsberg (Ler) (selenate sensitive) (Zhang et al. 2006a). Among them, the QTL on chromosome 3 may have the potential for selenate resistance function because several genes involved in S assimilation are located in the region. The following four genes are present: (1) *ATPS1*, encoding the main (chloroplast) form of ATP sulfurylase; (2) *SERAT*, encoding the chloroplast form of serine-*O*-acetyl-transferase; (3) *HMT1*, an ortholog of the SeCys methyltransferase gene in Se hyper-accumulator *Astragalus* species; and (4) a predicted Se-binding protein (SBP); these could be candidates for the acquisition of selenate resistance (Neuhierl et al. 1999). However, gene expression analysis showed no evidence for the involvement of *ATPS1* (At3g22890), *SERAT* (At1g55920), and *SMT/HMT2* (At3g63250) in the acquisition of Se resistance in Col because no difference in transcript levels was found between selenate-treated Col and Ler (Zhang et al. 2006b). Another genetic and physiologic study using a different combination of *A. thaliana* accessions provided insight into selenite resistance and accumulation mechanisms in this species. *Arabidopsis* accessions grown with different concentrations of selenite demonstrated that Wassilewskija (Ws) is more susceptible to selenite than Col (Fig. 3.1) (Zhang et al. 2006b; Tamaoki et al. 2008a, b). A QTL analysis using F_2 population that was prepared from a cross between Ws and Col indicated that a selenite-resistance gene is located on chromosome 4 and that the resistance allele may inherit in a recessive manner (Zhang et al. 2006b). From all of the above-mentioned studies, selenate- and selenite-resistance in *Arabidopsis* species appear to be controlled, at least in part, by different loci. In addition, selenate-sensitive accessions Ler accumulate more Se than selenate-resistant accession Col (Zhang et al. 2006b). Although this may support the existence of negative correlation between Se-resistance and -accumulation in non-accumulator plants, such correlation was not observed from a comprehensive study using 19 Arabidopsis accessions by Zhang et al. (2007). Further studies are necessary to understand the correlation between Se-resistance and Se-accumulation in non-accumulator plants. Alternatively,

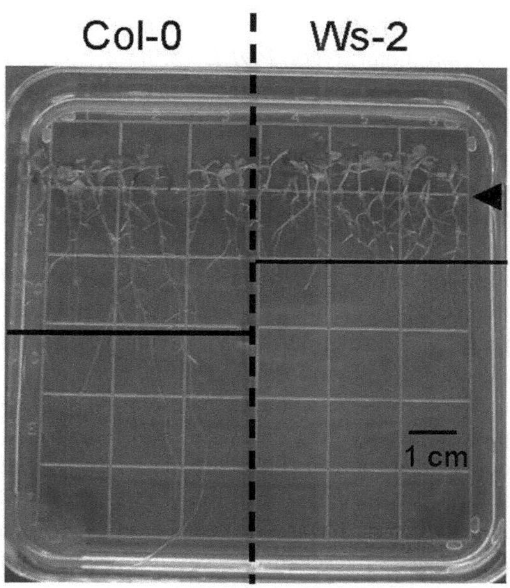

Fig. 3.1 Difference between *Arabidopsis* accessions in root growth inhibition by selenite treatment. *Arabidopsis* accessions Col-0 (selenite resistant) and Ws-2 (selenite sensitive) were vertically grown on 15 μM selenite for a week. The *arrowhead* indicates the position of seeds. Horizontal lines represent average root length in each accession

studies of fine mapping of recombinant inbred lines or F_2 population are also needed to identify genes involved in Se resistance in this non-accumulator species.

3.3 Genomic and Biochemical Analysis of Se Resistance in Non-accumulator Plants

Comprehensive analysis of Se-responsive genes in the non-accumulator species *A. thaliana* showed evidence of the involvement of defense-related phytohormones in the acquisition of Se resistance in plants. Indeed, a transcriptome study carried out to identify selenate-responsive genes revealed that selenate treatment induced expression of many ethylene- and/or jasmonic acid (JA)-responsive genes (Van Hoewyk et al. 2008). Moreover, induction of many of these ethylene- and/or JA-responsive genes was also observed in selenite-treated plants by comprehensive gene expression analysis (Tamaoki et al. 2008a). In fact, induction of ethylene response factor 1 (*ERF1*, At3g23240), pathogenesis-related 4 (*PR4*, At3g04720), plant defensin 1.2 (*PDF1.2*, At5g44420), vegetative storage protein 1 (*VSP1*, At5g24780), proteinase inhibitor 2 (*PIN2*, At2g02100), and JA-responsive gene (*JR*, At3g16470), which are responsive to ethylene and/or JA, was observed in Se-treated plants (Tamaoki et al. 2008a, b; Van Hoewyk et al. 2008). Moreover, Se treatment induced the genes encoding the key enzymes in ethylene biosynthesis (Zarembinski and Theologis 1994), *S*-adenosyl-Met synthase (*SAM*, At1g02500) and 1-aminocyclopropane-1-carboxylate synthase 6 (*ACS6*, At4g11280), as well as lipoxygenase 2 (*LOX2*, At3g45140) and allene oxide synthase (*AOS*, At5g42650) genes, which encode key enzymes in JA biosynthesis (Stenzel et al. 2004; Tamaoki et al. 2008a). Selenium regulation of these genes was confirmed using transgenic *AOS* promoter::*GUS*, *PDF1.2* promoter::*GUS* and *ACS8* promoter::*GUS* plants, in which GUS activities were only detected in leaves or roots of selenite-treated plants (Tamaoki et al. 2008b; Lehotai et al. 2012). All these results suggest that Se treatment triggers ethylene and JA production and responses. It is noteworthy that enhancement of ethylene generation and JA accumulation with selenite treatment was more pronounced in the selenite-resistant accession Col-0 than in selenite-sensitive accession Ws-2 (Tamaoki et al. 2008a). This indicates that ethylene and JA may play important roles in the acquisition of Se resistance in non-accumulator plants. The importance of ethylene and JA for Se resistance in non-accumulator plants was further investigated using mutants impaired in the biosynthesis or signaling of these phytohormones in the Se-resistant accession Col-0 background: *acs6* (lacking ethylene owing to the mutation of the biosynthetic *ACS6* gene), *ein2* (completely lacking ethylene signaling) (Guzmán and Ecker 1990), and *jar1* (lacking active form of JA) (Staswick et al. 1992). These mutants were more sensitive to selenite and selenate than the wild-type mutant Col-0 (Tamaoki et al. 2008a; Van Hoewyk et al. 2008). Conversely, an experiment to determine whether the selenite resistance of the sensitive accession Ws-2 is limited by its lower ethylene generation or JA accumulation showed that supply of 1-aminocyclopropane-1-carboxylic acid (ACC, precursor of ethylene) or MeJA enhanced selenite resistance in Ws-2 (Tamaoki et al. 2008a). These results further suggest that ethylene and JA play an important role in Se resistance in plants.

In selenite-treated *A. thaliana*, accumulation of another defense-related phytohormone, salicylic acid (SA), was also observed (Tamaoki et al. 2008a). SA is a major phenylpropanoid compound, whose biosynthesis is triggered by various biotic and abiotic stresses (Durner et al. 1997; Yalpani et al. 1994; Overmyer et al. 2003; Tamaoki 2008a). Isochorismate synthase 1 (*ICS1*) encodes a rate-limiting enzyme of SA biosynthetic pathway in *A. thaliana* (Wildermuth et al. 2001). Selenite induced both accumulation of SA as well as *ICS1* expression, in both Se-resistance accession Col-0 and Se-sensitive accession Ws-2 (Tamaoki et al. 2008a). However, mutants deficient in SA production (*sid2*) (Wildermuth et al. 2001) or signaling (*npr1*) (Cao et al. 1997) showed no difference in selenite resistance in comparison with their wild-type, Col-0, and treatment with SA increased selenite susceptibility in a Se-resistant *A. thaliana* accession, Col-0 (Tamaoki et al. 2008a). In contrast to the function of ethylene and JA, these results suggest that SA accumulation suppresses the acquisition of Se resistance in plants rather than enhancing it. The underlying mechanism of the negative effect of SA on Se resistance is still unclear, but one possible explanation is crosstalk among ethylene, JA, and SA signaling pathways. Indeed, many studies have shown that these phytohormones act in a mutually antagonistic or coordinated manner in plants exposed to biotic or abiotic stresses. Several recent genetic studies provide evidence for an antagonistic effect of SA on JA signaling in *A. thaliana*. For instance, the *eds4* and *pad4* mutants, which are impaired in SA accumulation, exhibit enhanced responses to inducers of JA-dependent gene expression (Clarke et al. 2000; Gupta et al. 2000; Lorenzo and Solano 2005). Conversely, characterization of three JA-signaling mutants, *mitogen-activated protein kinase4* (*mpk4*), *suppressor of SA insensitivity2* (*ssi2*) and *coronatine insensitive1* (*coi1*), provided genetic evidence that JA signaling also negatively regulates the expression of SA-mediated defenses in *A. thaliana* (Petersen et al. 2000; Kachroo et al. 2001; Kloek et al. 2001). Similarly, SA is known to inhibit the activity of the last step in the ethylene biosynthesis pathway, ACC oxidase (Leslie and Romani 1988), and inhibition of the SA pathway with ethylene signaling was also demonstrated using ethylene insensitive mutant or transcriptome analysis (Lawton et al. 1994; Tamaoki et al. 2003). Together, the observed increased Se sensitivity in the presence of SA might act through inhibition of ethylene- and/or JA-signaling pathways. In addition to ethylene, JA, and SA, other phytohormones such as abscisic acid (ABA), auxins, brassinosteroids, cytokinins (CKs), and gibberellins have been shown to affect defense signaling; however, their roles in plant defense, including Se resistance, are not characterized sufficiently (Pieterse et al. 2009; Kazan and Lyons 2014). Involvement of auxins and CKs in the acquisition of Se resistance has been discussed in part in the section below.

As described above, ethylene and JA may play important roles in Se resistance in *A. thaliana*. In general, production of defense-related phytohoromones, such as ethylene and JA, is often preceded by the generation of reactive oxygen species (ROS)

(Dong 1998; Overmyer et al. 2003; Tamaoki 2008). In the case of Se, ROS accumulation was also detected in both non-accumulator and Se hyper-accumulator plants (Gomes-Jr et al. 2007; Tamaoki et al. 2008a; Freeman et al. 2010). Interestingly, selenite-inducible ROS production was more pronounced in Se-resistant accession Col-0 than in Se-sensitive accession Ws-2 (Tamaoki et al. 2008a), suggesting that ROS generation has potential to provide Se resistance in non-accumulator plants. However, very high levels of ROS production decreased Se resistance in plants. Such plants showed signs of high levels of SA production, which may have attenuated ethylene and/or JA function, as described above. Thus, optimal levels of ROS production appear to be necessary for the acquisition of Se resistance in non-accumulator plants. Although the source of ROS in Se-treated plants is still unclear, it is expected to be triggered by cytosolic calcium as discussed in more detail below.

3.4 An Overview of Se Responses Contributing to Se Resistance in Non-Accumulator Plants Through the "Stress Pathway"

A simple model for Se resistance in non-accumulator plants, designated as "stress pathway," is proposed based on the studies described in the previous section (Fig. 3.2). According to the model, selenate or selenite is absorbed from soil through sulfate or phosphate transporters and induce ROS generation in the plant. The Se-induced ROS mimics an oxidative burst in plant cells, and the perception of this change triggers a wide array of signaling cascades similar to those induced by biotic or abiotic stresses. In parallel, an increase in cytosolic calcium concentration is also expected in these plants, as many genes related to calcium signaling, such as a calcium transporter (At5g26220), calmodulin-related proteins (At1g76640, At1g76650 and At2g26530), and a calcium-binding protein (At4g27280), were identified in the transcriptome analysis of selenate-treated *A. thaliana* (Van Hoewyk et al. 2008). Changes in the concentration of free cytosolic calcium may directly enhance NADPH oxidase activity, which is known to generate ROS, or it may phosphorylate one of the subunits of NADPH oxidase because this enzyme has an N-terminal sequence with two calcium-binding EF-hand motifs (Keller et al. 1998; Torres et al. 1998). NADPH-dependent ROS generation activates the production of ethylene, JA, and SA. Ethylene signaling up-regulates stress-responsive genes and JA signaling up-regulates both stress-responsive and S-uptake/metabolism genes (Lorenzo et al. 2005; Sasaki-Sekimoto et al. 2005). Up-regulation of S-uptake/metabolism genes by JA is quite important for the acquisition of Se resistance. Several transgenic plants with enhanced Se accumulation and resistance have already been generated through overexpression of genes involved in S/Se assimilation and volatilization. For instance, overexpression of the *ATPS1* gene in *Brassica juncea* (Indian mustard) enhanced Se accumulation two to threefold relative to the wild-type, and the plants also had enhanced selenate resistance (Pilon-Smits et al. 1999). Over-production of

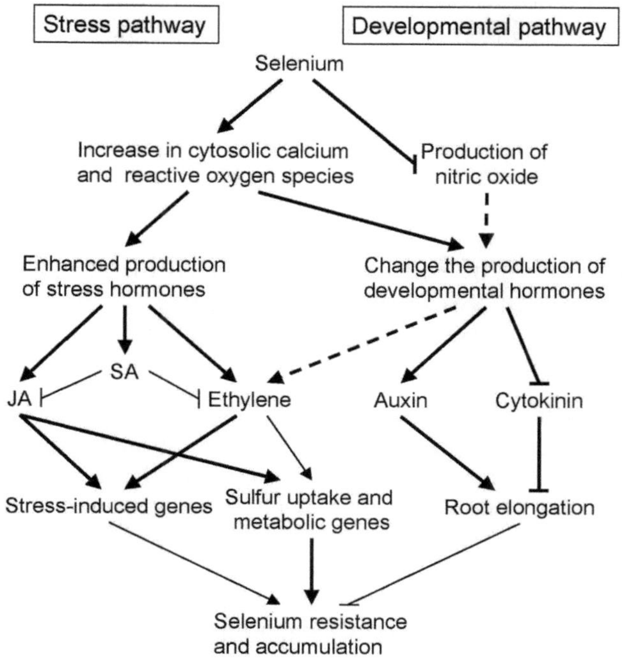

Fig. 3.2 Schematic model of the acquisition of Se resistance and accumulation in non-accumulator plants

SeCys methyltransferase from hyper-accumulator *Astragalus bisulcatus* in *A. thaliana* and Indian mustard enhanced Se resistance and Se volatilization (Ellis et al. 2004; LeDec et al. 2004). Higher S levels in plants likely help prevent incorporation of Se into S compounds, particularly proteins. Moreover, higher levels of the reduced S compound glutathione (GSH) may help alleviate Se-induced oxidative stress. Thus, JA likely plays an important role in Se resistance through the up-regulation of S metabolism genes.

In contrast to JA, the contribution of ethylene-induced stress-responsive genes in the acquisition of Se resistance of plants is still unclear. Any involvement of ethylene signaling in up-regulation of S-uptake/metabolism genes has not been shown until now. However, transgenic *A. thaliana* that overproduce *Arabidopsis halleri* plant defensin (*AhPDF1.1*) showed a slight but significant increase in tolerance to selenite compared with wild-type plants (Tamaoki et al. 2008b). *PDF1.2* gene and orthologue of *AhPDF1.1* in *A. thaliana* are known to be induced with the ethylene pathway (Penninckx et al. 1998), and the enhancement of *PDF1.2* expression was observed in Se-treated plants (Tamaoki et al. 2008a, b; Van Hoewyk et al. 2008). Moreover, this same gene was shown earlier to confer zinc (Zn) tolerance when overexpressed in *A. thaliana* (Mirouze et al. 2006) and yeast (Shahzad et al. 2013). As the function of PDF proteins is still largely unknown, the mechanism of these positive effects of PDF1.1 on Zn and Se resistance is unclear. However, *PDF1* transcripts were constitutively more abundant in Zn-tolerant *A. halleri* than in Zn-sensitive *A. thaliana* (Shahzad et al. 2013). As for SA, at low levels, it might have no effect on these processes, and thus on Se resistance, but at higher levels, SA decreases *A. thaliana* Se resistance, perhaps through inhibition of ethylene and/or JA synthesis or signaling.

3.5 An Overview of Se Responses Contributing to Se Resistance in Non-accumulator Plants Through the "Developmental Pathway"

Apart from the defense responses controlled with ethylene, JA, and SA, Se is also considered to affect root morphogenesis. Such a viewpoint is also important for discussion of Se resistance in plants, as Se resistance in higher plants is usually tested by inhibition of root elongation. A recent study showed that Se treatment inhibits root elongation processes in terms of plant development (Lehotai et al. 2012); thus, it seems difficult to distinguish it from Se-induced defense responses. However, Se-induced inhibition of root elongation can be considered to be an acclimation process by the plant, as retardation of root development reduces Se absorbance and this may ensure better survival of plants from Se toxicity. New insights indicate that plant defense networking involves more than just ethylene, JA, and SA, with more integrative models implicating a coordinated range of phytohormones in configuring the plant's response to biotic and abiotic stressors. These include ABA, auxins, brassinosteroids, CKs, and gibberellins (De Bruyne et al. 2014). Among these, auxins and CKs are known to be involved in root elongation processes. Hence, Se-dependent root growth regulation is designated as a "developmental pathway" in Fig. 3.2. Auxins are a class of essential plant hormones that control almost every aspect of plant growth and development (Woodward and Bartel 2005; Vanneste and Friml 2009). In roots, the most well-characterized auxin-associated phenotype is the dose-dependent increase of primary root length (Overvoorde et al. 2010). CKs are also known to play an important role in plant growth and development, and the reduction of CK levels in mutants lead to increased meristem size and primary root elongation compared with the wild-type (Werner et al. 2010). *In situ* expression analysis of phytohormone-associated genes showed that selenite treatment inhibits auxin accumulation, but increases the CK level in primary roots (Lehotai et al. 2012). These results indicate that Se affects auxin and CK levels in primary root, which inhibit elongation. However, the impact of these phytohormones on primary root growth is not always constant, as demonstrated by the inhibition of root growth with the application of high levels of auxin (Mähönen et al. 2014). Together, these apparently conflicting findings clearly illustrate the importance of auxin and CK homeostasis in the establishment of root development. Given the importance of auxins and CKs in regulating defense responses, these observations are compatible with the idea that auxins and CKs merge normal growth and developmental programs with plant defense functions, thus serving as important regulators of the innate trade-off between defense responses and plant growth. Signaling and/or production of auxins and CKs are known to be regulated by nitric oxide (NO), a multifunctional gaseous signaling molecule that plays a regulatory role in developmental processes (Lehotai et al. 2012). Nitric oxide positively regulates auxin signaling in root development (Correa-Aragunde et al. 2006) and represses CK signaling by inhibiting the phosphorylation activity through *S*-nitrosylation (Feng et al. 2013). Thus, NO regulates root elongation positively through activation of auxin signaling and/or inhibition of CK signaling. Selenite treatment decreases NO level in roots (Lehotai et al. 2012). Further studies are needed of the potential function of this Se-induced NO suppression in acquisition of Se resistance in plants. However, it is noteworthy that high levels of NO in GSNO reductase-deficient mutant (*gsnor1*) (Feechan et al. 2005) was reported to lead to selenite resistance because the viability of cells in root meristem was not affected, whereas reduction of NO production in NO-lacking double mutant (*nia1nia2*) (Wilkinson and Crawford 1993) resulted in selenite sensitivity (Lehotai et al. 2012).

3.6 Selenium Tolerance in Hyper-accumulator Plants

From an application perspective, it appears that increasing ethylene, JA, and/or SA levels in plants may be a useful approach to develop plants with enhanced Se resistance and/or accumulation. In this context, it is interesting to note that providing plants with ethylene or JA precursors resulted in both higher Se resistance and accumulation. This finding may provide insight into the Se resistance and accumulation mechanisms in Se hyper-accumulator plants. Comparative studies using a Se hyper-accumulator *Stanleya pinnata* and a non-hyper-accumulator *Stanleya albescens* showed that Se-resistance mechanisms were similar to those found in non-accumulator plants. Generation of ROS in leaves was highly induced in *S. albescens* rather than in *S. pinnata* with selenite treatment (Freeman et al. 2010). ROS generation in *Stanleya* species was different to that in *A. thaliana*, as a higher level of ROS generation was observed in the Se-resistant *A. thaliana* accession than in the Se-sensitive one (Tamaoki et al. 2008a). *S. pinnata* showed constitutive high levels of JA, MeJA, and free SA. Ethylene was generated at lower levels in *S. pinnata* in the absence of Se, but it appeared to be induced more strongly by Se treatment in the plant. Selenium responses of phytohormones in the non-hyper-accumulator *S. albescens* showed similar trends to those in the non-accumulator plant *A. thaliana*. Moreover, transcriptome analyses of *S. pinnata* showed a constitutively higher expression of genes involved in S assimilation, antioxidant activities, defense, and responses to ethylene, JA, and SA (Freeman et al. 2010). Accordingly, Se accumulation was slightly enhanced in both species when these phytohormones were supplied. These results together indicate that the stress pathway (Fig. 3.2) is constitutively activated in hyper-accumulators, which may provide Se resistance and accumulation in the plants.

Previous studies showed that hyper-accumulator plants accumulate Se predominantly as methylselenocysteine (MeSeCys), whereas non-accumulator plants store mainly selenate (de Souza et al. 1998; Freeman et al. 2006), indicating differences in Se metabolic pathways. Selenium hyper-accumulator plants also appear to have specialized Se sequestrating cells in the leaf epidermis or leaf hairs, and about 90% of the accumulated Se is present as MeSeCys in these specialized cells (Freeman et al. 2006). This may indicate that the hyper-accumulators have special transport pathways for MeSeCys into the specialized cells. Although the special transport pathways in hyper-accumulator plants are still unclear, the finding that higher levels of ethylene and JA correlate with higher Se resistance and accumulation may give insight into the evolution of Se hyper-accumulators from non-accumulator plants.

3.7 Application of Se-Responsive Genes to Plant-Based System for Detecting Selenate in the Environment

To prevent the release of toxic levels of Se to the environment, monitoring systems as well as efficient remediation techniques need to be developed (Salt et al. 1998; Kovalchuk et al. 2001; Krizek et al. 2003; Peuke and Rennenberg 2005; Pilon-Smits 2005; Pilon-Smits et al. 2009). The common analytical methods for determining trace Se in environmental samples include inductively-coupled plasma mass spectrometry (ICP-MS) or hydride generation atomic absorption spectrometry (HG-AAS). However, a simpler and less expensive analytical method is required to determine the risk of Se contamination in the environment. Application of plant-based biomarkers for Se contamination will take advantage of the sedentary nature of plants, providing an inexpensive and low-tech means of environmental analysis and management (Kovalchuk et al. 2001; Krizek et al. 2003).

Knowledge about Se-responsive genes has been accumulated mainly from the transcriptome analysis of Se-treated *A. thaliana*, as described above (Tamaoki et al. 2008a; Van Hoewyk et al. 2008). The positive relationships between Se treatment and the expression of these Se-responsive genes enable us to construct a system that traces their expression for quantitative monitoring of Se in the environment. Construction of the monitoring system includes selection of Se-responsive genes, determination of the Se-responsive regions in their promoters, design and construction of the fusion genes, generation of transgenic plants, and detection of Se content using those transgenic plants. The sensitivity of the system relies on the Se-responsive regions and the reporter genes selected for the fusion gene constructs.

In the environment, Se exists mainly as selenate. Selenate is imported into plant cells through the activities of sulfate transporters and competitively inhibits the influx of sulfate (Shibagaki et al. 2002; Kassis et al. 2007). Based on this physiology, studies in *Arabidopsis* revealed that several sulfate transporters (SULTRs), including

SULTR1;2 and *SULTR2;1*, are highly responsive to selenate treatment as well as to S deficiency (Takahashi et al. 2000; Yoshimoto et al. 2002; Zhang et al. 2006b; Kassis et al. 2007; Van Hoewyk et al. 2008). *SULTR1;2* is a high-affinity sulfate transporter localized to the root epidermis and cortex (Shibagaki et al. 2002; Yoshimoto et al. 2002) and is responsible for the initial uptake of sulfate from the environment (Shibagaki et al. 2002; Yoshimoto et al. 2002; Maruyama-Nakashita et al. 2003). Induction of *SULTR1;2* expression under S deficiency is controlled by the promoter activities of the 5′-region, which contributes to plant survival by maximizing sulfate uptake (Shibagaki et al. 2002; Yoshimoto et al. 2002, 2007; Maruyama-Nakashita et al. 2004). In contrast to *SULTR1;2*, the S deficiency-induced expression of *SULTR2;1* in roots is controlled by the *cis*-acting elements in the 3′-non-transcribed intergenic region (Maruyama-Nakashita et al. 2015). *SULTR2;1* is a low-affinity sulfate transporter expressed in the vascular tissues of plants and contributes to the root-to-shoot transport of sulfate (Takahashi et al. 1997, 2000; Kawashima et al. 2011).

By using the regulatory sequences of *SULTR1;2* and *SULTR2;1*, the model systems for detecting and quantifying Se levels by measuring the accumulation of green fluorescent protein (GFP) in transgenic plants were developed (Maruyama-Nakashita et al. 2007, 2016). Two transgenic lines expressing GFP under the control of the S-responsive promoter region of *SULTR1;2* ($P_{SULTR1;2}$) in combination with nopaline synthase terminator (T_{NOS}) or another S-responsive region of the sulfate transporter *SULTR2;1* located in the 3′-non-transcribed intergenic region ($T_{SULTR2;1}$) were generated and named as $P_{SULTR1;2}$-*GFP*-T_{NOS} and $P_{SULTR1;2}$-*GFP*-$T_{SULTR2;1}$, respectively. In this system, plants were vertically grown on an agar medium for 6 days and then transferred to media containing various concentrations of selenate. Two days after the transfer, the expression of GFP in plant roots was quantified using plant images obtained by GFP imaging/using a fluorescence image analyzer (Fig. 3.3).

The detection limit of selenate was 10 μmol/L when $P_{SULTR1;2}$-*GFP*-T_{NOS} was used (Maruyama-Nakashita et al. 2007), whereas the detection limit was reduced to 1 μmol/L by the use of $P_{SULTR1;2}$-*GFP*-$T_{SULTR2;1}$ (Maruyama-Nakashita et al. 2016). A concentration-dependent increase in the GFP levels was observed between 1 and 30 μmol/L in $P_{SULTR1;2}$-*GFP*-T_{NOS} and between 0.3 and 10 μmol/L in $P_{SULTR1;2}$-*GFP*-$T_{SULTR2;1}$ (Maruyama-Nakashita et al. 2007, 2016), respectively, which allowed us to estimate the external concentration of selenate. These results suggest their potential use for quantifying selenate in the environment. The minimum effluent standards of Se is determined as 0.1 mg/L in most countries, which is equivalent to 1.3 μmol/L of selenate, indicating that this method can be used for monitoring the Se concentration in the effluent. However, in the case of drinking water, a current limitation of this method is still the detection limit: the recommended maximum allowable concentration of Se in drinking water is 0.01 mg/L (WHO 2011), which is equivalent to 0.13 μmol/L of selenate, i.e. higher than the detection limit of this system. Perhaps more sensitive indicators could be designed by arranging the promoter regions and reporter proteins.

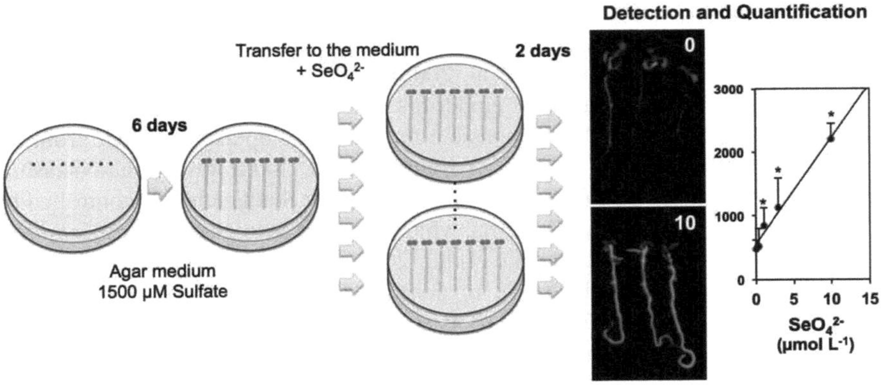

Fig. 3.3 Procedure of Se detection using transgenic plants. Transgenic *Arabidopsis* plants containing the fusion gene constructs were vertically grown on mineral nutrient media supplied with sufficient sulfate. Six-day-old seedlings were transferred to the media with or without selenate and grown for a further 2 days. The GFP signal in intact *Arabidopsis* roots was increased in 10 μM selenate-treated plants relative to control plants. Then, a part of root GFP image was surrounded by the same size of rectangles and the value of GFP intensities within the rectangles were subtracted by the intensity value obtained with a same size of rectangle put on the media without plant and used as the relative fluorescence intensities of each plant

In addition to the detection limit, there are several problems that we need to overcome in future. One is the specificity of this system. $P_{SULTR1;2}$ and $T_{SULTR2;1}$ are induced not only by selenate treatment but also by S deficiency, chromate treatment, and other possible stresses (Maruyama-Nakashita et al. 2007, 2016; Rouached et al. 2008, 2009; Yamaguchi et al. 2016). Another problem is the root-specific expression of GFP in transgenic plants. Furthermore, for easier bio-monitoring of selenate contamination in soil, use of shoot tissues is more convenient than using roots. The above-mentioned problems can be overcome by the selection of sequences that are specifically induced by selenate in shoots. The accumulated knowledge from earlier transcriptome studies allows the mining of potential candidates of sequences most suitable for bio-monitoring of selenate (Tamaoki et al. 2008a; Van Hoewyk et al. 2008). As the plant-based monitoring system depends on protein synthesis of the reporter proteins as well as their transcript levels, this system would not quantify selenate levels so high as to inhibit protein synthesis.

References

Cao H, Glazebrook J, Clarke JD, Volko S, Dong X (1997) The Arabidopsis *NPR1* gene that control systemic acquired resistance encodes a novel protein containing ankyrin repeats. Cell 88:57–63

Clarke JD, Volko SM, Ledford H, Ausubel FM, Dong X (2000) Roles of salicylic acid, jasmonic acid, and ethylene in *cpr*-induced resistance in *Arabidopsis*. Plant Cell 12:2175–2190

Correa-Aragunde N, Graziano M, Chevalier C, Lamattina L (2006) Nitric oxide modulates the expression of cell cycle regulatory genes during lateral root formation in tomato. J Exp Bot 57:581–588

De Bruyne L, Höfte M, De Vleesschauwer D (2014) Connecting growth and defense: the emerging roles of brassinosteroids and gibberellins in plant innate immunity. Mol Plant 7:943–959

Denancé N, Sánchez-Vallet A, Goffner D, Molina A (2013) Disease resistance or growth: the role of plant hormones in balancing immune responses and fitness costs. Front Plant Sci 4:155

de Souza MP, Pilon-Smits EAH, Lytle CM, Hwang S, Tai JC, Honma TSU, Yeh L, Terry N (1998) Rate-limiting steps in selenium volatilization by *Brassica juncea*. Plant Physiol 117:1487–1494

Dong X (1998) SA, JA, ethylene, and disease resistance in plants. Curr Opin Plant Biol 1:316–323

Durner J, Shah J, Klessig DF (1997) Salicylic acid and disease resistance in plants. Trends Plant Sci 2:162–165

Ellis DR, Sors TG, Brunk DG, Albrecht C, Orser C, Lahner B, Wood KV, Harris HH, Pickering IJ, Salt DE (2004) Production of Se-methylselenocysteine in transgenic plants expressing selenocysteine methyltransferase. BMC Plant Biol 4:1–11

Eustice DC, Kull FJ, Shrift A (1981) Selenium toxicity: aminoacylation and peptide bond formation with selenomethionine. Plant Physiol 67:1054–1058

Feechan A, Kwon E, Yun B-W, Wang Y, Pallas JA, Loake GJ (2005) A central role for *S*-nitrosothiols in plant disease resistance. Proc Nat Acad Sci USA 102:8054–8059

Feng J, Wang C, Chen Q, Chen H, Ren B, Li X, Zuo J (2013) *S*-nitrosylation of phosphotransfer proteins represses cytokinin signaling. Nat Commun 4:152910

Freeman JL, Zhang LH, Marcus MA, Fakra S, McGrath SP, Pilon-Smits EAH (2006) Spatial imaging, speciation and quantification of selenium in the hyperaccumulator plants *Astragalus bisulcatus* and *Stanleya pinnata*. Plant Physiol 142:124–134

Freeman JL, Tamaoki M, Stushnoff C, Quinn CF, Cappa JJ, Devonshire J, Fakra SC, Marcus MA, McGrath SP, Van Hoewyk D, Pilon-Smits EAH (2010) Molecular mechanisms of selenium tolerance and hyperaccumulation in *Stanleya pinnata*. Plant Physiol 153:1630–1652

Gomes-Jr RA, Gratão PL, Gaziola SA, Mazzafera PM, Lea PJ, Azevedo RA (2007) Selenium-induced oxidative stress in coffee cell suspension cultures. Funct Plant Biol 34:449–456

Gromer S, Gross JH (2002) Methylseleninate is a substrate rather than an inhibitor of mammalian thioredoxin reductase. Implications for the antitumor effects of selenium. J Biol Chem 277:9701–9706

Gupta V, Willits MG, Glazebrook J (2000) *Arabidopsis thaliana* EDS4 contributes to salicylic acid (SA)-dependent expression of defense responses: evidence for inhibition of jasmonic acid signaling by SA. Mol Plant-Microbe Interact 13:503–511

Guzmán P, Ecker JR (1990) Exploiting the triple response of Arabidopsis to identify ethylene-related mutants. Plant Cell 2:513–523

Hamilton SJ (2004) Review of selenium toxicity in the aquatic food chain. Sci Total Environ 326:1–31

Hansen D, Duda PJ, Zayed AM, Terry N (1998) Selenium removal by constructed wetlands: role of biological volatilization. Environ Sci Technol 32:591–597

Hira CK, Partal K, Dhillon K (2004) Dietary selenium intake by men and women in high and low selenium areas of Punjab. Public Health Nutr 7:39–43

Kachroo P, Shanklin J, Shah J, Whittle EJ, Klessig DF (2001) A fatty acid desaturase modulates the activation of defense signaling pathways in plants. Proc Natl Acad Sci USA 98:9448–9453

Kassis EE, Cathala N, Rouached H, Fourcroy P, Berthomieu P, Terry N, Davidian JC (2007) Characterization of a selenate-resistant Arabidopsis mutant. Root growth as a potential target for selenate toxicity. Plant Physiol 143:1231–1241

Kawashima CG, Matthewman CA, Huang S, Lee BR, Yoshimoto N, Koprivova A, Rubio-Somoza I, Todesco M, Rathjen T, Saito K, Takahashi H, Dalamay T, Kopriva S (2011) Interplay of SLIM1 and miR395 in the regulation of sulfate assimilation in *Arabidopsis*. Plant J 66:863–876

Kazan K, Lyons R (2014) Intervention of phytohormone pathways by pathogen effectors. Plant Cell 26:2285–2309

Keller T, Damude HG, Werner D, Doerner P, Dixon RA, Lamb C (1998) A plant homologue of the neutrophil NADPH oxidase $gp91^{phox}$ subunit gene encodes a plasma membrane protein with Ca^{2+} binding motifs. Plant Cell 10:255–266

Kloek AP, Verbsky ML, Sharma SB, Schoelz JE, Vogel J, Klessig DF, Kunkel BN (2001) Resistance to *Pseudomonas syringae* conferred by an *Arabidopsis thaliana* coronatine-insensitive (*coi1*) mutation occurs through two distinct mechanisms. Plant J 26:509–522

Kovalchuk O, Titov V, Hohn B, Kovalchuk I (2001) A sensitive transgenic plant system to detect toxic inorganic compounds in the environment. Nat Biotechnol 19:568–572

Krizek BA, Prost V, Joshi RM, Stoming T, Glenn TC (2003) Developing transgenic Arabidopsis plants to be metal-specific bioindicators. Environ Toxicol Chem 22:175–181

Läuchli A (1993) Selenium in plants: uptake, functions, and environmental toxicity. Bot Acta 106:455–468

Lawton KA, Potter SL, Uknes S, Ryals J (1994) Acquired resistance signal transduction in *Arabidopsis* is ethylene independent. Plant Cell 6:581–588

LeDuc DL, Tarun AS, Montes-Bayon M, Meija J, Malit MF, Wu CP, AbdelSamie M, Chiang C-Y, Tagmount A, deSouza MP, Neuhierl B, Böck A, Caruso JA, Terry N (2004) Overexpression of selenocysteine methyltransferase in Arabidopsis and indian mustard increases selenium tolerance and accumulation. Plant Physiol 135:377–383

Lehotai N, Kolbert Z, Petõ A, Feigl G, Ördög A, Kumar D, Tari I, Erdei L (2012) Selenite-induced hormonal and signalling mechanisms during root growth of *Arabidopsis thaliana* L. J Exp Bot 63:5677–5687

Leslie CA, Romani RJ (1988) Inhibition of ethylene biosynthesis by salicylic acid. Plant Physiol 88:833–837

Lorenzo O, Solano R (2005) Molecular players regulating the jasmonate signaling network. Curr Opin Plant Biol 8:532–540

Mähönen AP, ten Tusscher K, Siligato R, Smetana O, Díaz-Triviño S, Salojärvi J, Wachsman G, Prasad K, Heidstra R, Scheres B (2014) PLETHORA gradient formation mechanism separates auxin responses. Nature 515:125–129

Maruyama-Nakashita A (2016) Combinatorial use of sulfur-responsive regions of sulfate transporters provides a highly sensitive plant-based system for detecting selenate and chromate in the environment. Soil Sci Plant Nutr 62:386–391

Maruyama-Nakashita A, Inoue E, Watanabe-Takahashi A, Yamaya T, Takahashi H (2003) Transcriptome profiling of sulfur-responsive genes in *Arabidopsis* reveals global effect on sulfur nutrition on multiple metabolic pathways. Plant Physiol 132:597–605

Maruyama-Nakashita A, Nakamura Y, Yamaya T, Takahashi H (2004) A novel regulatory pathway of sulfate uptake in *Arabidopsis* roots: implication of CRE1/WOL/AHK4-mediated cytokinin-dependent regulation. Plant J 38:779–789

Maruyama-Nakashita A, Inoue E, Saito K, Takahashi H (2007) Sulfur-responsive promoter of sulfate transporter gene is potentially useful to detect and quantify selenate and chromate. Plant Biotechnol 24:261–263

Maruyama-Nakashita A, Watanabe-Takahashi A, Inoue E, Yamaya T, Saito K, Takahashi H (2015) Sulfur-responsive elements in the 3′-nontranscribed intergenic region are essential for the induction of SULFATE TRANSPORTER 2;1 gene expression in Arabidopsis roots under sulfur deficiency. Plant Cell 27:1279–1296

McNeal JM, Balisteri LS (1989) Geochemistry and occurrence of selenium: an overview. Selenium in agriculture and the environment, Special publication no. 23. Soil Science Society of America, Madison

Mirouze M, Sels J, Richard O, Czernic P, Loubet S, Jacquier A, François IEJA, Cammue BPA, Lebrun M, Berthomieu P, Marquès L (2006) A putative novel role for plant defensins: a defensin from the zinc hyper-accumulating plant, *Arabidopsis halleri*, confers zinc tolerance. Plant J 47:329–342

Neuhierl B, Thanbichler M, Lottspeich F, Böck A (1999) A family of S-methylmethionine-dependent thiol/selenol methyltransferases. Role in selenium tolerance and evolutionary relation. J Biol Inorg Chem 274:5407–5414

Overmyer K, Broschè M, Kangasjärvi J (2003) Reactive oxygen species and hormonal control of cell death. Trends Plant Sci 8:335–342

Overvoorde P, Fukaki H, Beeckman T (2010) Auxin control of root development. Cold Spring Harb Perspect Biol 2:a001537

Penninckx IAMA, Thomma BPHJ, Buchla A, Métraux J-P, Broekaert WF (1998) Concomitant activation of jasmonate and ethylene response pathway is required for induction of a plant defensin gene in *Arabidopsis*. Plant Cell 10:2103–2114

Petersen M, Brodersen P, Naested H, Andreasson E, Lindhart U, Johansen B, Nielsen HB, Lacy M, Austin MJ, Parker JE, Sharma SB, Klessig DF, Martienssen R, Mattsson O, Jensen AB, Mundy J (2000) *Arabidopsis* map kinase 4 negatively regulates systemic acquired resistance. Cell 103:1111–1120

Peuke AD, Rennenberg H (2005) Phytoremediation: molecular biology, requirements for application, environmental protection, public attention and feasibility. EMBO Rep 6:497–501

Pieterse CMJ, Leon-Reyes A, Van der Ent S, Van Wees SCM (2009) Networking by small-molecules hormones in plant immunity. Nat Chem Biol 5:308–316

Pilon-Smits EAH (2005) Phytoremediation. Annu Rev Plant Biol 56:15–39

Pilon-Smits EAH, Hwang S, Lytle CM, Zhu YL, Tai JC, Bravo RC, Chen Y, Leustek T, Terry N (1999) Overexpression of ATP sulfurylase in Indian mustard leads to increased selenate uptake, reduction, and tolerance. Plant Physiol 119:123–132

Pilon-Smits EAH, Quinn CF, Tapken W, Malagoli M, Schiavon M (2009) Physiological functions of beneficial elements. Curr Opin Plant Biol 12:267–274

Rouached H, Wirtz M, Alary R, Hell R, Arpat AB, Davidian JC, Fourcroy P, Berthomieu P (2008) Differential regulation of the expression of two high-affinity sulfate transporters, SULTR1.1 and SULTR1.2, in Arabidopsis. Plant Physiol 147:897–911

Rouached H, Secco D, Arpat AB (2009) Getting the most sulfate from soil: regulation of sulfate uptake transporters in Arabidopsis. J Plant Physiol 166:893–902

Salt DE, Smith RD, Raskin I (1998) Phytoremediation. Annu Rev Plant Physiol Plant Mol Biol 49:643–668

Sasaki-Sekimoto Y, Taki N, Obayashi T, Aono M, Matsumoto F, Sakurai N, Suzuki H, Yokota-Hirai M, Noji M, Saito K, Masuda T, Takamiya K, Shibara D, Ohta H (2005) Coordinated activation of metabolic pathway for antioxidants and defence compounds by jasmonates and their roles in stress tolerance in Arabidopsis. Plant J 44:653–668

Shahzad Z, Ranwez V, Fizames C, Marquès L, Le Martret B, Alassimone J, Godé C, Lacombe E, Castillo T, Saumitou-Laprade P, Berthomieu P, Gosti F (2013) Plant Defensin type 1 (PDF1): protein promiscuity and expression variation within the Arabidopsis genus shed light on zinc tolerance acquisition in *Arabidopsis halleri*. New Phytol 200:820–833

Shibagaki N, Rose A, McDermott JP, Fujiwara T, Hayashi H, Yoneyama T, Davies JP (2002) Selenate-resistant mutants of *Arabidopsis thaliana* identify *Sultr1;2*, a sulfate transporter required for efficient transport of sulfate into roots. Plant J 29:475–486

Sors TG, Ellis DR, Salt DE (2005) Selenium uptake, translocation, assimilation and metabolic fate in plants. Photosynth Res 86:373–389

Stadtman TC (1990) Selenium biochemistry. Annu Rev Biochem 59:111–127

Staswick PE, Su W, Howell SH (1992) Methyl jasmonate inhibition of root growth and induction of a leaf protein are decreased in an *Arabidopsis thaliana* mutant. Proc Natl Acad Sci USA 89:6837–6840

Stenzel I, Hause B, Miersch O, Kurz T, Mauzher H, Weichert H, Ziegler J, Feussner I, Wasternack C (2004) Jasmonate biosynthesis and the allene oxide cyclase family of *Arabidopsis thaliana*. Plant Mol Biol 51:895–911

Takahashi H, Yamazaki M, Sasakura N, Watanabe A, Leustek T, Engler JA, Engler G, Van Montagu M, Saito K (1997) Regulation of sulfur assimilation in higher plants: a sulfate transporter induced in sulfate starved roots plays a central role in *Arabidopsis thaliana*. Proc Natl Acad Sci USA 94:11102–11107

Takahashi H, Watanabe-Takahashi A, Smith FW, Blake-Kalff M, Hawkesford MJ, Saito K (2000) The roles of three functional sulfate transporters involved in uptake and translocation of sulfate in *Arabidopsis thaliana*. Plant J 23:171–182

Tamaoki M (2008) The role of phytohormone signaling in ozone-induced cell death in plants. Plant Signal Behav 3:166–174

Tamaoki M, Nakajima N, Kubo A, Aono M, Matsuyama T, Saji H (2003) Transcriptome analysis of O_3-exposed Arabidopsis reveals that multiple signal pathways act mutually antagonistically to induce gene expression. Plant Mol Biol 53:443–456

Tamaoki M, Freeman JL, Pilon-Smits EAH (2008a) Cooperative ethylene and jasmonic acid signaling regulates selenite resistance in *Arabidopsis thaliana*. Plant Physiol 146:1219–1230

Tamaoki M, Freeman JL, Marquès L, Pilon-Smits EAH (2008b) New insights into the roles of ethylene and jasmonic acid in the acquisition of selenium resistance in plants. Plant Signal Behav 3:865–867

Terry N, Zayed AM, de Souza MP, Tarun AS (2000) Selenium in higher plants. Annu Rev Plant Physiol Plant Mol Biol 51:401–432

Torres MA, Onouchi H, Hamada S, Machida C, Hammond-Kosack KE, Jones JD (1998) Six *Arabidopsis thaliana* homologues of the human respiratory burst oxidase (gp^{91phox}). Plant J 14:365–370

Van Hoewyk D, Takahashi H, Inoue E, Hess A, Tamaoki M, Pilon-Smits EAH (2008) Transcriptome analyses give insights into selenium-stress responses and selenium tolerance mechanisms in *Arabidopsis*. Physiol Plant 132:236–253

Vanneste S, Friml J (2009) Auxin: a trigger for change in plant development. Cell 136:1005–1016

Werner T, Nehnevajova E, Köllmer I, Novák O, Strnad M, Krämer U, Schmülling T (2010) Root-specific reduction of cytokinin causes enhanced root growth, drought tolerance, and leaf mineral enrichment in Arabidopsis and tobacco. Plant Cell 22:3905–3920

WHO (2011) Guidelines for drinking-water quality: forth edition. ISBN: 978 92 4 154815 1.

Wilber CG (1980) Toxicology of selenium: a review. Clin Toxicol 17:171–230

Wildermuth MC, Dewdney J, Wu G, Ausubel FM (2001) Isochorismate synthase is required to synthesize salicylic acid for plant defense. Nature 414:562–565

Wilkinson JQ, Crawford NM (1993) Identification and characterization of a chlorate resistant mutant of Arabidopsis with mutations in both NIA1 and NIA2 nitrate reductase structural genes. Mol Gen Genet 239:289–297

Woodward AW, Bartel B (2005) Auxin: regulation, action, and interaction. Ann Bot 95:707–735

Yalpani N, Enyedi AJ, León J, Raskin I (1994) Ultraviolet light and ozone stimulate accumulation of salicylic acid, pathogenesis-related proteins and virus resistance in tobacco. Planta 193:372–376

Yamaguchi C, Takimoto Y, Ohkama-Ohtsu N, Hokura A, Shinano T, Nakamura T, Suyama A, Maruyama-Nakashita A (2016) Effects of cadmium treatment on the uptake and translocation of sulfate in *Arabidopsis thaliana*. Plant Cell Physiol 57:2353–2366

Yoshimoto N, Takahashi H, Smith FW, Yamaya T, Saito K (2002) Two distinct high-affinity sulfate transporters with different inducibilities mediate uptake of sulfate in *Arabidopsis* roots. Plant J 29:465–473

Yoshimoto N, Inoue E, Watanabe-Takahashi A, Saito K, Takahashi H (2007) Posttranscriptional regulation of high-affinity sulfate tansporters in Arabidopsis by sulfur nutrition. Plant Physiol 145:378–388

Zarembinski TI, Theologis A (1994) Ethylene biosynthesis and action: a case of conservation. Plant Mol Biol 26:1579–1597

Zhang LH, Byrne PF, Pilon-Smits EAH (2006a) Mapping quantitative trait loci associated with selenate tolerance in *Arabidopsis thaliana*. New Phytol 170:33–42

Zhang LH, Abdel-Ghany SE, Freeman JL, Ackley AR, Schiavon M, Pilon-Smits EAH (2006b) Investigation of selenium tolerance mechanism in *Arabidopsis thaliana*. Physiol Plant 128:212–223

Zhang LH, Ackley AR, Pilon-Smits EAH (2007) Variation in selenium tolerance and accumulation among 19 *Arabidopsis thaliana* accessions. J Plant Physiol 164:327–336

Chapter 4
Mechanisms of Plant Selenium Hyperaccumulation

Elizabeth A.H. Pilon-Smits

Abstract Selenium hyperaccumulator plants can accumulate Se to at least 0.1% of dry weight while growing on naturally seleniferous soil. Selenium hyperaccumulation has been reported for 45 taxa from six dicot families; they are perennials native to seleniferous areas, predominantly in western North America. Compared to other plants, hyperaccumulators are characterized by 10–100× higher Se levels and higher Se to sulfur (S) ratios, suggestive of a transporter with a preference for Se over S. Furthermore, hyperaccumulators have higher organic/inorganic Se ratios (i.e. enhanced selenate assimilation). Hyperaccumulators also have higher shoot/root Se ratios (i.e. higher xylem translocation), higher source/sink Se ratios (i.e. higher phloem translocation), and their patterns of spatial and temporal Se sequestration are different from non-accumulators, and different from S patterns. Transcriptomic and biochemical investigations into the mechanisms of Se hyperaccumulation indicate that hyperaccumulators have constitutive high expression of several sulfate/selenate transporters that likely mediate Se uptake and translocation. They also have enhanced transcript levels of several enzymes in the sulfate/selenate assimilation pathway. Hyperaccumulators also have elevated selenocysteine methyltransferase (SMT) levels, whose product is the main form accumulated, methyl-selenocysteine. This form is sequestered in hyperaccumulators mainly in epidermis and reproductive tissues. Transcriptomic and biochemical analyses indicate constitutively elevated levels of the hormones jasmonic acid, salicylic acid and ethylene, which may explain the constitutive upregulation of sulfate uptake and assimilation. Hyperaccumulators also have higher transcript levels of genes involved in oxidative stress resistance and defense against biotic stress, which may contribute to Se tolerance and are upregulated by the same stress/defense hormones.

Keywords Hyperaccumulation • Molecular mechanisms • Evolution • Ecology

E.A.H. Pilon-Smits (✉)
Biology Department, Colorado State University, Fort Collins, CO 80523-1878, USA
e-mail: epsmits@colostate.edu

E.A.H. Pilon-Smits et al. (eds.), *Selenium in plants*, Plant Ecophysiology 11,
DOI 10.1007/978-3-319-56249-0_4

4.1 Introduction to Se Hyperaccumulation

All plants can take up and metabolize Se via sulfur (S) transporters and pathways, due to the chemical similarity between Se and S (Anderson 1993). There is variation, however, in the degree to which they accumulate Se. Three classes can be distinguished: Se hyperaccumulators, (secondary) Se accumulators and non-accumulators (Brown and Shrift 1982; Läuchli 1993). Plant Se hyperaccumulation has been defined to have a threshold of 0.1% Se of plant dry weight (DW) or 1000 mg per kg DW (ppm) while growing in a natural setting, typically on seleniferous soil containing 0.1–10 mg Se per kg. Thus, hyperaccumulators bioconcentrate Se around 1000-fold. Non-accumulator vegetation on seleniferous soils contain <100 mg/kg Se, and plants that accumulate between 100 and 1000 mg/kg Se in nature are considered Se accumulators or secondary Se accumulators (Anderson 1993; Läuchli 1993).

The general mechanisms of Se uptake, assimilation and accumulation in plants are the same processes used for S. Species differ in the extent to which these processes occur, resulting in differences in uptake into the root, translocation to the shoot, remobilization to sink organs, assimilation to organic forms, sequestration in particular tissues, and volatilization. Secondary Se accumulators such as *Brassica* and *Allium* species often have elevated levels of both Se and S, as compared to non-accumulators, but the Se/S ratio is the same for both groups and reflects the Se/S ratio in the growth substrate (White et al. 2007). The differences between both groups appear to be mainly quantitative, and the elevated Se levels a side effect of the elevated S levels. Selenium hyperaccumulators have several qualities that make them truly different from non-hyperaccumulator plants. Not only do they take up more Se but they also take up Se preferentially over S, assimilate most of the Se into organic forms, mobilize Se more between organs, and store Se in special tissues. Interesting questions related to Se hyperaccumulation are: How do plants hyperaccumulate? Why do plants hyperaccumulate? What implications does hyperaccumulation have for the local ecology and for Se fluxes in ecosystems? This chapter will focus on the *how*, i.e. the mechanisms of Se hyperaccumulation. As such, it sets the stage for later chapters that further explore the evolutionary and ecological questions.

Selenium hyperaccumulation was discovered in the 1930s by Orville Beath and coworkers, who considered the hyperaccumulators indicator species for seleniferous soils (Rosenfeld and Beath 1964). Hyperaccumulation has meanwhile been reported in 45 taxa from 14 genera and 6 families (Cappa and Pilon-Smits 2014; White 2016). Most Se hyperaccumulators (25 taxa) have been reported in two North American clades within the large genus *Astragalus* (Fabaceae); other well-documented hyperaccumulators include *Stanleya pinnata* and *S. bipinnata* (Brassicaceae) and the Asteraceae genera *Oonopsis*, *Xylorhiza* and *Symphyotrichum* (Rosenfeld and Beath 1964; El Mehdawi et al. 2014). These species all occur on naturally seleniferous soils in the Western United States of America. In addition, *Neptunia amplexicaulis* (Mimosaceae) has been reported to hyperaccumulate Se when growing on seleniferous soil in Queensland, Australia (Peterson and Butler 1967). It is worth noting that individuals from hyperaccumulator species do not always contain hyperaccumulator levels of Se in nature; sometimes their levels are quite low (Beath et al. 1939). The reason for this variability remains to be determined, and may be due to local soil Se availability, genetic differences between individuals, or microbiome differences. For this reason, it is reasonable to consider a species a Se hyperaccumulator once it has been reliably documented to have tissue Se levels upwards of 0.1% Se in at least one location in a natural setting. Within a hyperaccumulator species, different populations may also vary genetically in their Se accumulation capacity (Feist and Parker 2001; El Mehdawi et al. 2015a), which may be adaptive. The Se levels in hyperaccumulator species can be as high as 1.5% of DW and are high in all organs (Galeas et al. 2007). Ingestion of such high-Se plant material causes toxicity in both vertebrate and invertebrate herbivores (Wilber, 1980). Grazing on Se hyperaccumulator species has been known since the 1930s to cause toxicity in livestock in the Western USA, and was termed blind staggers and alkali disease (Wilber 1980). Seleniferous soils also occur in other areas of the world including Australia, Brazil, China and India, where the local vegetation also can accumulate toxic levels of Se to not only grazers, but to the human population (Peterson and Butler 1967; Dhillon and Dhillon 2003).

Based on taxonomic distribution, it is likely that Se hyperaccumulation has evolved independently in different clades, perhaps under similar ecological and physiological selection pressures. The benefits to plants of having elevated Se levels may be both physiological and ecological. Selenium is a beneficial element for plants (Pilon-Smits et al. 2009); low tissue Se levels are associated with enhanced growth and enhanced antioxidant activity (Hartikainen 2005), thus offering a physiological advantage. Similar low tissue Se levels (around 10 mg/kg DW) also already offer the ecological advantage of protection from certain herbivores, such as aphids (Hanson et al. 2004). At more elevated tissue levels, Se can protect plants against a wide variety of invertebrate and vertebrate herbivores as well as some fungal pathogens, and may also offer allelopathic benefits (El Mehdawi and Pilon-Smits 2012). There is no evidence of an evolutionary constraint on Se hyperaccumulation, since hyperaccumulators are tolerant to their extreme Se levels, and their mutualistic ecological partners also appear to be Se tolerant (El Mehdawi and Pilon-Smits 2012). In the convergent evolution of Se hyperaccumulation and hypertolerance in different clades, the same or different selection pressures may have driven the evolution of these traits, and may have acted on the same or on different loci. Thus, Se hyperaccumulators from different clades may to some extent share molecular mechanisms, but may also differ. Most studies so far have focused on *Stanleya pinnata* (Brassicaceae) and *Astragalus bisulcatus* (Fabaceae), and more recently *Symphyotrichum ericoides* (Asteraceae). These Se hyperaccumulators from different families show many similarities in physiology and biochemistry, and also some differences. We will review and discuss these in the next sections.

4.2 Plant Se Uptake and Metabolism – How Are Hyperaccumulators Different?

4.2.1 Uptake, Translocation and Remobilization of Selenate

Selenate is the most prevalent bioavailable form of Se under oxidizing conditions like aerated soils; under more reducing conditions such as in wetlands, the most common available form is selenite (Terry et al. 2000; Sors et al. 2005a). Plants are thought to take up selenate via sulfate transporters, and to take up selenite via phosphate transporters or anion channels (Shibagaki et al. 2002; El Kassis et al. 2007; Gigolashvili and Kopriva 2014; Zhang et al. 2014). Non-hyperaccumulator plants do not appear to be able to distinguish between selenate and sulfate. They take them up proportional to their abundance in the growth medium, and incorporate Se and S proportionally into all S compounds. A special feature of hyperaccumulator species is that they appear to be able to discriminate between Se and S and take up Se preferentially over S. As a result, they have higher Se/S ratios in their tissues as compared to their growth medium and to surrounding vegetation (White et al. 2007; Harris et al. 2014). The mechanism of this Se/S discrimination is not known, but may involve a transporter that preferentially transports selenate over sulfate, or that may even be selenate-specific.

In addition to this qualitative difference in selenate transport, there is a quantitative difference. Hyperaccumulators obviously have much higher levels of Se accumulation, which may be due to higher selenate influx rates and/or lack of feed-back inhibition by high tissue S levels. Results from mRNA abundance analysis (macroarrays, RNA sequencing) indicate that the major gene involved in root sulfate/selenate uptake (a type 1 high-affinity sulfate transporter or *Sultr*) is constitutively overexpressed in hyperaccumulators relative to non-hyperaccumulator sister taxa (Freeman et al. 2010; Cabannes et al. 2011; Schiavon et al. 2015). In view of the elevated expression levels of *Sultr* genes, it is not surprising that hyperaccumulators not only have higher levels of Se, but also higher S levels than comparable non-accumulators (Galeas et al. 2007). The molecular mechanisms for the enhanced transcript levels of *Sultr* genes are not known, and may involve mutations in regulatory regions or rather gene duplication, as was found for metal transporters in metal hyperaccumulators (Cappa and Pilon-Smits 2014). This will be an interesting area of further study.

After uptake into the root, selenate and sulfate can be moved across membranes within and between cells, tissues and organs using a variety of high- and low-affinity SULTR proteins (Takahashi et al. 2011). Accumulation of selenate and selenite in metabolically active compartments like the cytosol likely causes oxidative stress (Van Hoewyk 2013). Within cells, selenate/sulfate can be transported to the vacuoles; tonoplast transport is mediated by group 4 SULTR but these are thought to promote efflux. Selenate/sulfate can also be transported to the plastids via a group 3 SULTR (*Sultr3;1* in non-accumulator *Arabidopsis thaliana*) for reductive assimilation (Takahashi et al. 2011; Cao et al. 2013). The first step in reductive assimilation is selenate activation by coupling it to ATP. Some of the ATP sulfurylases that catalyze this step were found to be highly overexpressed in hyperaccumulator *S. pinnata* relative to non-hyperaccumulator *B. juncea* (Schiavon et al. 2015). This likely enhances the rate of selenate reduction to organic forms, as was indeed found when an ATP sulfurylase from *A. thaliana (APS1)* was overexpressed in *B. juncea* (Pilon-Smits et al. 1999). This may explain why *S. pinnata* accumulates organic forms of Se (Freeman et al. 2006) while *B. juncea* accumulates selenate (Pilon-Smits et al. 1999). In a comparison of *Astragalus* Se hyperaccumulators and non-accumulators, no difference was found in activity levels of sulfate reductive assimilation enzymes (Sors et al. 2005a), leading the authors to hypothesize that while these enzymes likely are important for selenate reduction, the flux through this pathway is probably controlled by sink depletion (conversion of Cys to other forms of organic Se).

Enhanced conversion of inorganic to organic Se explains in part the enhanced Se tolerance of hyperaccumulators, since inorganic forms of Se are thought to cause more oxidative stress than organic forms of Se (van Hoewyk 2013). The seleno-aminoacid SeCys (and perhaps also SeMet) can, however, cause toxicity if it is non-specifically incorporated into proteins, replacing Cys (or Met) (Brown and Shrift 1982; Stadtman 1990). Hyperaccumulators avoid this type of toxicity by methylating SeCys via the enzyme SeCys methyltransferase (SMT) (Neuhierl et al. 1999; Sors et al. 2009). The resulting methyl-SeCys can be safely accumulated and constitutes the main (>80%) Se fraction in hyperaccumulators (Pickering et al. 2003; Freeman et al. 2006). Some secondary Se accumulators (*Brassica spp., Allium spp.*) also are able to produce methyl-SeCys (Lyi et al. 2005), although this may be restricted to certain plant parts such as in the flower in *B. juncea* (Quinn et al. 2011). In addition to methyl-SeCys, hyperaccumulator *S. pinnata* accumulates up to 20% selenocystathionine, an intermediate in the pathway between SeCys and SeMet. The Australian Se hyperaccumulator *Neptunia amplexicaulis* also accumulates selenocystathionine (Peterson and Butler 1967). Furthermore, *A. bisulcatus* can accumulate up to 50% γ-glutamyl-methylSeCys in its seeds (Freeman et al. 2006), an intermediate in glutathione biosynthesis. Methyl-SeCys may be further metabolized in hyperaccumulators to volatile dimethyldiselenide, or DMDSe (Terry et al. 2000), which is malodorous and may contribute to herbivore deterrence. Finally, Se tolerance may be mediated in plants by breakdown of SeCys into elemental Se (Se^0) and alanine via a plastid-localized SeCys lyase (Van Hoewyk et al. 2005). This does not appear to be a main Se tolerance mechanism in Se hyperaccumulator species, since they do not consistently accumulate elemental Se. This form of Se has been found in hyperaccumulators while growing in the field, even up to 30% of total Se, but was not found in greenhouse-grown plants and may be due to the activity of fungal or bacterial endophytes (Valdez et al. 2012; Lindblom et al. 2013).

Between organs, Se can be transported via the xylem and phloem. For selenate, this is mediated by various low- and high-affinity SULTR proteins including SULTR1;3, SULTR2;1, SULTR2;2 and SULTR3;5 (Takahashi et al. 2011). Organic forms of Se/S can also move through these long-distance transport routes; for many of these compounds the transporters are unknown. In most plants, Se movement in the xylem is likely in the form of selenate, since reductive sulfate/selenate assimilation is thought to happen predominantly in the shoot, where it depends on chloroplast reducing power (Leustek 1996; White 2016). In hyperaccumulators, there is some evidence that Se reduction may happen to a larger part in the root: first, the form of Se in guttation (xylem) fluid of hyperaccumulator *A. bisulcatus* was found to be methyl-SeCys (Freeman et al. 2006), and second, the transcript levels of ATP sulfurylase and other S assimilation genes were particularly elevated relative to non-accumulator counterparts in the roots of hyperaccumulators, and not so much in the shoots (Freeman et al. 2010; Schiavon et al. 2015).

Hyperaccumulators show more root-to-shoot Se translocation than non-hyperaccumulators (i.e. higher shoot/root Se ratios) and also higher phloem remobilization (i.e. reproductive/vegetative organ Se ratios) (Quinn et al. 2011; Cappa et al. 2014). Moreover, patterns of Se and S translocation via xylem and phloem in hyperaccumulators appear to be independent from patterns of S translocation, while in non-hyperaccumulators Se and S movement show identical patterns (Galeas et al. 2007; Cappa et al. 2014). Further research is needed to determine the underlying mechanisms. Perhaps hyperaccumulators have long-distance transporters that can discriminate between Se and S substrates; alternatively, Se and S may be translocated in different forms (e.g. inorganic vs. organic) in hyperaccumulators, but in the same forms in non-hyperaccumulators. It is also possible that hyperaccumulators and non-hyperaccumulators have different levels of positive or negative S feedback signaling compounds (e.g. sulfate, O-acetylserine, glutatione, miRNA395).

4.2.2 Sequestration of Se

Hyperaccumulators store their (organic) Se in what may be specialized organs and tissues. Their Se is found particularly in young leaves and reproductive organs, and within these the Se is concentrated in the epidermis, in leaf hairs of *A. bisulcatus* and in the vacuoles of epidermal cells in *S. pinnata*, especially along the leaf margins (Freeman et al. 2006, 2010). In flowers, hyperaccumulator *S. pinnata* was found to sequester the highest Se levels in the pistil and anthers, especially in ovules and pollen, while non-hyperaccumulator relative *B. juncea* showed uniform Se distribution throughout this organ (Quinn et al. 2011). If Se serves a protective function in hyperaccumulators, these Se hyperaccumulation patterns may have been selected for because they optimally protect the plant's most sensitive and precious organs (young leaves, gamete-producing structures, seeds, seedlings) and tissues (mesophyll) against Se toxicity and biological attack. The mechanisms involved in the tissue-specific Se sequestration in hyperaccumulators remain to be elucidated, and may involve elevated expression of a seleno-aminoacid transporter in certain phloem loading/unloading tissues and epidermal tissues.

4.2.3 Regulation of Se/S Metabolism

Because Se is metabolized via S transporters and enzymes, the elements Se and S tend to influence each other's uptake and conversion. Selenium and sulfur analogues tend to competitively inhibit each other as enzyme substrates and during plant uptake (Wilson and Bandurski 1958; Harris et al. 2014). However, in most plant species selenate also upregulates genes involved in sulfate uptake and assimilation in plants, which may lead to higher S levels when plants are grown in the presence of Se (Harris et al. 2014). Indeed, transcriptome responses of *A. thaliana* to selenate are similar to the S starvation transcriptomic response (van Hoewyk et al. 2008). Selenium hyperaccumulator species stand out from other plants with respect to Se-S interactions. Selenate more effectively inhibits sulfate uptake, while sulfate is much less capable of inhibiting selenate uptake, and Se supply does not boost plant S levels in hyperaccumulators (Harris et al. 2014). Also, the fluxes between plant organs and across the seasons are different for Se and S in hyperaccumulators (Cappa et al. 2014; Galeas et al. 2007). Furthermore, in most plants, sulfate –and with that, selenate- uptake is regulated by plant S status, upregulated under S starvation and downregulated at S repletion (Lappartient and Touraine 1996). In hyperaccumulators, however, genes involved in sulfate uptake and assimilation appear to be constitutively expressed at high levels, as if the plant senses S starvation continuously (Freeman et al. 2010; Cabannes et al. 2011).

If Se hyperaccumulators are perpetually in S starvation mode, could it be that a signaling pathway that turns on selenate/sulfate transport and assimilation is constitutively activated in these plants? Results so far from transcriptome analyses, in combination with genetic and biochemical/physiological studies, point to the stress/defense plant hormones jasmonic acid (JA), salicylic acid (SA) and ethylene (Freeman et al. 2010). Genes involved in biosynthesis and signaling of these hormones appear constitutively upregulated in *S. pinnata* relative to non-hyperaccumulator relatives. This may lead to enhanced sulfate/selenate uptake and assimilation (Sasaki-Sekimoto et al. 2005). Pathways involved in antioxidant activities and defense were also constitutively upregulated in *S. pinnata* (Freeman et al. 2010), and have also been reported to be upregulated by these hormones (Sasaki-Sekimoto et al. 2005). Higher antioxidant activity may contribute to Se tolerance in the hyperaccumulator. Whether upregulated defense-related proteins may also confer Se tolerance remains to be investigated, but it is interesting to note that overexpression of a plant defensin protein (PDF) was shown to enhance tolerance to selenate (Tamaoki et al. 2008b) and zinc (Shahzad et al. 2013). It is intriguing why hyperaccumulators have constitutively upregulated defense pathways and hormones. There may be a trigger for these pathways that is switched on even in the absence of stress. Further studies are needed to explore the importance of these receptors for Se hyperaccumulation. Based on the findings so far, a model for Se hyperaccumulation is shown in Fig. 4.1.

Further support for the involvement of the hormones JA and ethylene in Se accumulation and tolerance in plants was obtained from studies using the model

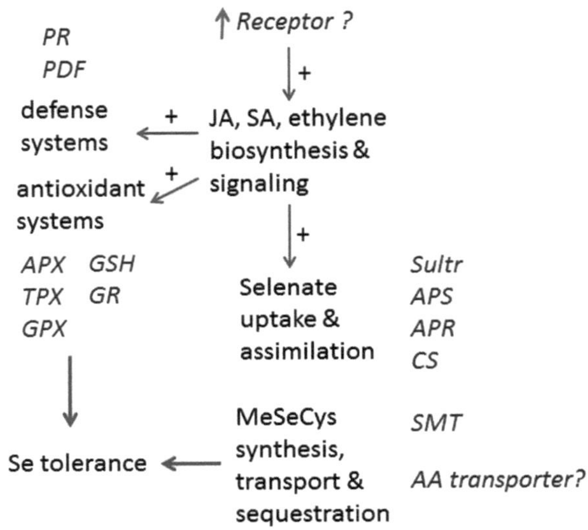

Fig. 4.1 Schematic model of key pathways and genes proposed to mediate Se hyperaccumulation in *S. pinnata*. Constitutive upregulation of a putative defense-related receptor may trigger the defense signaling pathways, leading to increased hormone synthesis and an increase in overall antioxidant activity and S/Se accumulation. SeCys is further methylated by SMT (SeCys methyltransferase), and sequestered by an unknown aminoacid transporter. The MeSeCys production and sequestration, together with the enhanced antioxidant activity likely confer Se tolerance. The enhanded selenate uptake and assiimilation likely confer Se hyperaccumulation. The upregulation of defense proteins may be a side effect, or may contributte to the hyperaccumulation via an unknown mechanism. Abbreviations: *AA* transporter: unknown aminoacid transporter; *APS*: ATP sulfurylase; *APR*: APS reductase; *APX*: ascorbate peroxidase; *CS*: cysteine synthase; *DMDSe*: dimethyldiselenide (volatile); *GPX*: glutathione peroxidase; *GR*: glutathione reductase; *GSH*: glutatione synthetase; *JA*: jasmonic acid (hormone) *PDF*: plant defensin; *PR*: pathogenesis related protein; *SA*: salicylic acid (hormone); *SMT*: selenocysteine synthase; *Sultr*: sulfate transporter; *TRX*: thioredoxin reductase; *MeSeCys*: methylselenocysteine

Brassicaceae species *A. thaliana*. Mutants or transgenics that had reduced or elevated levels or signaling of these hormones showed reduced or elevated Se resistance, and external supplementation with these hormones led to elevated Se resistance (Tamaoki et al. 2008a). Thus, hyperaccumulator and non-accumulator relatives within the Brassicaceae share Se response and tolerance mechanisms. These overall similarities should facilitate manipulation of Se responses in non-accumulator Brassicaceae crops using genes and insight gained from hyperaccumulator relative *S. pinnata*. In the next section the similarities and differences in Se biochemistry and physiology among Brassiceae species is integrated into a model for the evolution of Se hyperaccumulation in this clade.

Hyperaccumulators were shown to have differing seasonal fluxes for Se and S, while these were the same in non-hyperaccumulator species (Galeas et al. 2007). Selenium levels in hyperaccumulators peaked in the spring and decreased over the course of the growth season, while their S levels peaked in the summer; both Se and

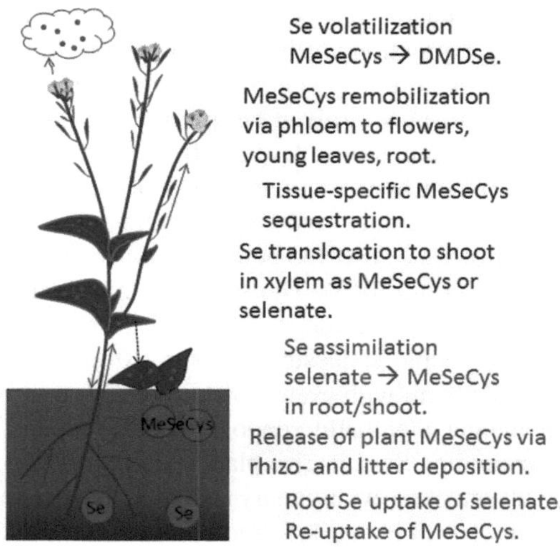

Fig. 4.2 Selenium fluxes and metabolic transformation in Se hyperaccumulators. *Red circles* represent Se. *MeSeCys* methylselenocysteine, *DMDSe* dimethyldiselenide (volatile)

S peaked in summer in non-hyperaccumulators. Based on their seasonal fluctuations in Se levels in different organs over the course of the seasons, it appears that the -perennial- hyperaccumulators store Se in their roots and leaf buds in the winter, and mobilize this stored Se in the spring (in addition to newly taken up Se from the soil) from the root to the growing shoot via the xylem. In the summer, this Se is remobilized via the phloem from source leaves to young leaves and reproductive structures, particularly seeds. In the fall, Se is remobilized to some extent back to the root via the phloem, and some is deposited back to the soil in leaf litter. Some Se is also deposited from live roots or via root turnover (El Mehdawi et al. 2012). In the process of sequestration and redeposition of Se by hyperaccumulators, Se is converted from inorganic to organic form. This may affect soil Se speciation, as evidenced in several studies (Beath et al. 1946; El Mehdawi et al. 2015b).

Figure 4.2 summarizes Se cycling (form, distribution) at the whole plant level. These processes are relevant when considering the ecological importance of Se hyperaccumulators in Se cycling in their local ecosystem.

4.3 How May Hyperaccumulation Have Evolved?

Hyperaccumulation of Se likely has evolved independently in at least three different plant lineages (Brassicaceae, Fabaceae, Asteraceae) as a new trait (Cappa et al. 2015). Individuals with gene-based elevated Se levels may have been selected for in evolution through physiological and ecological benefits, particularly protection from Se-sensitive herbivores. Selenium accumulator species such as *B. juncea* may be evolutionary intermediates between Se non-accumulators such as *A. thaliana* and true hyperaccumulators such as *S. pinnata*. The first step from non-accumulator to

Se accumulator may be quantitative rather than qualitative, and mediated simply by higher expression levels of certain genes, in particular sulfate transporters. For the second evolutionary step, from Se accumulator to hyperaccumulator, Se hypertolerance must have evolved before or concomitant with Se hyperaccumulation, since hyperaccumulator Se levels (>0.1% of DW) impair growth and reproductive functions in non-hyperaccumulators (Prins et al. 2011). Indeed, analysis of evolution of Se hypertolerance and hyperaccumulation in the genus *Stanleya* indicated that hypertolerance evolved before hyperaccumulation: it is more prevalent in the genus (Cappa et al. 2015). The tolerance mechanisms that evolved in this step include metabolic conversion to less toxic forms (particularly methyl-SeCys) as well as tissue-specific sequestration. The extreme Se levels in hyperaccumulators likely provide them with even broader protection from herbivores, and may also mediate elemental allelopathy (deposition of Se around hyperaccumulators may be toxic to surrounding Se-sensitive plant species, El Mehdawi et al. 2011). So far, there is no evidence of an evolutionary cost of Se hyperaccumulation in Se hyperaccumulators: plant growth is not impaired but rather promoted by Se in hyperaccumulators, and reproductive functions and pollinator visitation are not impaired, nor is nodulation or endophyte colonization (Prins et al. 2011; Quinn et al. 2011; El Mehdawi et al. 2012; Alford et al. 2012; Sura-de Jong et al. 2015).

4.4 Outlook: Can We Transfer Hyperaccumulation to a Crop Species?

Selenium hyperaccumulators have several traits that would be attractive to transfer to crop species: the ability to take up selenate uninhibited by sulfate (so high S levels do not impede Se accumulation), high Se accumulation capacity, the capacity to accumulate anticarcinogenic methyl-SeCys, and high capacity to volatilize Se. These can all be attractive traits to transfer to crop species, if we have knowledge of the underlying mechanisms and genes. Once a selenate-specific transporter has been identified, for instance, its gene can be expressed in non-hyperaccumulator species, in the organs/tissues/intracellular compartments of choice, under control of a suitable promoter. The ability to rapidly convert selenate to organic Se may be transferred by overexpressing a limiting enzyme of the selenate assimilation pathway (as was done successfully using an ATP sulfurylase) (Pilon-Smits et al. 1999). To achieve accumulation of methyl-SeCys, the enzyme SMT from hyperaccumulators can be expressed in crop species. This has been done successfully, resulting in enhanced total Se accumulation, in the form of methyl-SeCys (Ellis et al. 2004; LeDuc et al. 2004). Tissue-specific sequestration of the organic selenocompound methyl-SeCys may be achieved once the responsible transporter has been identified. It is even possible that the Se hyperaccumulation "syndrome", i.e. all of the above features, can be triggered by a single master switch, for instance a receptor whose expression turns on the synthesis of defense hormones that in turn upregulate S/Se uptake and assimilation. Further analysis of transcriptomic, metabolomic and genomic data will likely provide a wealth of new information about differentially expressed genes in hyperaccumulators and related non-accumulators, as well as sequence differences that may underlie the observed differential expression or differences in kinetic properties. Hyperaccumulator genes emerging from these studies may be expressed in non-accumulators, or may be knocked out in the hyperaccumulators to test their importance. This will be an exciting topic of research in future years.

Acknowledgements The author was supported by National Science Foundation grant IOS-1456361.

References

Anderson JW (1993) Selenium interactions in sulfur metabolism. In: de Kok LJ (ed) Sulfur nutrition and assimilation in higher plants: regulatory, agricultural and environmental aspects. SPB Academic Publishing, The Hague, pp 49–60

Alford ER, Pilon-Smits EAH, Marcus MA et al (2012) No evidence for a cost of tolerance: selenium hyperaccumulation by Astragalus does not inhibit root nodule symbiosis. Am J Bot 99:1930–1941

Beath OA, Gilbert CS, Eppson HF (1939) The use of indicator plants in locating seleniferous areas in the western United States. I. General. Am J Bot 26:257–269

Beath OA, Hagner AF, Gilbert CS (1946) Some rocks and soils of high selenium content. The Geological Survey of Wyoming. Univ Wyo Agric Exp Stat Bull 36

Brown TA, Shrift A (1982) Selenium: toxicity and tolerance in higher plants. Biol Rev 57:59–84

Cabannes E, Buchner P, Broadley MR et al (2011) A comparison of sulfate and selenium accumulation in relation to the expression of sulfate transporter genes in *Astragalus* species. Plant Physiol 157:2227–2239

Cao MJ, Wang Z, Wirtz M et al (2013) SULTR3;1 is a chloroplast-localized sulphate transporter in *Arabidopsis thaliana*. Plant J 73:607–616

Cappa JJ, Pilon-Smits EAH (2014) Evolutionary aspects of elemental hyperaccumulation. Planta 239:267–275

Cappa JJ, Cappa PJ, El Mehdawi AF et al (2014) Characterization of selenium and sulfur accumulation across the genus *Stanleya* (Brassicaceae): a field survey and common-garden experiment. Am J Bot 101:830–839

Cappa JJ, Yetter C, Fakra S et al (2015) Evolution of selenium hyperaccumulation in *Stanleya* (Brassicaceae) as inferred from phylogeny, physiology and X-ray microprobe analysis. New Phytol 205:583–595

Dhillon KS, Dhillon SK (2003) Distribution and management of seleniferous soils. Adv Agron 79:119–184

El Kassis E, Cathala N, Rouached H et al (2007) Characterization of a selenate-resistant Arabidopsis mutant. Root growth as a potential target for selenite toxicity. Plant Physiol 143:1231–1241

El Mehdawi AF, Pilon-Smits EAH (2012) Ecological aspects of plant selenium hyperaccumulation. Plant Biol 14:1–10

El Mehdawi AF, Quinn CF, Pilon-Smits EAH (2011) Effects of selenium hyperaccumulation on plant–plant interactions: evidence for elemental allelopathy? New Phytol 191:120–131

El Mehdawi AF, Cappa JJ, Fakra SC et al (2012) Interactions of selenium and non-accumulators during co-cultivation on seleniferous or non-seleniferous soil – the importance of having good neighbors. New Phytol 194:264–277
El Mehdawi AF, Reynolds RJB, Prins CN et al (2014) Analysis of selenium accumulation, speciation and tolerance of potential selenium hyperaccumulator *Symphyotrichum ericoides*. Physiol Plant 152:70–83
El Mehdawi AF, Paschke MW, Pilon-Smits EAH (2015a) *Symphyotrichum ericoides* populations from seleniferous and nonseleniferous soil display striking variation in selenium accumulation. New Phytol 206:231–242
El Mehdawi AF, Lindblom SD, Cappa JJ et al (2015b) Do selenium hyperaccumulators affect selenium speciation in neighboring plants and soil? An x-ray microprobe analysis. Int J Phytoremediation 17:753–765
Ellis DR, Sors TG, Brunk DG et al (2004) Production of Se-methylselenocysteine in transgenic plants expressing selenocysteine methyltransferase. BMC Plant Biol 4:1–11
Feist LJ, Parker DR (2001) Ecotypic variation in selenium accumulation among populations of *Stanleya pinnata*. New Phytol 149:61–69
Freeman JL, Zhang LH, Marcus MA et al (2006) Spatial imaging, speciation and quantification of selenium in the hyperaccumulator plants *Astragalus bisulcatus* and *Stanleya pinnata*. Plant Physiol 142:124–134
Freeman JL, Tamaoki M, Stushnoff C et al (2010) Molecular mechanisms of selenium tolerance and hyperaccumulation in *Stanleya pinnata*. Plant Physiol 153:1630–1652
Galeas ML, Zhang LH, Freeman JL et al (2007) Seasonal fluctuations of selenium and sulfur accumulation in selenium hyperaccumulators and related non-accumulators. New Phytol 173:517–525
Gigolashvili T, Kopriva S (2014) Transporters in plant sulfur metabolism. Front Plant Sci 5:422
Hanson BR, Lindblom SD, Loeffler ML et al (2004) Selenium protects plants from phloem-feeding aphids due to both deterrence and toxicity. New Phytol 162:655–662
Harris J, Schneberg KA, Pilon-Smits EAH (2014) Sulfur–selenium–molybdenum interactions distinguish selenium hyperaccumulator *Stanleya pinnata* from non-hyperaccumulator *Brassica juncea* (Brassicaceae). Planta 239:479–491
Hartikainen H (2005) Biogeochemistry of selenium and its impact on food chain quality and human health. J Trace Elem Med Biol 18:309–318
Lappartient AG, Touraine B (1996) Demand-driven control of root ATP sulfurylase activity and sulfate uptake in intact canola (the role of phloem-translocated glutathione). Plant Physiol 111:147–157
Läuchli A (1993) Selenium in plants: uptake, functions, and environmental toxicity. Bot Acta 106:455–468
LeDuc DL, Tarun AS, Montes-Bayon M et al (2004) Overexpression of selenocysteine methyltransferase in Arabidopsis and indian mustard increases selenium tolerance and accumulation. Plant Physiol 135:377–383
Leustek T (1996) Molecular genetics of sulfate assimilation in plants. Physiol Plant 97:411–419
Lindblom SD, Valdez-Barillas JR, Fakra SC et al (2013) Influence of microbial associations on selenium localization and speciation in roots of *Astragalus* and *Stanleya* hyperaccumulators. Environ Exp Bot 88:33–42
Lyi SM, Heller LI, Rutzke M et al (2005) Molecular and biochemical characterization of the selenocysteine Se-methyltransferase gene and Se-methylselenocysteine synthesis in broccoli. Plant Physiol 138:409–420
Neuhierl B, Thanbichler M, Lottspeich F, Böck A (1999) A family of S-methylmethionine-dependent thiol/selenol methyltransferases. Role in selenium tolerance and evolutionary relation. J Biol Inorg Chem 274:5407–5414
Peterson PJ, Butler GW (1967) Significance of selenocystathionine in an Australian selenium-accumulating plant, *Neptunia amplexicaulis*. Nature 213:599–600

Pickering IJ, Wright C, Bubner B et al (2003) Chemical form and distribution of selenium and sulfur in the selenium hyperaccumulator *Astragalus bisulcatus*. Plant Physiol 131:1460–1467

Pilon-Smits EAH, Hwang S, Lytle CM et al (1999) Overexpression of ATP sulfurylase in Indian mustard leads to increased selenate uptake, reduction, and tolerance. Plant Physiol 119:123–132

Pilon-Smits EAH, Quinn CF, Tapken W et al (2009) Physiological functions of beneficial elements. Curr Opin Plant Biol 12:267–274

Prins CN, Hantzis LJ, Quinn CF et al (2011) Effects of selenium accumulation on reproductive functions in *Brassica juncea* and *Stanleya pinnata*. J Exp Bot 62:5633–5640

Quinn CF, Prins CN, Gross AM, Hantzis L, Reynolds RJB, Freeman JL, Yang SI, Covey PA, Bañuelos GS, Pickering IJ, Fakra SF, Marcus MA, Arathi HS, Pilon-Smits EAH (2011) Selenium accumulation in flowers and its effects on pollination. New Phytol 192:727–737

Rosenfeld I, Beath OA (1964) Selenium: geobotany, biochemistry, toxicity and nutrition. New York, Acad Press

Sasaki-Sekimoto Y, Taki N, Obayashi T et al (2005) Coordinated activation of metabolic pathway for antioxidants and defence compounds by jasmonates and their roles in stress tolerance in Arabidopsis. Plant J 44:653–668

Schiavon M, Pilon M, Malagoli M et al (2015) Exploring the importance of sulphate transporters and ATPsulphurylases for selenium hyperaccumulation – comparison of *Stanleya pinnata* and *Brassica juncea* (Brassicaceae). Front Plant Sci 6:2

Shahzad Z, Ranwez V, Fizames C et al (2013) Plant Defensin type 1 (PDF1): protein promiscuity and expression variation within the Arabidopsis genus shed light on zinc tolerance acquisition in *Arabidopsis halleri*. New Phytol 200:820–833

Shibagaki N, Rose A, McDermott JP et al (2002) Selenate-resistant mutants of *Arabidopsis thaliana* identify *Sultr1;2*, a sulfate transporter required for efficient transport of sulfate into roots. Plant J 29:475–486

Sors TG, Ellis DR, Salt DE (2005a) Selenium uptake, translocation, assimilation and metabolic fate in plants. Photosynth Res 86:373–389

Sors TG, Ellis DR, Na GN et al (2005b) Analysis of sulfur and selenium assimilation in *Astragalus* plants with varying capacities to accumulate selenium. Plant J 42:785–797

Sors TG, Martin CP, Salt DE (2009) Characterization of selenocysteine methyltransferases from Astragalus species with contrasting selenium accumulation capacity. Plant J 59:110–122

Stadtman TC (1990) Selenium biochemistry. Annu Rev Biochem 59:111–127

Sura-de Jong M, Reynolds RJ, Richterova K et al (2015) Selenium hyperaccumulators harbor a diverse endophytic bacterial community characterized by extreme selenium tolerance and plant growth promoting properties. Front Plant Sci 6:113

Takahashi H, Kopriva S, Giordano M et al (2011) Sulfur assimilation in photosynthetic organisms: molecular functions and regulations of transporters and assimilatory enzymes. Annu Rev Plant Biol 62:157–184

Tamaoki M, Freeman JL, Pilon-Smits EAH (2008a) Cooperative ethylene and jasmonic acid signaling regulates selenite resistance in *Arabidopsis thaliana*. Plant Physiol 146:1219–1230

Tamaoki M, Freeman JL, Marquès L, Pilon-Smits EAH (2008b) New insights into the roles of ethylene and jasmonic acid in the acquisition of selenium resistance in plants. Plant Signal Behav 3:865–867

Terry N, Zayed AM, de Souza MP, Tarun AS (2000) Selenium in higher plants. Annu Rev Plant Physiol Plant Mol Biol 51:401–432

Valdez JR, Quinn CF, Freeman JL et al (2012) Selenium distribution and speciation in hyperaccumulator *Astragalus bisulcatus* and associated ecological partners. Plant Physiol 159:1834–1844

Van Hoewyk D (2013) A tale of two toxicities: malformed selenoproteins and oxidative stress both contribute to selenium stress in plants. Ann Bot 112:965–972

Van Hoewyk D, Garifullina GF, Ackley AR et al (2005) Overexpression of AtCpNifS enhances selenium tolerance and accumulation in Arabidopsis. Plant Physiol 139:1518–1528

Van Hoewyk D, Takahashi H, Hess A et al (2008) Transcriptome and biochemical analyses give insights into selenium-stress responses and selenium tolerance mechanisms in Arabidopsis. Physiol Plant 132:236–253

White PJ (2016) Selenium accumulation by plants. Ann Bot 117:217–235

White PJ, Bowen HC, Marshall B et al (2007) Extraordinarily high leaf selenium to sulphur ratios define 'Se-accumulator' plants. Ann Bot 100:111–118

Wilber CG (1980) Toxicology of selenium: a review. Clin Toxicol 17:171–230

Wilson LG, Bandurski RS (1958) Enzymatic reactions involving sulfate, sulfite, selenate, and molybdate. J Biol Chem 233:975–981

Zhang L, Hu B, Li W et al (2014) OsPT2, a phosphate transporter, is involved in the active uptake of selenite in rice. New Phytol 201:1183–1191

Part II
Selenium Metabolism in Non-plant Organisms – Influence on Se Fluxes in Ecosystems and Relevance for Human Health

oxidative toxicity. Such a pro-oxidative effect of selenite has also been found for plants (Van Hoewyk 2013).

5.3 Selenium Accumulation in Algae

Microalgae Se accumulation depends on the form and the dosage of given Se (Umysová et al. 2009). In *C. reinhardtii*, uptake of selenite was about tenfold higher than selenate uptake at similar exposure concentrations, with intracellular Se concentrations up to 55 mM and 4 mM when the microalga was exposed for 24 h–100 μM selenite or selenate, respectively (Vriens et al. 2016).

In *S. quadricauda* wild type strains, increasing selenite or selenate concentration in the media caused a dose-dependent increase of the total content of Se and selenomethionine (SeMet) (Umysová et al. 2009). The concentration of total Se when cells where treated with 100 mg/L selenite or selenate was 4 mg/L and 33 mg/L, respectively (Vítová et al. 2011), while the content of SeMet in the same microalga after incubation with selenate or selenite accounted for 29% and 41% of the accumulated Se, respectively (Umysová et al. 2009). These findings were comparable to those obtained in *Chlorella vulgaris* (24% and 39%) (Neumann et al. 2003).

Selenium accumulation is also influenced by the ambient S concentration (Fournier et al. 2010; Vriens et al. 2016). In *C. reinhardtii*, Se uptake was strongly affected by S concentration in the media (Fournier et al. 2010). Indeed, for the same selenate concentrations, Se bioaccumulation was significantly higher in the presence of 8 μmol/L than 80 μmol/L of sulfate anions. In the former case, a plateau was reached at 0.60 μmol/L of selenate in the culture medium, in the latter one a linear increase in Se was observed.

In the case of macroalgae, the literature focusing on Se absorption by these organisms is scarce. The investigations conducted so far suggest a role for them as Se bioindicators and/or removers in Se phytoremediation technologies, even though they accumulate Se less efficiently than microalgae. A study carried out on the macroalga *Chara canescens* highlighted the capacity of this species to accumulate Se coming from the drainage water of a farmland (Lin et al. 2002). In another study, *Cladophora hutchinsiae* was identified as an alternative biosorbent organism for the treatment of wastewater containing Se(IV) ions, due to it being low-cost biomass and having a considerably high sorption capacity (Tuzen and Sari 2010).

Differences in Se accumulation have been reported among different seaweed classes, with values within 0.014–0.135 mg/kg for Phaeophyceae, 1.153–0.434 mg/kg for Rhodophyceae and 0.053–0.264 mg/kg for Chlorophyceae (Maher et al. 1992). Sànchez-Rodrìguez et al. (2001) measured Se concentration in 14 different species of macroalgae, and values varied from 0.078 to 0.86 μg/g. In other macroalgae, like *Ulva pertusa* and *Dyctiopters divaricate*, Se concentration was 0.549 μg/g and 0.289 μg/g, respectively (Hou and Yan 1998), while it was 0.53–0.75 μg/g in *Ulva spp.* and 1.12–1.73 μg/g in *Porphyra colombina* (Pèrez et al. 2007). The Se concentration in *Fucus vesciculosus* and *Fucus ceranoides* was 0.05–0.31 μg/g and

0.05–0.51 μg/g, respectively (Turner 2013), and in different red, green and brown macroalgae species from Mexico the Se concentration varied from 0.10 to 0.32 mg/kg (Tenorio Rodriguez et al. 2013).

5.4 Nutritional Aspects of Se Accumulation in Algae

Algae may represent a potential source of Se and other essential minerals required for the healthy metabolism of humans and animals. The content of these nutrients depends on the alga species, location and season of harvest (Garcia-Vaquero and Hayes 2016).

Selenium content and speciation in algae may impact the bioavailability of this element in human diet if algae represent one of the main dietary components (Yan et al. 2004). For humans, the concentration range for Se required as a nutrient is 50–70 μg Se per day, and Se becomes toxic at around tenfold higher levels (Pilon-Smits and Le Duc 2009; Zhu et al. 2009; USDA 2012). Low Se intake can result in dysfunction of the immune system, cardiovascular diseases, decreased fertility and hypothyroidism, whereas excessive dietary Se can induce adverse cardio-metabolic effects, as well as chronic Se poisoning symptoms such as loss of hair and nails, generally known as "selenosis" (Rayman 2012).

Selenium supplementation via Se-enriched algae may be used to correct Se deficiency-related diseases that affect people living in low Se-regions worldwide. In East Asian countries, fresh marine macroalgae are commonly used as food by the local populations because they represent a valuable source of proteins, polysaccharides, fiber, vitamins and trace elements (Chapman and Chapman 1980; Dawczynsky et al. 2007; Cornish and Garbary 2010). For instance, the average intake for the Japanese population is about 1.6 kg dry weight of sea weed per person per year (Chandini et al. 2008).

The form of Se accumulated in algae is also crucial for the establishment of their role as powerful supplementary source of this element in human and animal nutrition. The macroalga *Laminaria japonica*, for instance, possesses high capability to accumulate and convert inorganic Se into organic Se-metabolites that function as anticarcinogens and promoting agents of immune system and thyroid metabolism (Yan et al. 2004). The unicellular alga *C. vulgaris* was also reported to produce higher antioxidants and organic selenocompounds when supplied with selenite concentrations lower than 75 mg/L, and it has been proposed as antioxidative food for aquaculture and human health (Sun et al. 2014). Together, these studies show beneficial effects on human and animal health related to the consumption of edible seaweeds rich in selenocompounds. Further studies focusing on better understanding the algal mechanisms involved will help optimize their potential.

References

Abdel-Hamid M, Skulberg OM (2006) Effect of selenium on the growth of some selected green and blue-green algae. Lakes Reserv Manag 1:205–211

Araie H, Shiraiwa Y (2009) Selenium utilization strategy by microalgae. Molecules 14(12):4880–4891

Araie H, Suzuki I, Shiraiwa Y (2008) Identification and characterization of a selenoprotein, thioredoxin reductase, in a unicellular marine haptophyte lga, *Emiliania huxleyi*. J Biol Chem 283(51):35329–35336

Chandini SK, Ganesan P, Bhaskar N (2008) In vitro antioxidant activities of three selected brown seaweeds of India. Food Chem 107:707–713

Chapman VJ, Chapman DJ (1980) Seaweeds and their uses, 3rd edn. Springer/Chapman and Hall, London/New York

Chapman PM, Adams WJ, Brooks ML, Delos CG, Luoma SN, Maher WA, Ohlendorf HM, Presser TS, Shaw DP (2010a) Ecological assessment of selenium in the aquatic environment. CRC Press, Boca Raton

Chapman PM, Adams WJ, Brooks ML, Delos CG, Luoma SN, Maher WA, Ohlendorf HM, Presser TS, Shaw DP (eds) (2010b) Ecological assessment of selenium in the aquatic environment. SETAC, Pensacola, p 339

Cornish M, Garbary DJ (2010) Antioxidant from macroalgae: potential applications in human health and nutrition. Algae 25(4):155–171

Cutter GA, Brulan KW (1984) The marine biogeochemistry of selenium: a re-evaluation. Limnol Oceanogr 29:1179–1192

Dawczynsky C, Schubert R, Jahreis G (2007) Amino acids, fatty acids and dietary fibre in edible seaweed products. Food Chem 103:891–899

Dazhi W, Zhaodi C, Shaojing L, Yahui G (2003) Toxicity and accumulation of selenite in four microalgae. Chin J Oceanol Limnol 21:280–285

Fournier E, Adam-Guillermin C, Potin-Gautier M, Pannier F (2010) Selenate bioaccumulation and toxicity in *Chlamidomonas reinhardtii*: influence of ambient sulphate ion concentration. Aquat Toxicol 97:51–57

Garcia-Vaquero M, Hayes M (2016) Red and green macroalgae for fish and animal feed and human functional food development. Food Rev Int 32(1):15–45

Geering HR, Cary EE, Jones LHP, Allaway WH (1968) Solubility and redox criteria for the possible forms of selenium in soils. Soil Sci Soc Am Proc 32:35–40

Geoffroy L, Gilbin R, Simon O, Floriani M, Adam C, Pradines C, Cournac L, Garnier-Laplace J (2007) Effect of selenate on growth and photosynthesis of *Chlamydomonas reinhardtii*. Aquat Toxicol 83(2):149–158

Grossman AR, Croft M, Gladyshev VN, Merchant SS, Posewitz MC, Prochnik S, Spalding MH (2007) Novel metabolism in *Chlamydomonas* through the lens of genomics. Curr Opin Plant Biol 10(2):190–198

Hartikainen H (2005) Biogeochemistry of selenium and its impact on food chain quality and human health. J Trace Elem Med Biol 18:309–318

Hou X, Yan X (1998) Study on the concentration and seasonal variation of inorganic elements in 35 species of marine algae. Sci Total Environ 222:141–156

Lemly AD (2004) Aquatic selenium pollution is a global environmental safety issue. Ecotoxicol Environ Saf 59:44–56

Lin ZQ, De Souza M, Pickering IJ, Terry N (2002) Evaluation of the macroalga, muskgrass, for the phytoremediation of selenium-contaminated agricultural drainage water by microcosms. J Environ Qual 31:2104–2110

Lobanov AV, Fomenko DE, Zhang Y, Sengupta A, Hatfield DL, Gladyshev VN (2007) Evolutionary dynamics of eukaryotic selenoproteomes: large selenoproteomes may associate with aquatic life and small with terrestrial life. Genome Biol 8(9):R198

Maher W, Baldwin S, Deader M, Irving M (1992) Characteristics of selenium in Australian marine biota. Appl Organomet Chem 6:103–11.2

Maruyama S, Misumi O, Ishii Y, Asakawa S, Shimizu A, Sasaki T, Matsuzaki M, Shin-I T, Nozaki H, Kohara Y et al (2004) The minimal eukaryotic ribosomal DNA units in the primitive red alga *Cyanidioschyzon merolae*. DNA Res 11(2):83–91

Mitchell K, Mason PRD, Van Cappellen P, Johnson TM, Gill BC, Owens JD (2012) Selenium as paleo-oceanographic proxy: a first assessment. Geochim Cosmochim Acta 89:302–317

Morlon H, Fortin C, Floriani M, Adam C, Garnier-Laplace J, Boudou A (2005) Toxicity of selenite in the unicellular green alga *Chlamydomonas reinhardtii*: comparison between effects at the population and sub-cellular level. Aquat Toxicol 73(1):65–78

Morlon H, Fortin C, Adam C, Garnier-Laplace J (2006) Selenite transport and its inhibition in the unicellular green alga *Chlamydomonas reinhardtii*. Environ Toxicol Chem 25:1408–1417

Neumann PM, De Souza MP, Pickering IJ, Terry N (2003) Rapid microalgal metabolism of selenate to volatile dimethylselenide. Plant Cell Environ 26(6):897–905

Novoselov S, Rao M, Onoshko N, Zhi H, Kryukov G, Xiang Y, Weeks D, Hatfield D, Gladyshev V (2002) Selenoproteins and selenocysteine insertion system in the model plant cell system, *Chlamydomonas reinhardtii*. EMBO J 21:3681–3693

Obata T, Shiraiwa Y (2005) A novel eukaryotic selenoprotein in the haptophyte alga *Emiliania huxleyi*. J Biol Chem 280:18462–18468

Obata T, Araie H, Shiraiwa Y (2004) Bioconcentration mechanism of selenium by a coccolithophorid, *Emiliania huxleyi*. Plant Cell Physiol 45(10):1434–1441

Palenik B, Grimwood J, Aerts A, Rouze P, Salamov A, Putnam N, Dupont C, Jorgensen R, Derelle E, Rombauts S et al (2007) The tiny eukaryote *Ostreococcus* provides genomic insights into the paradox of plankton speciation. Proc Natl Acad Sci U S A 104(18):7705–7710

Papp LV, Holmgren A, Khanna KK (2010) Selenium and selenoproteins in health and disease. Antioxid Redox Signal 12:793–795

Pastierova J, Kramarova Z, Molnarova M, Fargasova A (2009) Comparison of the sensitivity of four freshwater macroalgae to selenate and selenite. Fresenius Environ Bull 18(11):2029–2033

Pérez AA, Farìas SS, Strobl AM, Pérez LB, Lòpez CM, Piñero A, Roses O, Fajardo MA (2007) Levels of essential and toxic elements in *Porphyra columbina* and *Ulva* sp. from San Jorge Gulf, Patagonia Argentina. Sci Total Environ 376:51–59

Pilon-Smits EAH, Le Duc DL (2009) Phytoremediation of selenium using transgenic plants. Curr Opin Biotechnol 20(2):207–212

Plant JA, Kinniburgh DG, Smedley PL, Fordyce FM, Klinck BA (2004) Arsenic and selenium. In: Holland HD, Turekian KK (eds) Treatise on geochemistry, vol 9. Elsevier, Amsterdam, pp 17–66

Price NM, Harrison PJ (1988) Specific selenium-containing macromolecules in the marine diatom *Thalassiosira pseudonana*. Plant Physiol 86(1):192–199

Rayman MP (2000) The importance of selenium to human health. Lancet 356(9225):233–241

Rayman MP (2012) Selenium and human health. Lancet 379(9822):1256–1268

Riedel GF, Sanders GJ (1996) The influence of pH and media composition on the uptake of inorganic selenium by *Chlamidomonas reinhardtii*. Environ Toxicol Chem 15:1577–1583

Robberecht H, van Grieken RV (1982) Selenium in environmental waters: determination, speciation and concentration levels. Talanta 29(10):823–844

Sànchez-Rodrìguez I, Huerta-Dia MA, Choumiline E, Holguin-Quiñones O, Zertuche-Gonzàlez JA (2001) Elemental concentrations in different species of seaweeds from Loreto Bay, Baja California Sur, Mexico: implication for the geochemical control of metals in algal tissue. Environ Pollut 114:145–160

Schiavon M, Moro I, Pilon-Smits EAH, Matozzo V, Malagoli M, Dalla Vecchia F (2012) Acumulation of selenium in *Ulva* sp. and effects on morphology, ultrastructure and antioxidant enzymes and metabolites. Aquat Toxicol 122–123:222–231

Schiavon M, Pilon-Smits EAH, Citta A, Folda A, Rigobello M, Dalla Vecchia F (2016) Comparative effects of selenate and selenite on selenium accumulation, morpho-physiology and glutathione synthesis in *Ulva australis*. Environ Sci Pollut Res 15:15023–15032

Simmons DBD, Emery RJN (2011) Phytochelatin induction by selenite in *Chlorella vulgaris* and regulation on effect by sulfate levels. Environ Toxicol Chem 30(2):469–476

Sors TG, Ellis DR, Salt DE (2005) Selenium uptake, translocation, assimilation and metabolic fate in plants. Photosynth Res 86(3):373–389

Sun X, Zhong Y, Huang Z, Yang Y (2014) Selenium accumulation in unicellular green alga *Chlorella vulgaris* and its effects on antioxidant enzymes and content of photosynthetic pigments. PLoS One 9(11):e112270

Tenorio Rodriguez PA, Méndez-Rodrìguez LC, Serviere-Saragoza E, O'Hara T, Zenteno-Savìn T (2013) Antioxidant substances and trace element content in macroalgae from a subtropical lagoon in the west of the Baja California Peninsula. Vitamin Trace Elem 2(1):1–5

Turner A (2013) Selenium in sediments and biota from estuaries of Southwest England. Mar Pollut Bull 73:192–198

Tuzen M, Sari A (2010) Biosorption of selenium from aqueous solution by green algae (*Cladophora hutchinsiae*) biomass: equilibrium, thermodynamic and kinetic studies. Chem Eng J 158:200–206

Umysová D, Vìtovà M, Douskovà I, Bisovà K, Hlavovà M, Cizkovà M, Machàt J, Doucha J, Zachleder V (2009) Bioaccumulation and toxicity of selenium compounds in the green alga *Scenedesmus quadricauda*. BMC Plant Biol 9:58–74

USDA (2012) USDA National Nutrient Database for Standard Reference, Release 25. Nutrient Data Laboratory Home Page. U.S. Department of Agriculture, Agricultural Research Service, Washington, DC http://www.ars.usda.gov/research/publications/Publications.htm?seq_no_115=285841

Van Hoewyk D (2013) A tale of two toxicities: malformed selenoproteins and oxidative stress both contribute to selenium stress in plants. Ann Bot 112:965–972

Vítová M, Bišová K, Hlavová M, Zachleder V, Rucki M, Čížková M (2011) Glutathione peroxidase activity in the selenium-treated alga *Scenedesmus quadricauda*. Aquat Toxicol 102(1–2):87–94

Vriens B, Behra R, Voegelin A, Zupanic A, Winkel LH (2016) Selenium uptake and methylation by the microalga *Chlamydomonas reinhardtii*. Environ Sci Technol 50(2):711–720

Wang WX, Dei RCH (2001) Influences of phosphate and silicate on Cr(VI) and Se(IV) accumulation in marine phytoplankton. Aquat Toxicol 52(1):39–47

Wheeler AE, Zingaro RA, Irgolic K (1982) The effect of selenite, selenite and sulphate on the growth of six unicellular marine algae. J Exp Mar Biol Ecol 57:181–194

Yan X, Zheng L, Chen H, Lin W, Zhang W (2004) Enriched accumulation and biotransformation of selenium in the edible seaweed *Laminaria japonica*. J Agric Food Chem 52(21):6460–6464

Zhong Y, Chen T, Zheng W, Yang Y (2015) Selenium enhances antioxidant activity and photosynthesis in *Ulva fasciata*. J Appl Phycol 27:555–562

Zhu Y-G, Pilon-Smits EAH, Zhao F-J, Williams PN, Meharg AA (2009) Selenium in higher plants: understanding mechanisms for biofortification and phytoremediation. Trends Plant Sci 14(8):436–442

Chapter 6
Bacteria Versus Selenium: A View from the Inside Out

Lucian C. Staicu, Ronald S. Oremland, Ryuta Tobe, and Hisaaki Mihara

Abstract Bacteria and selenium (Se) are closely interlinked as the element serves both essential nutrient requirements and energy generation functions. However, Se can also behave as a powerful toxicant for bacterial homeostasis. Conversely, bacteria play a tremendous role in the cycling of Se between different environmental compartments, and bacterial metabolism has been shown to participate to all valence state transformations undergone by Se in nature. Bacteria possess an extensive molecular repertoire for Se metabolism. At the end of the 1980s, a novel mode of anaerobic respiration based on Se oxyanions was experimentally documented for the first time. Following this discovery, specific enzymes capable of reducing Se oxyanions and harvesting energy were found in a number of anaerobic bacteria. The genes involved in the expression of these enzymes have later been identified and cloned. This iterative approach undertaken *outside-in* led to the understanding of the molecular mechanisms of Se transformations in bacteria. Based on the extensive knowledge accumulated over the years, we now have a full(er) view from the *inside out*, from DNA-encoding genes to enzymes and thermodynamics. Bacterial transformations of Se for assimilatory purposes have been the object of numerous studies predating the investigation of Se respiration. Remarkable contributions related to the understating of the molecular picture underlying seleno-amino acid biosynthesis are reviewed herein. Under certain circumstances, Se is a toxicant for bacterial metabolism and bacteria have evolved strategies to counteract this toxicity, most notably by the formation of elemental Se (nano)particles. Several biotechnological

L.C. Staicu
Faculty of Applied Chemistry and Materials Science, University Politehnica of Bucharest, Bucharest, Romania
e-mail: lucian.staicu@upb.ro

R.S. Oremland
US Geological Survey, 345 Middlefield Road, MS 480, Menlo Park, CA 94025, USA
e-mail: roremlan@usgs.gov

R. Tobe • H. Mihara (✉)
Laboratory of Applied Molecular Microbiology, Department of Biotechnology, College of Life Sciences, Ritsumeikan University, 1-1-1 Noji-Higashi, Kusatsu, Shiga 525-8577, Japan
e-mail: tober@fc.ritsumei.ac.jp; mihara@fc.ritsumei.ac.jp

E.A.H. Pilon-Smits et al. (eds.), *Selenium in plants*, Plant Ecophysiology 11,
DOI 10.1007/978-3-319-56249-0_6

applications, such as the production of functional materials and the biofortification of crop species using Se-utilizing bacteria, are presented in this chapter.

Keywords Selenium • Bacteria • Anaerobic respiration • Selenium detoxification • Selenoenzymes

6.1 Introduction

Bacteria are involved in the cycling of selenium (Se) through different compartments of the environment. Se has several oxidation states: (VI), (IV), (0), and (-II), that display variable solubility, bioavailability, and toxicological profiles (Chapman et al. 2010). Se oxyanions, SeO_x, i.e. selenate (SeO_4^{2-}) and selenite (SeO_3^{2-}), are water-soluble, bioavailable, and toxic to aquatic life (Simmons and Wallschlaeger 2005). Elemental Se (Se^0) is considered practically nontoxic in view of its solid state and negligible water solubility. However, a number of reports documented adverse effects exhibited by Se^0 against ecological receptors such as filter-feeding mollusks and fish (Chapman et al. 2010). The bioremediation approach employed by various bioreactor systems relies on the microbial conversion of SeO_x to Se^0 (Staicu et al. 2015a). Selenide, Se(−II), the most reduced valence state of Se, is present in strongly reducing conditions. Both inorganic Se, e.g. hydrogen selenide (H_2Se and HSe^-), metal selenides ($M^{n+}Se^{2-}$), and selenocyanate ($SeCN^-$), and organic selenides (e.g. methylated species such as dimethylselenide (DMSe), aminoacids such as selenocysteine (Sec or SeCys) and selenomethionine (SeMet), and metabolic products (e.g. trimethylselenonium) have been described (Fernandez-Martinez and Charlet 2009).

A biogeochemical cycle of inorganic and organic forms of Se was first proposed in a seminal article by Shrift (1964). Following this article, bacteria were later found to participate in most transformations undergone by Se in aquatic and terrestrial ecosystems. A major finding that came out at the end of 1980s was the capacity of some anaerobic bacteria to use Se as terminal electron acceptor for cellular respiration (Macy et al. 1989; Oremland et al. 1989). The first bacteria described that carry out anaerobic respiration on selenate were *Thauera selenatis*, belonging to the beta subclass (Macy et al. 1993), and *Sulfurospirillum barnesii* and *S. arsenophilus* of the epsilon subclass of Proteobacteria (Oremland et al. 1994; Laverman et al. 1995; Stolz and Oremland 1999). *T. selenatis* was isolated in California by Joan Macy and coworkers from a bioreactor setting treating agricultural wastewater rich in Se-oxyanion effluents. *S. barnesii* was isolated from a Se-rich drainage slough located near Fallon, NV (Oremland et al. 1994). Apart from their use in cellular respiration, Se compounds can also behave as powerful toxicants due to the production of dysfunctional biomolecules when Se is mis-incorporated into sulfur-rich proteins (Stadtman 1974). Additionally, the metabolism of SeO_x was linked to oxidative stress (Hoffman 2002). However, bacteria have evolved different strategies to counteract this toxicity and a major one is the production of Se^0 nanoparticles with significantly reduced toxicity.

This chapter discusses the bacterial transformations of Se from a molecular biology perspective. We briefly examine the main strategies employed by anaerobic and aerobic bacteria to transform Se, and then we expound upon the genetics and the enzymes that underlie these transformations. The chapter also presents the potential use of Se-transforming bacteria for the production of functional materials and the biofortification of plant and crop species. The reader is referred to some recent reviews for further details on this re-emerging subject of scientific interest (Santos et al. 2015; Winkel et al. 2015).

6.2 Bacterial Metabolism of Selenium

6.2.1 *Selenium Respiration*

Under anaerobic conditions, various electron acceptors (e.g. NO_3^-, SO_4^{2-}, S^0, Fe^{3+} or Mn^{4+}) can be utilized by bacteria for respiration as the terminal step of their electron transport chain. Macy et al. (1989) showed that selenate can be used by bacteria for cellular respiration. This strategy is termed *dissimilatory* reduction. From a thermodynamic point of view, the reduction of selenate to selenite coupled with the oxidation of an electron donor, such as formate, acetate (Eq. 6.1) or lactate (Eq. 6.2) provides energy to support bacterial growth (Macy et al. 1989; Oremland et al. 1994):

$$\mathrm{C_2H_3O_2}^- + 4SeO_4^{\,2} + \mathrm{H}^+ \rightarrow 4SeO_3^{\,2} + 2CO_2 + 2\mathrm{H_2O} \tag{6.1}$$

$$\Delta G_f{\circ} = -556kJ\,/\,mol\,\mathrm{acetate}\,(\mathrm{C_2H_3O_2}^-)$$

$$\mathrm{C_3H_6O_3}^- + 2SeO_4^{\,2} \rightarrow 2SeO_3^{\,2} + \mathrm{C_2H_3O_2}^- + HCO_3^- + \mathrm{H}^+ \tag{6.2}$$

$$\Delta G_f{\circ} = -343.1kJ\,/\,mol\,\mathrm{lactate}\,(\mathrm{C_3H_6O_3}^-)$$

Provided the bacteria involved are capable of metabolizing both selenate and selenite, usually sequentially, then elemental Se is the end product of selenate reduction according to Eq. 6.3 (Oremland et al. 1994):

$$3\mathrm{C_3H_6O_3}^- + 2SeO_4^{\,2} + \mathrm{H}^+ \rightarrow 2Se^0 + 3\mathrm{C_2H_3O_2}^- + 3HCO_3^- + 2\mathrm{H_2O} \tag{6.3}$$

$$\Delta G_f{\circ} = -467.4kJ\,/\,mol\,\mathrm{lactate}$$

Some species can respire either selenate or selenite, but not both. For example, *Bacillus selenitireducens*, a halo-alkaliphile isolated from Mono Lake, California,

Fig. 6.1 Biogenic Se^0 and HSe^-: (**a**) Formation of HSe^- from accumulated Se^0 over a time sequence by *B. selenitireducens* incubated with an excess of the electron donor lactate. The black precipitate at the bottom of the bottle was the hexagonal crystalline allotrope of Se^0, while the red/orange was the amorphous/monoclinic allotrope of Se^0 (Photos courtesy of M. Herbel); (**b**) Scanning electron micrograph of a black hexagonal crystal of Se^0 taken at the end of the above time course. The hair-like threads on the crystal consist of Se^0 formed by auto-oxidation of HSe^- when exposed to air (M. Herbel, unpublished)

can carry out the reductive dissimilation of selenite (Eq. 6.4) that yields energy for growth (Switzer Blum et al. 1998):

$$\mathrm{C_3H_6O_3}^- + SeO_3^{\,2} + \mathrm{H}^+ \rightarrow Se^0 + \mathrm{C_2H_3O_2}^- + HCO_3^- + \mathrm{H_2O} \qquad (6.4)$$

$$\Delta G_f \circ = -241.6 kJ\,/\,mol\,\mathrm{lactate}\,(pH = 10)$$

If provided with an excess of electron donor (lactate) over the available selenite supplied, then this microorganism can carry out a further reduction of the accumulated mass of extracellular Se^0 to HSe^-, according to Eq. 6.5 (Herbel et al. 2003):

$$\mathrm{C_3H_6O_3}^- + 2Se^0 + 2\mathrm{H_2O} \rightarrow 2HSe^- + \mathrm{C_2H_3O_2}^- + HCO_3^- + 3\mathrm{H}^+ \qquad (6.5)$$

$$\Delta G_f \circ = -98.8\, kJ\,/\,mol\,\mathrm{lactate}\,(pH = 10)$$

The time course progression of this reaction is visually quite striking. It starts out with the mass accumulation of bright orange amorphous Se^0, which grows darker and separates into layers, and eventually clears into a tawny-tinged fluid with a dense, black precipitate of hexagonal Se^0 at the bottom of the large serum bottle (Fig. 6.1a). If one removes the rubber stopper allowing air to exchange with the N_2 headspace, a rapid exothermic reaction occurs whereby the accumulated HSe^- autoxidizes back to Se^0, and the bottle turns back to the thick orange color. A scanning electron micrograph of the black hexagonal Se^0 allotrope is shown in Fig. 1b, where the thread-like hairs seen on its surface are strands of Se^0 formed by the oxidation of HSe^- upon its exposure to the oxygen in air. A cell of *B. selenitireducens* also adorns the surface of the crystal. Se respiration in Archaea was only marginally reported to date (Huber et al. 2000).

6.2.2 *Selenium Assimilation*

In contrast to the dissimilatory reduction of SeO_x used to energize the bacterial cell, the *assimilatory* reduction of Se oxyanions is employed by both aerobes and anaerobes for the synthesis of Se amino acids, namely Sec and SeMet. These amino acids are incorporated into selenoproteins having essential roles in the proper functioning of bacterial metabolism. In selenoproteins, Se has structural and enzymatic roles, serving oxidoreductase functions against reactive oxygen species (ROS) (Labunskyy et al. 2014). Sects. 6.3.3, 6.3.4, 6.3.5 from this chapter provide an in-depth presentation of the seleno-amino acids metabolism.

6.2.3 *Selenium Detoxification*

Se exerts toxic effects on bacteria and several mechanisms have been proposed for the reduction of selenite to Se^0 in microorganisms, including a glutathione (GSH) system, thioredoxin system, siderophore-mediated reduction, sulfide-mediated reduction, and dissimilatory reduction (Zannoni et al. 2008).

The reduction sequence of selenite to Se^0 by GSH occurs according to the following reactions (Eqs. 6.6, 6.7, 6.8, 6.9, and 6.10) (Ganther 1968):

$$\mathrm{H_2}SeO_3 + GSH \rightarrow \mathrm{GSSeO_2H} + \mathrm{H_2O} \tag{6.6}$$

$$\mathrm{GSSeO_2H} + GSH \rightarrow \mathrm{GSSeOSG} + \mathrm{H_2O} \tag{6.7}$$

$$\mathrm{GSSeOSG} + GSH \rightarrow GS - Se - SG + \mathrm{GSOH} \tag{6.8}$$

$$GS\text{–}Se\text{–}SG + \mathrm{NADPH} + \mathrm{H^+} \rightarrow GS\text{–}SeH + GSH + \mathrm{NADP^+} \tag{6.9}$$

$$\mathrm{GSSeH} + GSH \rightarrow GSH + Se^0 \tag{6.10}$$

where $GSSeO_2H$, glutathione selenone; GSSeOSG, diglutathione selenone; GS-Se-SG, selenodiglutathione; and GSSeH, L-γ-glutamyl-*S*-selanyl-L-cysteinylglycine.

Reduced thioredoxin and thioredoxin reductase were hypothesized to be involved in the reduction of selenite and GS-Se-SG (Björnstedt et al. 1992). Reduced thioredoxin reacts with GS-Se-SG to form oxidized thioredoxin, reduced GSH, and selenopersulfide anion, and then Se^0 is released from the reactive selenopersulfide anion. In addition, selenite can react with the reactive biogenic sulfide abiotically, yielding Se^0 and elemental sulfur, S^0 (Hockin and Gadd 2003; Pettine et al. 2013). An iron siderophore, pyridine-2,6-bis(thiocarboxylic acid), produced by

Pseudomonas stutzeri KC, has also been proposed to detoxify selenite through reduction and formation of insoluble Se^0 precipitates (Zawadzka et al. 2006).

The exposure of bacteria to growth media containing selenite resulted in phenotypic changes and altered cell morphology in *Wolinella succinogenes* (Tomei et al. 1992) and *Desulfovibrio desulfuricans* DSM 1924 (Tomei et al. 1995). *P. moraviensis* stanleyae showed impaired growth (40% less bacterial cell density during stationary phase) and extended lag time when it was exposed to 10 mM sodium selenite (Staicu et al. 2015b). Figure 6.2 presents Se^0 nanoparticles with a diameter around 60 nm that were found intracellularly, following exposure of *P. moraviensis* stanleyae to sodium selenite. Additionally, the Se^0 enzyme assay identified GSH, nitrite, and sulfite reductases as candidate enzymes involved in selenite reduction, suggestive of a detoxification mechanism at play (Ni et al. 2015). Aerobic reduction of selenite is ubiquitous amongst phylogenetically diverse bacterial groups, indicating shared metabolic pathways used for the reduction of other oxyanions such as nitrate or sulfate (Sura-de Jong et al. 2015).

6.2.4 Metabolic Explorations Using Se Isotopic Techniques

Further detailed investigations into the actual scope and rates of Se biotransformations occurring within Se-impacted and pristine environments were facilitated by the use of both radioisotopes (i.e., ^{75}Se-selenate) and the fact that this element also displays six naturally-occurring stable isotopes (i.e. ^{74}Se, ^{76}Se, ^{77}Se, ^{78}Se, and ^{80}Se). A radioisotopic procedure was devised whereby ^{75}Se-selenate was injected into subcores recovered from anoxic sediments, and after incubation and washing, the amount of $^{75}Se^0$ was quantified so as to yield rate constants (Oremland et al. 1990). Multiplication of the rate constants by the concentration of selenate in pore waters ($\leq$ 40 μM) yielded *in situ* rates of dissimilatory selenate reduction. The rates determined for a large agricultural evaporation pond located in the Se-impacted San Joaquin Valley (California) were calculated to be 300 μM SeO_4^{2-} per m^2 per day, which was sufficient to sequester all the pond water Se oxyanions as Se^0 in the bottom sediments within ~90 days. Similar results were obtained for an agricultural drainage slough located in western Nevada (Oremland et al. 1991). To answer the question of whether or not this phenomenon was widespread or confined to Se-contaminated regions, a broad survey was conducted to assay surficial aquatic sediments from a number of different locales and chemistries (e.g. freshwater, estuarine, soda lakes, contaminated ponds, and saturated salterns). The results were surprising in that rapid dissimilatory selenate reduction was found to be common to all of the 11 sediment types investigated, indicating that both the bacteria involved and their enzymes were constitutive and active in these diverse biomes (Steinberg and Oremland 1990). In stark contrast to these rapid Se(VI) reduction rates, the rates of oxidation of $^{75}Se^0$ back to soluble oxyanions by bacterial cultures, as well as

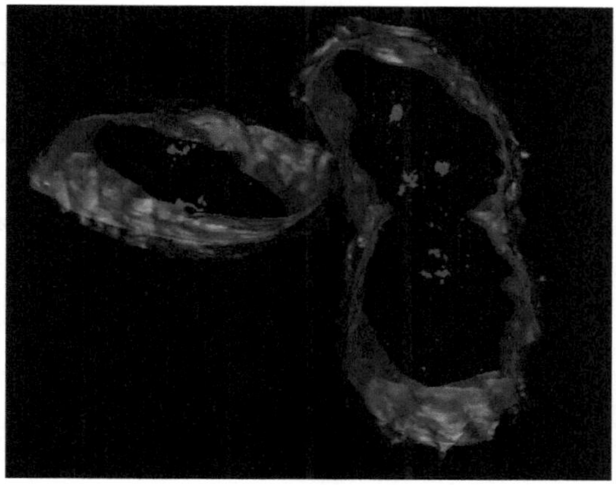

Fig. 6.2 Electron tomographic reconstruction with osmium staining of *P. moraviensis* stanleyae exposed to 10 mM of sodium selenite under aerobic conditions. Legend: The outer membrane (*in green*), inner membrane (*in yellow*), and Se^0 nanoparticles (*in pink*) (Adapted from Ni et al. 2015)

by oxic sediments, were very slow, with turnover rates measured in years rather than hours or days (Dowdle and Oremland 1998).

Selenate- and selenite-respiring bacteria are capable of a "classic" biological stable isotopic fractionation, selecting for the lighter isotopes, while leaving behind the heavier, as was demonstrated with pure cultures (Johnson et al. 1999). Cumulative reduction of Se(VI) and Se(IV) resulted in an enrichment of the $\delta^{80/76}$Se of ~11 per mil (Herbel et al. 2000). However, although fractionation was observed during incubation of live, manipulated sediment slurries (Ellis et al. 2003), little if any fractionation was observed in Se-contaminated drainages (Herbel et al. 2002). The absence of fractionation *in situ* was probably owing to the rapidity and completeness of the biological reactions involved. Nonetheless, Zhu et al. (2014) reported that at one particular site of an exposed outcrop in China, very broad ranges in the $\delta^{82/76}$Se ratios were observed (−14.2 to +11.37 per mil) reflecting on the alternating seasonal oxic vs. anoxic conditions in this subsurface aquifer. Analytical advances have been made which explore the use of stable oxygen isotopes ($^{18}O/^{16}O$) in Se oxyanions during dissimilatory reduction (Schellenger et al. 2015) which opens the possibility in the near future of employing multiple isotopes of Se and O (i.e. "clumped analyses") to better characterize the biogeochemical redox cycle of Se in nature.

6.3 Genes and Enzymes Involved in Bacterial Selenium Metabolism

6.3.1 Overview

As described above, inorganic and organic forms of Se are metabolized in a dissimilatory or assimilatory manner by Se-utilizing bacteria (Fig. 6.3). Some bacteria can use the Se oxyanions, selenate and selenite, as electron acceptors under anaerobic conditions, and other species can use these ions as substrates for producing

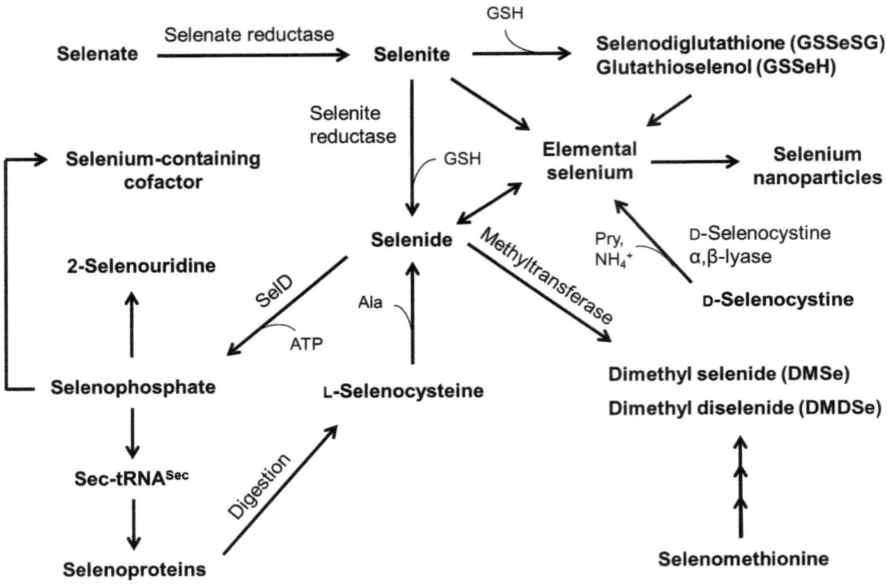

Fig. 6.3 Selenium metabolism in bacteria

biologically active Se compounds such as seleno-amino acids, selenouridine (SeU), and Se-containing cofactors. Until recently, only a few bacterial genes involved in Se utilization had been identified. Due to improvements in genome sequencing technology, genomic analysis has become a powerful way to predict if a bacterium can utilize Se. Previous studies identified genes involved in a specific Se-utilizing system. Based on the results of genomic screens using those already-identified genes as indicators, an increased number of Se-utilizing bacteria have been recognized.

Studies on the metabolism of Se oxyanions have identified genes responsible for selenate reduction, like *serABCD* in *Thauera selenatis* (Lowe et al. 2010), *srdBCA* in *Bacillus selenatarsenatis* SF-1 (Kuroda et al. 2011), and the *ygfKLMN* and *ynfEGHdmsD* operons in *Escherichia coli* (Bebien et al. 2002; Guymer et al. 2009). The microbial selenite reduction processes can be categorized broadly into either detoxification or anaerobic respiration. However, only a few selenite-respiring bacteria have been isolated (Stolz et al. 2006), and specific genes involved in selenite reduction have not as yet been identified. Various mechanisms have been proposed for the reduction of selenite to elemental Se, including the Painter-type reaction, a thioredoxin reductase system, siderophore-mediated reduction, sulfide-mediated reduction, and dissimilatory reduction (Zannoni et al. 2008).

Metabolic processes of biologically active Se molecules, such as seleno-amino acids, SeU, and Se-containing cofactors, have been characterized and their related genes have been identified. Such genes identified so far include those for biosynthesis of Sec (*selA*, *selB*, *selC*, and *selD* in bacteria and *PSTK*, *SepSecS*, *selB*, *selC*, and *selD* in archaea), 2-SeU (*ybbB*), and Se-containing cofactors (*yqeB* and *yqeC*). Since the *selD* gene encoding selenophosphate synthetase (SelD) is commonly required for the synthesis of Sec, SeU, and Se-containing cofactors, it is one of the major biomarkers for identifying bacterial species that can utilize Se (Zhang et al. 2006; Lin et al. 2015). Recently, Peng et al. (2016) applied comparative genomic approaches to more than 5200 sequenced bacterial genomes to investigate Se utilization in bacteria. Among the species examined, 1121 Sec-utilizing (21.5%), 980 SeU-utilizing (18.8%) and 312 Se-cofactor-utilizing (6.0%) organisms were identified. Se utilization is hypothesized to be an ancient trait that was once common to almost all bacterial species. However, this study only detected the presence of *selD* in 1754 organisms (33.7%), suggesting that most species have lost the ability to use Se over the long process of evolution.

6.3.2 Metabolism of Selenium Oxyanions

Bacteria interact with all valence states of Se, thus contributing to the biogeochemical cycle of this element (Shrift 1964). However, the reduction of high-valence states of Se is more often reported and occurs at a considerably faster pace than the oxidative side of the cycle.

6.3.2.1 Selenate Respiration and Reduction

As shown in Eq. 6.11 and 6.12, selenate reduction occurs by a two-step process. Selenate reduction to selenite is catalyzed by selenate reductases, and then selenite reduction to insoluble Se^0 is catalyzed by nonspecific selenite reductases. Such reductions can be observed under aerobic, anoxic, and anaerobic conditions. The genes and enzymes involved in selenate reduction were investigated especially in *Thauera selenatis*, *Enterobacter cloacae* SLD1a-1, *E. coli*, and *Bacillus selenatarsenatis* SF-1.

$$SeO_4^{2} + 2e + 2H^+ \rightarrow SeO_3^{2} + H_2O \tag{6.11}$$

$$SeO_3^{2} + 4e + 6H^+ \rightarrow Se^0 + 3H_2O \tag{6.12}$$

The Gram-negative bacterium *T. selenatis* can effectively reduce selenate to selenite anaerobically (Rech and Macy 1992). The selenate to selenite reduction reaction occurs in the periplasmic compartment. The first identified respiratory selenate reductase of the bacterium was purified and characterized (Schröder et al. 1997). This enzyme consists of a catalytic unit (SerA), an Fe-S protein (SerB), a heme *b* protein (SerC), and a molybdenum cofactor (Lowe et al. 2010). Complete inhibition of selenate reduction was achieved in the presence of both myxothiazol and

2-*n*-heptyl-4-hydroxyquinoline *N*-oxide, suggesting the involvement of both a quinol cytochrome *c* reductase and a quinol dehydrogenase in selenate reduction (Lowe et al. 2010). In addition, a novel 95-kDa protein, SefA (Se factor A), was isolated from the elemental Se secreted from *T. selenatis* cells into the extracellular medium, suggesting that the SefA protein aids in the secretion process by stabilizing Se nanospheres and preventing their aggregation (Debieux et al. 2011).

Enterobacter cloacae SLD1a-1 is a selenate-reducing bacterium isolated from the Se-rich waters of the San Luis Drain in California (Losi and Frankenberger 1997). Selenate reductase of *E. cloacae* SLD1a-1 is a membrane-bound trimeric complex with a catalytic subunit of 100 kDa, which may contain molybdenum as a cofactor (Ridley et al. 2006). In *E. coli*, at least three systems for selenate reduction have been identified. Selenate reductase encoded within the *ynfEGHdmsD* operon is dependent on the twin arginine translocation (Tat) system (Guymer et al. 2009). The catalytic subunit YnfE is predicted to bind a bis-molybdopterin guanine dinucleotide cofactor and a [4Fe-4S] cluster. The small subunit YnfG exhibits four [4Fe-4S]-binding motifs, with each motif containing four conserved cysteine residues. On the other hand, it has been demonstrated using gene deletion analyses that another *E. coli* selenate reductase is a structural complex including the proteins YgfK, YgfM, and YgfN, encoded by the *ygfKLMN* putative operon (Bebien et al. 2002). Although the specific activity is low, *E. coli* nitrate reductases A and Z (encoded by *narGHIJ* and *narZUWV*, respectively) and periplasmic nitrate reductase NapA also possess selenate reductase activity (Avazeri et al. 1997).

A Gram-positive bacterium, *B. selenatarsenatis* SF-1, was also isolated as a selenate-reducing bacterium (Fujita et al. 1997). The strain shows a stoichiometric relationship between cell growth, lactate consumption, and selenate reduction. It was demonstrated using transposon mutagenesis that the *srdBCA* operon encodes a putative oxidoreductase complex as a respiratory selenate reductase complex (Kuroda et al. 2011). The selenate reductase SrdBCA is a membrane-bound, trimeric molybdoenzyme. Electrons from the quinol pool are channeled to the catalytic subunit SrdA via SrdB, an Fe-S protein, and selenate receives the electrons from SrdA via the molybdenum cofactor (Kuroda et al. 2011).

6.3.2.2 Selenite and Elemental Selenium Respiration and Reduction

Certain selenate-reducing bacteria can also perform dissimilatory selenite reduction (Nancharaiah and Lens 2015). However, the investigation of selenite reduction via respiratory electron transport pathways is limited to a study using *Shewanella oneidensis* MR-1 (Li et al. 2014). Apart from respiration, selenite can also be reduced by bacteria as a dissimilatory strategy which includes detoxification.

To date, there are only a few studies on microbial reduction of Se^0 to selenide. Some of selenate- or selenite-respiring bacteria may have the capacity to reduce Se^0 as well. *Bacillus selenitireducens*, a selenite-respiring bacterium, produced significant amounts of selenide from Se^0 or selenite (Herbel et al. 2003). However, the reduction of Se^0 to selenide was not observed in the case of selenate-respiring bacteria, and the responsible catalytic enzymes have not been identified.

6.3.2.3 Selenium Nanoparticles

Many bacteria synthesize Se nanoparticles (SeNPs) as a mechanism of Se detoxification (Kessi et al. 1999). SeNPs have wide applications in medicine, therapeutics, biosensors, and environmental remediation (Wadhwani et al. 2016). Synthesis of these nanoparticles can be extracellular, intracellular, or membrane-bound. Se deposits were first observed on cell walls and cell membranes of *E. coli* under electron microscopy (Gerrard et al. 1974). After that, several bacterial species, both Gram-negative and Gram-positive, like *Veillonella atypica* and *Pseudomonas* sp. RB, have been demonstrated to synthesize quantum dots such as CdSe and ZnSe (Pearce et al. 2008; Ayano et al. 2014). After Se reduction based on the previously mentioned mechanisms, SeNPs accumulate in bacterial cells during mid- to late-exponential growth phases and are secreted into the surrounding medium in the stationary phase (Butler et al. 2012). Se factor A (SefA), a protein of approximately 95 kDa, accompanies SeNPs during their export from the cytoplasmic compartment, and aids in biomineralization and stabilization of the nanoparticles (Butler et al. 2012). Furthermore, the metalloid reductase RarA has the highest number of peptides with strong affinity for SeNPs, thereby conferring stability (Lenz et al. 2011). A reductase enzyme is responsible for the conversion of selenate and selenite to nano-Se in bacteria, and this phenomenon has mostly been studied in *T. selenatis*, *E. coli*, and *E. cloacae* (Wadhwani et al. 2016). However, the existence of multiple electron transport pathways has not been ruled out, and the mechanism by which these nanoparticles exhibit antimicrobial action is still unclear.

6.3.3 Selenoprotein Biosynthesis in Bacteria and Archaea

6.3.3.1 Selenocysteine Synthesis

The most important and best-characterized biological form of Se is the amino acid Sec. Most studies on selenoprotein biosynthesis in bacteria have been carried out with *E. coli*. Böck et al. revealed that at least four Sec-specific genes, *selA, selB, selC*, and *selD*, are required for bacterial Sec synthesis (Fig. 6.4) (Forchhammer et al. 1990; Leinfelder et al. 1990; Forchhammer et al. 1991). The gene *selC* encodes the Sec-specific tRNA ($tRNA^{Sec}$) and its anticodon UCA is complementary to the Sec codon UGA (Leinfelder et al. 1989). This tRNA is the largest tRNA in *E. coli* and has a unique modification pattern (Schon et al. 1989). However, the most obvious distinction between $tRNA^{Sec}$ and canonical elongator tRNAs is the eight-base-pair aminoacyl-acceptor stem; all other tRNA species have a seven base-pair stem. $tRNA^{Sec}$ is first aminoacylated with L-serine by seryl-tRNA synthetase and then the conversion of seryl-$tRNA^{Sec}$ into selenocysteyl-$tRNA^{Sec}$ (Sec-$tRNA^{Sec}$) is catalyzed by Sec synthase (the *selA* gene product) using selenophosphate. The Sec synthase SelA of *E. coli* binds pyridoxal 5′-phosphate (PLP) as a cofactor, the carbonyl of which forms an aldimine linkage with serine's α-amino group, and 2,3-elimination

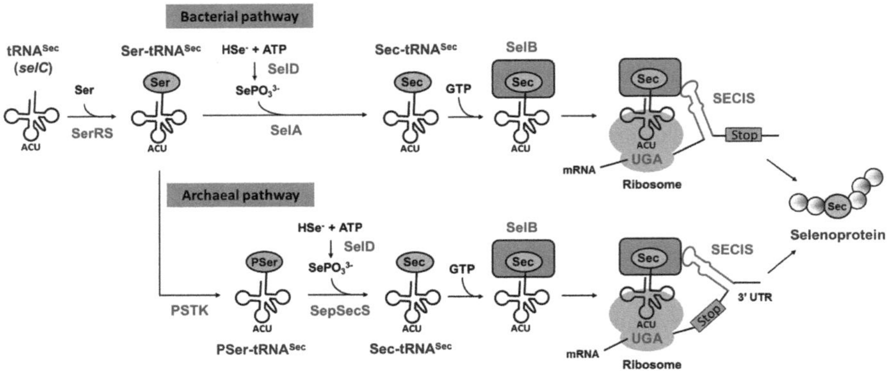

Fig. 6.4 Selenoprotein biosynthesis in bacteria and archaea. SerRS, seryl-tRNA synthetase; SelA, selenocysteine synthase; SelD, selenophosphate synthetase; GTP, guanosine-5′-triphosphate; SECIS, selenocysteine insertion sequence; SelB, SECIS-binding protein; PSKT, *O*-phosphoseryl-tRNA^{Sec} kinase; SepSecS, *O*-phosphoseryl-tRNA^{Sec}: selenocysteine synthase

of a water molecule generates enzyme-bound aminoacrylyl-tRNA^{Sec} (Forchhammer and Böck 1991). Nucleophilic addition of selenide to the aminoacrylyl double bond forms Sec-tRNA^{Sec}. The activated Se donor for the reaction, selenophosphate, is synthesized by selenophosphate synthetase, SelD (the product of the *selD* gene) (Veres et al. 1992). Due to the high K_m value of SelD for selenide (HSe^-), it is assumed that selenide is not the true substrate, but rather some activated, protein-bound Se species may specifically supply a substrate to selenophosphate synthetase.

Archaeal Sec-tRNA^{Sec} biosynthesis has some differences from the bacterial system (Fig. 6.4). Archaea and bacteria use a similar Sec-tRNA^{Sec} biosynthesis system, but archaea require an additional step, phosphorylation of Ser-tRNASec by *O*-phosphoseryl-tRNA^{Sec} kinase (PSTK) (Yuan et al. 2006). Investigating the process of Sec synthesis and incorporation in archaea started with the genome sequence of *Methanococcus jannaschii* (Bult et al. 1996). This genome encodes tRNA^{Sec} and its predicted structure resembles eukaryotic tRNA^{Sec} more closely than bacterial tRNA^{Sec} (Commans and Böck 1999). *In vitro* characterization of a *M. maripaludis* homolog of the eukaryotic *O*-phosphoseryl-tRNA^{Sec}:Sec synthase (SepSecS) showed its ability to catalyze the selenophosphate-dependent conversion of *O*-phosphoseryl-tRNA^{Sec} to Sec-tRNA^{Sec} (Yuan et al. 2006). Briefly, in the case of archaeal Sec-tRNA^{Sec} biosynthesis, after charging tRNA^{Sec} with serine, *O*-phosphoseryl-tRNA^{Sec} is synthesized as an intermediate upon phosphorylation of the seryl-tRNA^{Sec} by PSTK. Ultimately, *O*-phosphoseryl-tRNA^{Sec} is converted to Sec-tRNA^{Sec} by SepSecS using selenophosphate as the Se donor.

6.3.3.2 Selenocysteine Insertion

Because of its unique structural features, $tRNA^{Sec}$ is not recognized by the canonical elongation factor EF-Tu (Forster et al. 1990). Instead, a Sec-specific translation elongation factor, SelB, the product of the *selB* gene, is utilized for Sec insertion into a nascent polypeptide (Forchhammer et al. 1989). The N-terminus of SelB from *E. coli* shares significant homology to EF-Tu and it binds Sec-$tRNA^{Sec}$ and GTP stoichiometrically (Baron and Böck 1991). A unique property of bacterial SelB, crucial for its function, is its interaction with the selenoprotein mRNA. The segment responsible for this interaction, the SECIS element, has a stem-loop structure of approximately 40 nucleotides located immediately downstream of the UGA codon. Formation of the quaternary complex of SelB, Sec-$tRNA^{Sec}$, the SECIS, and GTP is cooperative and promotes Sec insertion into bacterial selenoproteins (Fig. 6.4). In addition, both *selA* and *selC* from the Gram-positive bacterium *Moorella thermoacetica* complement the corresponding genes in *E. coli* (Tormay et al. 1994; Kromayer et al. 1996). Thus, Gram-positive bacteria appear to utilize the same general strategy for Sec synthesis as Gram-negative bacteria.

In archaea, the Sec insertion system is still not fully proven. Analysis of Sec-coding genes in *M. voltae* first showed that Sec insertion in archaea is also directed by UGA (Halboth and Klein 1992). However, archaea do not have conserved SECIS sequences within the coding region of selenoprotein mRNAs (Fig. 6.4). Instead, conserved hairpin structures for different selenoprotein mRNAs were only found in untranslated regions of *M. jannaschii* transcripts (Wilting et al. 1997; Rother et al. 2001). In addition, inspection of the *M. jannaschii* genome revealed a putative archaeal SelB homolog, which was able to bind guanosine nucleotides and aminoacyl-$tRNA^{Sec}$ (Rother et al. 2000), suggesting that it is a key component of selenoprotein synthesis machinery in archaea.

6.3.4 *Selenoproteins in Bacteria and Archaea*

Most intracellular Se is found in selenoproteins in the form of Sec. Since selenol is highly nucleophilic and Sec is mostly deprotonated at physiological pH (pK_a: 5.2 for Sec vs. 8.3 for cysteine, Cys), Sec is more reactive than Cys (Zinoni et al. 1987; Axley et al. 1991). Therefore, due to the chemical properties of Se, almost all selenoproteins with Sec residues in their active site participate in intracellular redox systems (Table 6.1). The membrane-bound formate:hydrogen lyase-linked formate dehydrogenase H (FdhH encoded by *fdhF*) from *E. coli* was cloned, representing the first prokaryotic gene encoding a Sec residue, which is encoded by UGA (Zinoni et al. 1986). In *E. coli*, two other selenoprotein formate dehydrogenases, FdhO and FdhN, have been identified (Sawers et al. 2004). FDHs represent the most widespread selenoproteins in bacteria and archaea (Peng et al. 2016). FDH catalyzes the reversible oxidation of formate to CO_2 and is involved in energy metabolism, carbon

Table 6.1 Selenoproteins in bacteria

Selenoproteins	Gene	Characteristic organism	References
Formate dehydrogenase α subunit	*fdhA*	*E. coli, etc.*	Cox et al. (1981)
Selenophosphate synthetase	*selD*	*Eubacterium acidaminophilum, etc.*	Gursinsky et al. (2008)
Glycine reductase protein A	*grdA*	*Clostridium sticklandii, etc.*	Cone et al. (1976)
Glycine reductase protein B	*grdB*	*E. acidaminophilum, etc.*	Wagner et al. (1999)
Proline reductase	*pr*	*C. sticklandii, etc.*	Kabisch et al. (1999)
Sarcosine reductase	–	*E. acidaminophilum, etc.*	Hormann and Andreesen (1989)
Betaine reductase	–	*E. acidaminophilum, etc.*	Meyer et al. (1995)
Coenzyme F_{420}-reducing hydrogenase α subunit	*frhA*	*Syntrophobacter fumaroxidans, etc.*	Zhang et al. (2006)
Coenzyme F_{420}-reducing hydrogenase δ subunit	*frhD*	*S. fumaroxidans, etc.*	Zhang et al. (2006)
Heterodisulfide reductase subunit A	*hdrA*	*S. fumaroxidans, etc.*	Zhang et al. (2006)
Thioredoxin	*trx*	*Treponema denticola, etc.*	Kim et al. (2015)
Glutaredoxin	*grx*	*Clostridium* sp., *etc.*	Kim et al. (2011)
Peroxiredoxin	*prx*	*E. acidaminophilum, etc.*	Sohling et al. (2001)
Prx-like thiol:disulfide oxidoreductase	–	*Geobacter metallireducens, etc.*	Zhang et al. (2006)
Thiol:disulfide interchange protein	–	*Syntrophus aciditrophicus, etc.*	Zhang et al. (2006)
Fe-S oxidoreductase	*glpC*	*S. fumaroxidans, etc.*	Zhang et al. (2006)
NADH oxidase	–	*G. metallireducens*	Zhang et al. (2006)
Methionine sulfoxide reductase	*msrA*	*Clostridium* sp., *etc.*	Kim et al. (2009), etc.
Electron transfer protein	*prdC*	*C. sticklandii*	Fonknechten et al. (2010)
Glutathione peroxidase	*gpx*	*Treponema denticola*	Zhang et al. (2006)
HesB-like	–	*S. fumaroxidans, etc.*	Zhang et al. (2006)
SelW-like	–	*Desulfotalea psychrophila, etc.*	Zhang et al. (2006)
AhpD-like	–	*Alkaliphilus metalliredigenes*	Zhang et al. (2006)
ArsC-like	–	*D. psychrophila*	Zhang et al. (2006)
DsbA-like	–	*Anaeromyxobacter dehalogenans*	Zhang et al. (2006)
DsbG-like	–	*Symbiobacterium thermophilum*	Zhang et al. (2006)
DsrE-like	–	*Desulfovibrio vulgaris*	Zhang et al. (2006)
Homolog of AhpF	–	*Carboxydothermus hydrogenoformans*	Zhang et al. (2006)
Distant AhpD homolog	–	*Geobacter uraniumreducens*	Zhang et al. (2006)

fixation, and pH homeostasis (Ferry 1990). It contains an Fe-S cluster and either Mo or W (Andreesen and Makdessi 2008) coordinated by a pterin cofactor.

Besides *E. coli*, several selenoproteins have been characterized in *Eubacterium acidaminophilum* and *Clostridium sticklandii*, such as glycine reductase proteins A and B (Cone et al. 1976; Hormann and Andreesen 1989; Dietrichs et al. 1991; Garcia and Stadtman 1992; Meyer et al. 1995; Fonknechten et al. 2010). The glycine reductase system is essential for acetate formation via glycine; it comprises three proteins: glycine reductase proteins A, B, and C. The substrate-binding glycine reductase protein B is encoded by two genes, *grdB* and *grdE* (Wagner et al. 1999). Glycine reductase A (GrdA) is a small acidic, redox-active protein, which accepts the carboxymethyl group from GrdB. Another selenoprotein, D-proline reductase, appears to be similar to glycine reductase protein B and proline reductase B (PrdB) contains Sec in a motif similar to that found in GrdB (Kabisch et al. 1999).

In addition to experimentally verified selenoproteins, bioinformatics analyses have predicted the presence of additional selenoproteins from DNA sequence data (Zhang et al. 2006). A study shows that formate dehydrogenase α subunit (FdhA) and SelD are the most widespread selenoproteins in bacteria and that the bacterium *Syntrophobacter fumaroxidans* contains the largest number of selenoprotein genes among the bacteria that have been genome-sequenced. *S. fumaroxidans* has 31 selenoprotein-encoding genes including 1 for *selD*, 6 for *fdhA*, 3 for *frdA*, 8 for *frhD*, 7 for *hdrA*, 3 for *glpC*, *prx*, *hesB-like*, and *msrA*. Although those bioinformatics analyses provided a number of predicted selenoprotein genes, most gene products have not yet been experimentally characterized.

Several selenoproteins have been identified in methanogenic archaea (Table 6.2) (Jones et al. 1979; Yamazaki 1982; Halboth and Klein 1992; Vorholt et al. 1997; Wilting et al. 1997). The only archaea for which the presence of selenoproteins has been suggested, by either experimentation or prediction from genome sequence data, are methanogens dependent on the hydrogenotrophic methanogenesis pathway (Rother et al. 2001; Kryukov and Gladyshev 2004). However, not all hydrogenotrophic methanogens employ Sec. Within archaeal species, selenoproteins appear to be restricted to two genera, *Methanococcus* and *Methanopyrus*, according to an analysis of 56 available genome sequences representing 43 genera (Rother et al. 2001; Kryukov and Gladyshev 2004). Genome sequence analyses, radioactive *in vivo* labeling, and mutational studies identified at least six methanogenesis-related selenoproteins in *Methanococcus jannaschii*, *M. voltae*, *M. maripaludis*, *M. vannielii*, and *M. kandleri*. Selenophosphate synthetase (SelD) was identified as a selenoprotein, suggesting that selenoproteins synthesis is regulated by a selenoprotein itself (Wilting et al. 1997).

Table 6.2 Selenoproteins in archaea

Selenoproteins	Gene	Characteristic organism	References
Formate dehydrogenase	*fdhA*	*Methanococcus jannaschii*	Wilting et al. (1997)
			Jones et al. (1979)
		M. vannielii	
Selenophosphate synthetase	*selD*	*M. jannaschii*	Wilting et al. (1997)
Heterodisulfide reductase	*hdrA*	*M. jannaschii*	Wilting et al. (1997)
Formyl-methanofuran dehydrogenase	*fwuB*	*M. jannaschii*	Wilting et al. (1997)
		M. kandleri	Vorholt et al. (1997)
F_{420}-reducing hydrogenase	*fruA*	*M. jannaschii*	Wilting et al. (1997)
		M. voltae	Halboth and Klein (1992)
F_{420}-non-reducing hydrogenase	*vhuD/U*	*M. jannaschii*	Wilting et al. (1997)
		M. voltae	Halboth and Klein (1992)
		M. maripaludis	Sorgenfrei et al. (1993)
HesB-like protein	–	*M. jannaschii*	Kryukov and Gladyshev (2004)

6.3.5 Seleno-Amino Acids Metabolism

Three forms of seleno-amino acids have been identified in bacteria: Sec, selenocystine, and SeMet. The enzyme L-Sec lyase exists in various aerobic bacteria (Chocat et al. 1983) and was purified from *Citrobacter freundii* (Chocat et al. 1985). This enzyme exclusively catalyzes the pyridoxal 5′-phosphate (PLP)-dependent decomposition of L-Sec to L-alanine and elemental Se. On the other hand, *E. coli* possesses three cysteine desulfurases, IscS, SufS, and CsdB, which can utilize both L-Sec and L-Cys as a substrate (Mihara et al. 2000). Purified IscS from *E. coli* has been shown in an *in vitro* assay to supply Se from L-Sec for the synthesis of selenophosphate by SelD (Lacourciere et al. 2000). Another seleno-amino acid-acting enzyme, D-selenocystine α,β-lyase, which PLP-dependently decomposes D-selenocystine into pyruvate, ammonia, and elemental Se, was found in the anaerobic bacteria *C. sticklandii* and *C. sporogenes*. The enzyme was purified from *C. sticklandii* and characterized (Esaki et al. 1988). It consists of two subunits and has a molecular mass of approximately 74 kDa. In addition to D-selenocystine, other analogous amino acids can be used as substrates including D-cystine, D-lanthionine, *meso*-lanthionine, and D-cysteine. The biological roles of both L-Sec lyase and D-selenocystine α,β-lyase remain unknown. As Se has a very limited availability as a trace element for microorganisms, these enzymes are proposed to aid in recycling Se from degraded selenoproteins containing L-Sec to make new selenoproteins (Tamura et al. 2004).

L-Sec is the predominant seleno-amino acid in bacteria, but SeMet is also incorporated into proteins (Berntsson et al. 2009). Unlike L-Sec, SeMet is randomly incorporated into proteins instead of methionine because there is no specific machinery for selenomethionine. The incorporation of SeMet in place of Met is not expected to affect protein function. L-Methionine γ-lyase, which has been purified from *Pseudomonas putida*, decomposes SeMet into α-ketobutyrate, ammonia, and methaneselenol (Esaki et al. 1979). Recently, SeMet has been used in probiotic supplements (Krittaphol et al. 2011; Gojkovic et al. 2014) and in experimental analyses such as X-ray crystallography (Berntsson et al. 2009).

6.3.6 Other Selenium Compounds in Bacteria

Se is used for incorporation into selenoproteins as Sec and is also found in several bacterial tRNAs. The modified tRNA nucleoside, 5-methylaminomethyl-2-SeU, is located at the wobble position in the anticodons of $tRNA^{Lys}$, $tRNA^{Glu}$, $tRNA^{Pro}$, and $tRNA^{Gln}$. The modification likely contributes to tRNA recognition and translation efficiency (Chen and Stadtman 1980; Wittwer et al. 1984; Ching et al. 1985a, b; Wolfe et al. 2004). SeU is generated by the specific substitution of Se for sulfur in 2-thiouridine by tRNA 2-SeU synthase (YbbB) (Veres and Stadtman 1994), where selenophosphate serves as a Se donor (Veres et al. 1992; Glass et al. 1993). Mutants of *E. coli* and *S. typhimurium* containing a defective *selD* gene are unable to incorporate Se into proteins and tRNAs (Kramer and Ames 1988; Stadtman et al. 1989).

Comparative genomic and phylogenetic analyses showed the possibility that Se is used in Se-dependent molybdenum hydroxylases (SDMH) as a third pathway of Se utilization in bacteria and archaea (Haft and Self 2008; Zhang et al. 2008). In this pathway, the SelD protein may activate Se for SDMH maturation via two proteins, YqeB and YqeC, whose functions are still unknown. Three SDMHs have been characterized from two species: nicotinic acid hydroxylase and xanthine dehydrogenase from *Eubacterium barkeri* (Gladyshev et al. 1994; Schräder et al. 1999) and xanthine dehydrogenase and purine hydroxylase from *Clostridium purinolyticum* (Self and Stadtman 2000; Self et al. 2003).

Besides the above seleno-molecules, several minor Se-containing molecules have been identified in bacteria, such as Se exopolysaccharide (Ding et al. 2014) and Se-containing phycocyanin (Se-PC) (Chen and Wong 2008). Se-PC was identified and purified from Se-enriched *Spirulina platensis* (Chen and Wong 2008). Se-PC has stronger antioxidant activity than phycocyanin and it shows dose-dependent protective effects against H_2O_2-induced oxidative DNA damage in erythrocytes. Although it is artificial, Se-containing exopolysaccharides have been obtained using a *Rhizobium* sp. N6113 exopolysaccharide (Ding et al. 2014). These novel organic Se species have been proposed for use in antitumor chemoprevention applications.

6.3.7 Selenium Detoxification

As mentioned above, Se oxyanions exhibit toxicity to living organisms. Some bacteria relieve Se toxicity by glutathionylation and methylation of Se compounds. Glutathione is the most abundant low molecular weight thiol in the cell and the reduction of selenite with GSH, producing GS-Se-SG and glutathioselenol, was demonstrated (Ganther 1968; Ganther 1971). In addition to glutathionylation, methyl-selenides and methyl-selenoxides are common degradation and detoxification products for toxic Se oxyanions. The volatile forms, dimethyl selenide (DMSe) and dimethyl diselenide (DMDSe), are 500–700 times less toxic than other Se derivatives (Ganther et al. 1966). The conversion of inorganic and organic Se compounds to their volatile forms by microorganisms was first observed using lake water and sediment (Chau et al. 1976). *Rhodocyclus tenuis* and *Rhodospirillum rubrum* can produce both DMSe and DMDSe from selenate while growing phototrophically, and *R. tenuis* also produces DMSe from selenite (McCarty et al. 1993). One is the bacterial thiopurine methyltransferase (bTPMT) encoded by the *tpm* gene of *Pseudomonas syringae* (Ranjard et al. 2002). The enzyme is involved in the conversion of selenite and Se-methyl-Sec into DMSe and DMDSe. The other is a calicheamicin methyltransferase homolog encoded by the *mmtA* gene (Ranjard et al. 2004). Free SeMet is also converted to DMSe and DMDSe via the pathway including bTPMT (Ranjard et al. 2003). In addition, a recent report shows that *P. stutzeri* NT-I aerobically transform selenate, selenite, and biogenic elemental Se into DMSe and DMDSe; these volatile forms were temporarily accumulated in the aqueous phase and then transferred into the gaseous phase (Kagami et al. 2013). Demethylation of DMSe in anaerobic Se-contaminated sediments was reported to proceed via methanogenic pathways established for growth on dimethylsulfide (Oremland and Zehr 1986).

6.3.8 Transport of Selenium Compounds

Until now, two types of sulfate transporters have been demonstrated to transport Se oxyanions: the sulfate-thiosulfate permease type (Turner et al. 1998) and the SulP-type permease type (Zolotarev et al. 2008). Sulfate-thiosulfate permeases belong to the sulfate/tungstate uptake transporter (SulT) family of the ABC transporter superfamily. In *E. coli* and *Salmonella typhimurium*, two types of SulT sulfate-thiosulfate permeases were identified. They consist of: (1) periplasmic proteins Sbp, (sulfate-binding protein, Pflugrath and Quiocho 1985) and CysP (thiosulfate-binding protein, Hryniewicz et al. 1990); (2) membrane proteins CysT and CysW (Sirko et al. 1990); (3) ATP-binding protein CysA (Sirko et al. 1990). The SulP sulfate permease superfamily is a large and ubiquitous protein family with hundreds of sequenced members derived from all three domains of life. However, only a few proteins in this family have been functionally characterized (Kertesz 2001; Saier et al. 2006).

Selenate transport through the SulP sulfate permease has also been reported in *Mycobacterium tuberculosis* (Zolotarev et al. 2008) and in *Cupriavidus metallidurans* CH34 (Avoscan et al. 2009).

Recent genomic analyses suggest that the membrane protein YedE may also be involved in Se transport (Lin et al. 2015). This protein, exclusively found in Se-utilizing organisms, contains ten transmembrane domains and shows distant similarity to a sulfur transporter (Gristwood et al. 2011). The sulfur-related transporters contain several conserved glycines and an invariant Cys, which is probably an important functional residue. The *yedE* gene locus is often located next to known Se-related genes such as SirA-like, SelD, and Sec lyase in many bacteria of different phyla, implying that the protein may be involved in Se metabolism. A similar relationship between YedE and SelD was observed in archaea, and in this domain, the *yedE* gene is split into two adjacent genes.

6.4 Biotechnological Applications of Bacterial Selenium Metabolism

6.4.1 Production of Functional Materials by Microbial Biofactories

Microbial metabolism can be harnessed for the production of functional materials that find multiple industrial and domestic applications. Se^0 and metal selenide particles possess photo-optical and semiconducting physical properties required in devices such as photocopiers and microelectronic circuits (Oremland et al. 2004). The bulk of the Se^0 particles formed during dissimilatory reduction of Se oxyanions are not randomly shaped blobs of amorphous material, but instead are discretely shaped spheres of nano-sized dimensions (~200–400 nm). The spheres first accumulate on cell surfaces and then are sloughed off into the surrounding medium (Fig. 6.5) (Oremland et al. 2004). Washed nano-spheres recovered from 3 physiologically and

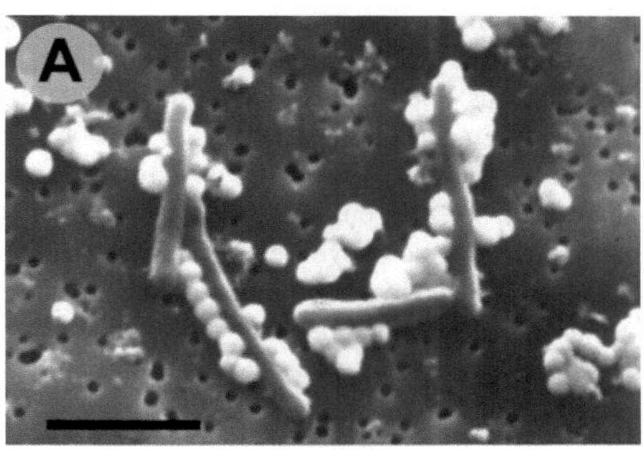

Fig. 6.5 Formation of Se^0 nanospheres on the surface of *B. selenitireducens* that slough off into the extracellular medium. Scale bar = 1 micrometer. (Adapted from Oremland et al. (2004) with permission)

phylogenetically different Se-respirers (*S. barnesii, B. selenitireducens*, and *Selenihalanaerobacter shriftii*) exhibited different spectral properties, which in turn were different from chemically precipitated Se^0. It was also found that washed Se-nanospheres were unable to undergo further reduction to HSe^-, whereas freshly-formed material could. Subsequent studies found that the nano-spheres were enveloped by a diaphanous layer, some of which consisted of peptides that are essential for bacterial adhesion to the materials and ultimately for their spherical shapes and sizes (Lenz et al. 2011; Debieux et al. 2011; Jain et al. 2015; Staicu et al. 2015c; Gonzalez-Gil et al. 2016). Over the past 12 years, considerable scientific interest has been piqued by the phenomenon of bacterial Se^0 nanospheres and many different species of prokaryotes (and microscopic eukaryotes like yeasts) have been found to be able to generate them, as summarized by Shirsat et al. (2015). Most of the described organisms carry out a reductive detoxification reaction when exposed to Se(IV) and do not use Se oxyanions as terminal electron acceptors.

One particular microorganism, *Veillonella atypica*, can form HSe^- by reduction of either Se(+IV) or Se(VI) oxyanions (Pearce et al. 2008, 2011). This organism is particularly attractive because reduction does not proceed through formation of Se^0 and thus the formed HSe^- anions can be precipitated with countering divalent cations of interest (e.g. Zn and Cd) and harvested as truly nanosized particles (~3–5 nm range). It remains to be determined whether or not they can somehow be placed uniformly in nano-arrays ("quantum-dots") that allows for quantum photonic effects to take place and make them suitable for practical applications in nano-technology (Fellowes et al. 2013; Mal et al. 2016). Nonetheless, use of microorganisms rather than harsh chemicals to form nanomaterials of Se and other Group 16 (G16) elements such as tellurium (Baesman et al. 2007, 2009) holds promise of a "green" technology that can produce nano-sized materials, perhaps with unique properties (Nancharaiah and Lens 2015).

6.4.2 Biofortification of Plant Species Using Selenium-Reducing Bacteria

Plant biofortification is a strategy which aims to increase the nutritional value and micronutrient levels (e.g. Se) in the edible parts of crop species (Wu et al. 2015). For this, various approaches can be used including conventional selective breeding, genetic engineering, Se fertilizers or microbial-mediated soil inoculation. Crops are the major source of dietary Se worldwide and may be employed to extract this element from seleniferous soils, thus providing dietary Se in low-Se areas. Se accumulator plant species have been shown to accumulate 100–1000 mg/kg dry weight (DW) Se and Se hyperaccumulators can even accumulate 1000–15,000 mg/kg DW Se on seleniferous soils, that is 0.1–1.5% (El Mehdawi and Pilon-Smits 2012). Endophytic bacteria were shown to have plant growth promoting properties and displayed extremely high tolerance to toxic Se oxyanions by forming red Se^0 particles (Fig. 6.6) (Sura-de Jong et al. 2015). In numerous studies, beneficial bacteria have been inoculated to plant growth medium in an attempt to increase the nutritional value and stimulate biomass production of the plants. Se accumulator *Brassica juncea* (Indian mustard) was inoculated with two Se-tolerant bacterial consortia (G1 and G2) and was shown to accumulate Se to 711 mg/kg DW in leaves, 276 mg/kg DW in pod husks, and 358 mg/kg DW in seeds (Yasin et al. 2015a). Plants inoculated with bacterial consortium G1 showed significantly increased growth (dry biomass and seed weight) as compared to control plants and G2-inoculated plants (Yasin et al. 2015a). Furthermore, the growth of *B. juncea* was stimulated by 1.7-fold using a novel Se-tolerant bacterium, *P. moraviensis* stanleyae, although no significant effect on Se accumulation was observed (Staicu et al. 2015b). In another study, the inoculation of wheat (*Triticum aestivum* L.) with YAM2 cultures, a bacterium with 99% similarity to *Bacillus pichinotyi*, resulted in enhanced plant growth. YAM2-inoculated wheat plants showed significantly higher dry weight, shoot length, and spike length compared to un-inoculated plants and significantly higher Se concentration in wheat kernels (167%) and stems (252%) (Yasin et al. 2015b). The significance of these studies stems mainly from the fact that wheat is a staple food for humans and animals and therefore a useful Se-delivery vehicle. In another study, Se biofortification of wheat plants using endophytic bacteria was shown to lead not only to enhanced plant growth, but could also act as a control strategy against pests such as *Gaeumannomyces graminis*, the principal soil-borne fungal pathogen in volcanic soils from southern Chile (Durán et al. 2014). Plant biofortification using Se-reducing bacterial inocula is an emerging research topic in need of further research and clarification. The exact contribution of bacteria to plant growth enhancement and micronutrient accumulation are open questions to be answered by future studies. In addition, it will be valuable to investigate to what extent Se-bacteria are colonizing plant organs and whether this colonization is an important factor for plant Se accumulation and speciation.

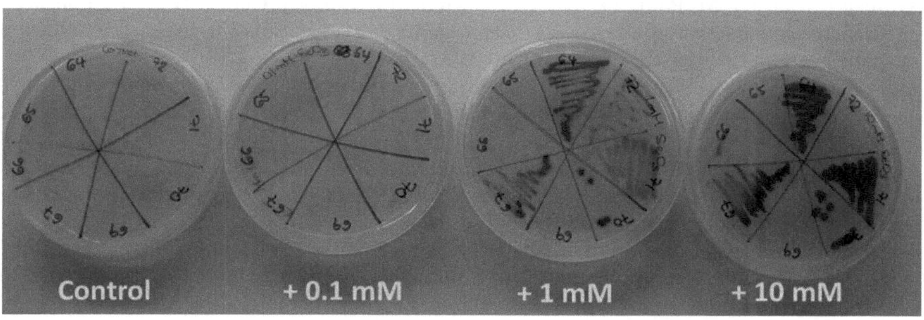

Fig. 6.6 Red Se^0 produced by endophytic bacteria isolated from the root tissue of *Stanleya pinnata* and *Astragalus bisulcatus* that were exposed to progressively higher concentrations [0.1, 1, and 10 mM] of SeO_3^{2-} (as Na_2SeO_3). Control represents a plate containing Luria Bertani growth medium without selenite

6.5 Conclusions

The relationship between bacteria and Se predates the Oxygen Revolution (~2.3 Ga). A large number of sequenced bacterial genomes indicate Se is an ancient trait once common for all bacterial species. Depending on several factors such as bacterial species, enzymatic repertoire, and geochemical context (e.g., oxygen profile, redox conditions, and nutrient availability) bacteria can exploit Se to energize its metabolic machinery, but it can also be affected by the toxicity exhibited by various forms of Se. On the other hand, being an essential micronutrient, a complex molecular mechanism is used by bacteria and archaea to incorporate Se into various cellular components (e.g. amino acids and proteins). Specifically, Se is involved in three major metabolic strategies employed by bacteria: *assimilatory* metabolism (biosynthesis), *dissimilatory* metabolism (energy generation), and detoxification. Several biotechnological applications using Se microbial specialists show high potential for the biofabrication of functional materials (e.g. Se^0 nanoparticles and quantum dots) and for the biofortification of crop species.

Acknowledgments R.S.O. is supported by the U.S. Geological Survey National Research Program. Mention of any brand name products does not constitute an endorsement by the USGS. The authors would like to acknowledge Elizabeth Pilon-Smits (Colorado State University) for helpful comments.

References

Andreesen JR, Makdessi K (2008) Tungsten, the surprisingly positively acting heavy metal element for prokaryotes. Ann N Y Acad Sci 1125:215–229

Avazeri C, Turner RJ, Pommier J, Weiner JH, Giordano G, Vermeglio A (1997) Tellurite reductase activity of nitrate reductase is responsible for the basal resistance of *Escherichia coli* to tellurite. Microbiology 143:1181–1189

Avoscan L, Carriere M, Proux O, Sarret G, Degrouard J, Coves J, Gouget B (2009) Enhanced selenate accumulation in *Cupriavidus metallidurans* CH34 does not trigger a detoxification pathway. Appl Environ Microbiol 75:2250–2252

Axley MJ, Böck A, Stadtman TC (1991) Catalytic properties of an *Escherichia coli* formate dehydrogenase mutant in which sulfur replaces selenium. Proc Natl Acad Sci U S A 88:8450–8454

Ayano H, Miyake M, Terasawa K, Kuroda M, Soda S, Sakaguchi T, Ike M (2014) Isolation of a selenite-reducing and cadmium-resistant bacterium *Pseudomonas* sp. strain RB for microbial synthesis of CdSe nanoparticles. J Biosci Bioeng 117:576–581

Baesman SM, Bullen TD, Dewald J, Zhang DH, Curran S, Islam FS, Beveridge TJ, Oremland RS (2007) Formation of tellurium nanocrystals during anaerobic growth of bacteria that use Te oxyanions as respiratory electron acceptors. Appl Environ Microbiol 73:2135–2143

Baesman SM, Stolz JF, Kulp TR, Oremland RS (2009) Enrichment and isolation of *Bacillus beveridgei* sp. nov., a facultative anaerobic haloalkaliphile from Mono Lake, California that respires oxyanions of tellurium, selenium, and arsenic. Extremophiles 13:695–705

Baron C, Böck A (1991) The length of the aminoacyl-acceptor stem of the selenocysteine-specific tRNA(Sec) of *Escherichia coli* is the determinant for binding to elongation factors SELB or Tu. J Biol Chem 266:20375–20379

Bebien M, Kirsch J, Mejean V, Vermeglio A (2002) Involvement of a putative molybdenum enzyme in the reduction of selenate by *Escherichia coli*. Microbiology 148:3865–3872

Berntsson RP, Alia Oktaviani N, Fusetti F, Thunnissen AM, Poolman B, Slotboom DJ (2009) Selenomethionine incorporation in proteins expressed in *Lactococcus lactis*. Protein Sci 18:1121–1127

Björnstedt M, Kumar S, Holmgren A (1992) Selenodiglutathione is a highly efficient oxidant of reduced thioredoxin and a substrate for mammalian thioredoxin reductase. J Biol Chem 267:8030–8034

Bult CJ, White O, Olsen GJ, Zhou L, Fleischmann RD, Sutton GG, Blake JA, FitzGerald LM, Clayton RA, Gocayne JD, Kerlavage AR, Dougherty BA, Tomb JF, Adams MD, Reich CI, Overbeek R, Kirkness EF, Weinstock KG, Merrick JM, Glodek A, Scott JL, Geoghagen NS, Venter JC et al (1996) Complete genome sequence of the methanogenic archaeon, Methanococcus jannaschii. Science 273:1058–1073

Butler CS, Debieux CM, Dridge EJ, Splatt P, Wright M (2012) Biomineralization of selenium by the selenate-respiring bacterium *Thauera selenatis*. Biochem Soc Trans 40:1239–1243

Chapman PM, Adams WJ, Brooks M, Delos CG, Luoma SN, Maher WA, Ohlendorf HM, Presser TS, Shaw P (2010) Ecological assessment of selenium in the aquatic environments. CRC Press, Boca Raton

Chau YK, Wong PT, Silverberg BA, Luxon PL, Bengert GA (1976) Methylation of selenium in the aquatic environment. Science 192:1130–1131

Chen CS, Stadtman TC (1980) Selenium-containing tRNAs from *Clostridium sticklandii*: cochromatography of one species with L-prolyl-tRNA. Proc Natl Acad Sci U S A 77:1403–1407

Chen T, Wong YS (2008) In vitro antioxidant and antiproliferative activities of selenium-containing phycocyanin from selenium-enriched *Spirulina platensis*. J Agric Food Chem 56:4352–4358

Ching WM, Alzner-DeWeerd B, Stadtman TC (1985a) A selenium-containing nucleoside at the first position of the anticodon in seleno-$tRNA^{Glu}$ from *Clostridium sticklandii*. Proc Natl Acad Sci U S A 82:347–350

Ching WM, Tsai L, Wittwer AJ (1985b) Selenium-containing transfer RNAs. Curr Top Cell Regul 27:497–507

Chocat P, Esaki N, Nakamura T, Tanaka H, Soda K (1983) Microbial distribution of selenocysteine lyase. J Bacteriol 156:455–457

Chocat P, Esaki N, Tanizawa K, Nakamura K, Tanaka H, Soda K (1985) Purification and characterization of selenocysteine β-lyase from *Citrobacter freundii*. J Bacteriol 163:669–676

Commans S, Böck A (1999) Selenocysteine inserting tRNAs: an overview. FEMS Microbiol Rev 23:335–351

Cone JE, Del Rio RM, Davis JN, Stadtman TC (1976) Chemical characterization of the selenoprotein component of clostridial glycine reductase: identification of selenocysteine as the organoselenium moiety. Proc Natl Acad Sci U S A 73:2659–2663

Cox JC, Edwards ES, DeMoss JA (1981) Resolution of distinct selenium-containing formate dehydrogenases from *Escherichia coli*. J Bacteriol 145:1317–1324

Debieux CM, Dridge EJ, Mueller CM, Splatt P, Paszkiewicz K, Knight I, Florance H, Love J, Titball RW, Lewis RJ, Richardson DJ, Butler CS (2011) A bacterial process for selenium nanosphere assembly. Proc Natl Acad Sci U S A 108:13480–13485

Dietrichs D, Meyer M, Rieth M, Andreesen JR (1991) Interaction of selenoprotein PA and the thioredoxin system, components of the NADPH-dependent reduction of glycine in *Eubacterium acidaminophilum* and *Clostridium litorale*. J Bacteriol 173:5983–5991

Ding GB, Nie RH, Lv LH, Wei GQ, Zhao LQ (2014) Preparation and biological evaluation of a novel selenium-containing exopolysaccharide from *Rhizobium* sp. N613. Carbohydr Polym 109:28–34

Dowdle PR, Oremland RS (1998) Microbial oxidation of elemental selenium in soils and bacterial cultures. Environ Sci Technol 32:3749–3755
Durán P, Acuña J, Jorquera M, Azcón R, Paredes C, Rengel Z, de la Luz Mora M (2014) Endophytic bacteria from selenium-supplemented wheat plants could be useful for plant-growth promotion, biofortification and *Gaeumannomyces graminis* biocontrol in wheat production. Biol Fertil Soils 50:983–990
El Mehdawi AF, Pilon-Smits EAH (2012) Ecological aspects of plant selenium hyperaccumulation. Plant Biol 14:1–10
Ellis AS, Johnson TM, Herbel MJ, Bullen TD (2003) Stable isotope fractionation by natural microbial consortia. Chem Geol 195:119–129
Esaki N, Tanaka H, Uemura S, Suzuki T, Soda K (1979) Catalytic action of L-methionine γ-lyase on selenomethionine and selenols. Biochemistry 18:407–410
Esaki N, Seraneeprakarn V, Tanaka H, Soda K (1988) Purification and characterization of *Clostridium sticklandii* D-selenocystine α, β-lyase. J Bacteriol 170:751–756
Fellowes JW, Pattrick RAD, Lloyd JR, Charnock JM, Mosselmans JFW, Weng T-C, Pearce CI (2013) *Ex situ* formation of metal selenide quantum dots using bacterially derived selenide precursors. Nanotech 24:145603
Fernandez-Martinez A, Charlet L (2009) Selenium environmental cycling and bioavailability: a structural chemist point of view. Rev Environ Sci Biotechnol 8:81–110
Ferry JG (1990) Formate dehydrogenase. FEMS Microbiol Rev 7:377–382
Fonknechten N, Chaussonnerie S, Tricot S, Lajus A, Andreesen JR, Perchat N, Pelletier E, Gouyvenoux M, Barbe V, Salanoubat M, Le Paslier D, Weissenbach J, Cohen GN, Kreimeyer A (2010) *Clostridium sticklandii*, a specialist in amino acid degradation: revisiting its metabolism through its genome sequence. BMC Genomics 11:555
Forchhammer K, Böck A (1991) Selenocysteine synthase from *Escherichia coli.* Analysis of the reaction sequence. J Biol Chem 266:6324–6328
Forchhammer K, Leinfelder W, Böck A (1989) Identification of a novel translation factor necessary for the incorporation of selenocysteine into protein. Nature 342:453–456
Forchhammer K, Rucknagel KP, Böck A (1990) Purification and biochemical characterization of SELB, a translation factor involved in selenoprotein synthesis. J Biol Chem 265:9346–9350
Forchhammer K, Leinfelder W, Boesmiller K, Veprek B, Böck A (1991) Selenocysteine synthase from *Escherichia coli.* Nucleotide sequence of the gene (*selA*) and purification of the protein. J Biol Chem 266:6318–6323
Forster C, Ott G, Forchhammer K, Sprinzl M (1990) Interaction of a selenocysteine-incorporating tRNA with elongation factor Tu from *E.coli.* Nucleic Acids Res 18:487–491
Fujita M, Ike M, Nisbimoto S, Takahasbi K, Kasbiwa M (1997) Isolation and characterization of a novel selenate-reducing bacterium, *Bacillus* sp. SF-1. J Ferment Bioeng 83:517–522
Ganther HE (1968) Selenotrisulfides. Formation by reaction of thiols with selenious acid. Biochem 7:2898–2905
Ganther HE (1971) Reduction of the selenotrisulfide derivative of glutathione to a persulfide analog by glutathione reductase. Biochemistry 10:4089–4098
Ganther HE, Levander OA, Baumann CA (1966) Dietary control of selenium volatilization in the rat. J Nutr 88:55–60
Garcia GE, Stadtman TC (1992) *Clostridium sticklandii* glycine reductase selenoprotein A gene: cloning, sequencing, and expression in *Escherichia coli.* J Bacteriol 174:7080–7089
Gerrard TL, Telford JN, Williams HH (1974) Detection of selenium deposits in *Escherichia coli* by electron microscopy. J Bacteriol 119:1057–1060
Gladyshev VN, Khangulov SV, Stadtman TC (1994) Nicotinic acid hydroxylase from *Clostridium barkeri*: electron paramagnetic resonance studies show that selenium is coordinated with molybdenum in the catalytically active selenium-dependent enzyme. Proc Natl Acad Sci U S A 91:232–236
Glass RS, Singh WP, Jung W, Veres Z, Scholz TD, Stadtman TC (1993) Monoselenophosphate: synthesis, characterization, and identity with the prokaryotic biological selenium donor, compound SePX. Biochemistry 32:12555–12559

Gojkovic Z, Vilchez C, Torronteras R, Vigara J, Gomez-Jacinto V, Janzer N, Gomez-Ariza JL, Marova I, Garbayo I (2014) Effect of selenate on viability and selenomethionine accumulation of *Chlorella sorokiniana* grown in batch culture. Sci World J 2014:401265

Gonzalez-Gil G, Lens PNL, Saikaly PE (2016) Selenite reduction by anaerobic microbial aggregates: microbial community structure, and proteins associated to the produced selenium spheres. Front Microbiol 7:571

Gristwood T, McNeil MB, Clulow JS, Salmond GP, Fineran PC (2011) PigS and PigP regulate prodigiosin biosynthesis in *Serratia* via differential control of divergent operons, which include predicted transporters of sulfur-containing molecules. J Bacteriol 193:1076–1085

Gursinsky T, Gröbe D, Schierhorn A, Jäger J, Andreesen JR, Söhling B (2008) Factors and selenocysteine insertion sequence requirements for the synthesis of selenoproteins from a gram-positive anaerobe in *Escherichia coli*. Appl Environ Microbiol 74:1385–1393

Guymer D, Maillard J, Sargent F (2009) A genetic analysis of in vivo selenate reduction by *Salmonella enterica* serovar Typhimurium LT2 and *Escherichia coli* K12. Arch Microbiol 191:519–528

Haft DH, Self WT (2008) Orphan SelD proteins and selenium-dependent molybdenum hydroxylases. Biol Direct 3:4

Halboth S, Klein A (1992) *Methanococcus voltae* harbors four gene clusters potentially encoding two [NiFe] and two [NiFeSe] hydrogenases, each of the cofactor F420-reducing or F420-non-reducing types. Mol Gen Genet 233:217–224

Herbel MJ, Johnson TM, Oremland RS, Bullen TD (2000) Fractionation of selenium isotopes during bacterial respiratory reduction of selenium oxyanions. Geochim Cosmochim Acta 64:3701–3709

Herbel MJ, Johnson TM, Tanji KK, Gao S, Bullen TD (2002) Selenium stable isotope ratios in California agricultural drainage water. J Environ Qual 31:1146–1156

Herbel MJ, Blum JS, Oremland RS, Borglin SE (2003) Reduction of elemental selenium to selenide: experiments with anoxic sediments and bacteria that respire Se-oxyanions. Geomicrobiol J 20:587–602

Hockin SL, Gadd GM (2003) Linked redox precipitation of sulfur and selenium under anaerobic conditions by sulfate-reducing bacterial biofilms. Appl Environ Microbiol 69:7063–7072

Hoffman DJ (2002) Role of selenium toxicity and oxidative stress in aquatic birds. Aquat Toxicol 57:11–26

Hormann K, Andreesen JR (1989) Reductive cleavage of sarcosine and betaine by *Eubacterium acidaminophilum* via enzyme systems different from glycine reductase. Arch Microbiol 153:50–59

Hryniewicz M, Sirko A, Pałucha A, Böck A, Hulanicka D (1990) Sulfate and thiosulfate transport in *Escherichia coli* K-12: identification of a gene encoding a novel protein involved in thiosulfate binding. J Bacteriol 172:3358–3366

Huber R, Sacher M, Vollmann A, Huber H, Rose D (2000) Respiration of arsenate and selenate by hyperthermophilic archaea. Syst Appl Microbiol 23:305–314

Jain R, Jordan N, Weiss S, Foerstendorf H, Heim K, Kacker R, Hübner R, Kramer H, van Hullebusch ED, Farges F, Lens PNL (2015) Extracellular polymeric substances (EPS) govern the surface charge of biogenic elemental selenium nanoparticles. Environ Sci Technol 49:1713–1720

Johnson TM, Herbel MJ, Bullen TD, Zawislanski PT (1999) Selenium isotope ratios as indicators of selenium sources and oxyanion reduction. Geochim Cosmochim Acta 63:2775–2783

Jones JB, Dilworth GL, Stadtman TC (1979) Occurrence of selenocysteine in the selenium-dependent formate dehydrogenase of *Methanococcus vannielii*. Arch Biochem Biophys 195:255–260

Kabisch UC, Grantzdorffer A, Schierhorn A, Rucknagel KP, Andreesen JR, Pich A (1999) Identification of D-proline reductase from *Clostridium sticklandii* as a selenoenzyme and indications for a catalytically active pyruvoyl group derived from a cysteine residue by cleavage of a proprotein. J Biol Chem 274:8445–8454

Kagami T, Narita T, Kuroda M, Notaguchi E, Yamashita M, Sei K, Soda S, Ike M (2013) Effective selenium volatilization under aerobic conditions and recovery from the aqueous phase by *Pseudomonas stutzeri* NT-I. Water Res 47:1361–1368

Kertesz MA (2001) Bacterial transporters for sulfate and organosulfur compounds. Res Microbiol 152:279–290

Kessi J, Ramuz M, Wehrli E, Spycher M, Bachofen R (1999) Reduction of selenite and detoxification of elemental selenium by the phototrophic bacterium *Rhodospirillum rubrum*. Appl Environ Microbiol 65:4734–4740

Kim HY, Zhang Y, Lee BC, Kim JR, Gladyshev VN (2009) The selenoproteome *of Clostridium* sp. OhILAs: characterization of anaerobic bacterial selenoprotein methionine sulfoxide reductase A. Proteins 74:1008–1017

Kim MJ, Lee BC, Jeong J, Lee KJ, Hwang KY, Gladyshev VN, Kim HY (2011) Tandem use of selenocysteine: adaptation of a selenoprotein glutaredoxin for reduction of selenoprotein methionine sulfoxide reductase. Mol Microbiol 79:1194–1203

Kim MJ, Lee BC, Hwang KY, Gladyshev VN, Kim HY (2015) Selenium utilization in thioredoxin and catalytic advantage provided by selenocysteine. Biochem Biophys Res Commun 461:648–652

Kramer GF, Ames BN (1988) Isolation and characterization of a selenium metabolism mutant of *Salmonella typhimurium*. J Bacteriol 170:736–743

Krittaphol W, Wescombe PA, Thomson CD, McDowell A, Tagg JR, Fawcett JP (2011) Metabolism of L-selenomethionine and selenite by probiotic bacteria: in vitro and in vivo studies. Biol Trace Elem Res 144:1358–1369

Kromayer M, Wilting R, Tormay P, Böck A (1996) Domain structure of the prokaryotic selenocysteine-specific elongation factor SelB. J Mol Biol 262:413–420

Kryukov GV, Gladyshev VN (2004) The prokaryotic selenoproteome. EMBO Rep 5:538–543

Kuroda M, Yamashita M, Miwa E, Imao K, Fujimoto N, Ono H, Nagano K, Sei K, Ike M (2011) Molecular cloning and characterization of the *srdBCA* operon, encoding the respiratory selenate reductase complex, from the selenate-reducing bacterium *Bacillus selenatarsenatis* SF-1. J Bacteriol 193:2141–2148

Labunskyy VM, Hatfield DL, Gladyshev VN (2014) Selenoproteins: molecular pathways and physiological roles. Physiol Rev 94:739–777

Lacourciere GM, Mihara H, Kurihara T, Esaki N, Stadtman TC (2000) *Escherichia coli* NifS-like proteins provide selenium in the pathway for the biosynthesis of selenophosphate. J Biol Chem 275:23769–23773

Laverman AM, Switzer Blum J, Schaefer JK, Phillips EJP, Lovley DR, Oremland RS (1995) Growth of strain SES-3 with arsenate and other diverse electron acceptors. Appl Environ Microbiol 61:3556–3561

Leinfelder W, Stadtman TC, Böck A (1989) Occurrence *in vivo* of selenocysteyl-tRNA(SERUCA) in *Escherichia coli*. Effect of *sel* mutations. J Biol Chem 264:9720–9723

Leinfelder W, Forchhammer K, Veprek B, Zehelein E, Böck A (1990) *In vitro* synthesis of selenocysteinyl-$tRNA_{UCA}$ from seryl-$tRNA_{UCA}$: involvement and characterization of the *selD* gene product. Proc Natl Acad Sci U S A 87:543–547

Lenz M, Kolvenbach B, Gygax B, Moes S, Corvini FX (2011) Shedding light on selenium biomineralization: proteins associated with bionanominerals. Appl Environ Microbiol 77:4676–4680

Li DB, Cheng YY, Wu C, Li WW, Li N, Yang ZC, Tong ZH, Yu HQ (2014) Selenite reduction by *Shewanella oneidensis* MR-1 is mediated by fumarate reductase in periplasm. Sci Rep 4:3735

Lin J, Peng T, Jiang L, Ni JZ, Liu Q, Chen L, Zhang Y (2015) Comparative genomics reveals new candidate genes involved in selenium metabolism in prokaryotes. Genome Biol Evol 7:664–676

Losi ME, Frankenberger WT (1997) Reduction of selenium oxyanions by *Enterobacter cloacae* SLD1a-1: isolation and growth of the bacterium and its expulsion of selenium particles. Appl Environ Microbiol 63:3079–3084

Lowe EC, Bydder S, Hartshorne RS, Tape HL, Dridge EJ, Debieux CM, Paszkiewicz K, Singleton I, Lewis RJ, Santini JM, Richardson DJ, Butler CS (2010) Quinol-cytochrome *c* oxidoreduc-

tase and cytochrome c_4 mediate electron transfer during selenate respiration in *Thauera selenatis*. J Biol Chem 285:18433–18442
Macy JM, Michel TA, Kirsch DG (1989) Selenate reduction by *Pseudomonas* species: a new mode of anaerobic respiration. FEMS Microbiol Lett 61:195–198
Macy JM, Rech S, Auling G, Dorsch M, Stackebrandt E, Sly LI (1993) *Thauera selenatis* gen. nov. a member of the beta subclass of Proteobacteria with a novel type of anaerobic respiration. Int J Syst Bacteriol 43:135–142
Mal J, Yarlagadda VN, Hullebusch EDV, Lens PNL (2016) Metal chalcogenide quantum dots: biotechnological synthesis and applications. RSC Adv 6:41477–41495
McCarty S, Chasteen T, Marshall M, Fall R, Bachofen R (1993) Phototrophic bacteria produce volatile, methylated sulfur and selenium compounds. FEMS Microbiol Lett 112:93–98
Meyer M, Granderath K, Andreesen JR (1995) Purification and characterization of protein PB of betaine reductase and its relationship to the corresponding proteins glycine reductase and sarcosine reductase from *Eubacterium acidaminophilum*. Eur J Biochem 234:184–191
Mihara H, Kurihara T, Yoshimura T, Esaki N (2000) Kinetic and mutational studies of three NifS homologs from *Escherichia coli*: mechanistic difference between L-cysteine desulfurase and L-selenocysteine lyase reactions. J Biochem 127:559–567
Nancharaiah YV, Lens PN (2015) Ecology and biotechnology of selenium-respiring bacteria. Microbiol Mol Biol Rev 79:61–80
Ni TW, Staicu LC, Nemeth R, Schwartz C, Crawford D, Seligman J, Hunter WJ, Pilon-Smits EAH, Ackerson CJ (2015) Progress toward clonable inorganic nanoparticles. Nanoscale 7:17320–17327
Oremland RS, Zehr JP (1986) Formation of methane and carbon dioxide from dimethylselenide in anoxic sediments and by a methanogenic bacterium. Appl Environ Microbiol 52:1031–1036
Oremland RS, Hollibaugh JT, Maest AS, Presser TS, Miller LG, Culberston CW (1989) Selenate reduction to elemental selenium by anaerobic bacteria in sediments and culture: biogeochemical significance of a novel, sulfate-independent respiration. Appl Environ Microbiol 55:2333–2343
Oremland RS, Steinberg NA, Maest AS, Miller LG, Hollibaugh JT (1990) Measurement of in situ rates of selenate removal by dissimilatory bacterial reduction in sediments. Environ Sci Technol 24:1157–1164
Oremland RS, Steinberg NA, Presser TS, Miller LG (1991) *In situ* bacterial selenate reduction in the agricultural drainage systems of western Nevada. Appl Environ Microbiol 57:615–617
Oremland RS, Switzer Blum J, Culberston CW, Visscher PT, Miller LG, Dowdle P, Strohmaier FE (1994) Isolation, growth and metabolism of an obligately anaerobic, selenate-respiring bacterium, strain SES-3. Appl Environ Microbiol 60:3011–3019
Oremland RS, Herbel MJ, Switzer Blum J, Langley S, Beveridge TJ, Sutto T, Ajayan PM, Ellis A, Curran S (2004) Structural and spectral features of selenium nanospheres formed by Se-respiring bacteria. Appl Environ Microbiol 70:52–60
Pearce CI, Coker VS, Charnock JM, Pattrick RA, Mosselmans JF, Law N, Beveridge TJ, Lloyd JR (2008) Microbial manufacture of chalcogenide-based nanoparticles via the reduction of selenite using *Veillonella atypica*: an in situ EXAFS study. Nanotechnology 19:155603
Pearce CI, Baesman SM, Blum JS, Fellowes JW, Oremland RS (2011) Nanoparticles formed from microbial oxyanion reduction of toxic group 15 and group 16 metalloids. In: Stolz JF, Oremland RS (eds) Microbial metal and metalloid metabolism: advances and applications. ASM Press, Washington, DC
Peng T, Lin J, Xu YZ, Zhang Y (2016) Comparative genomics reveals new evolutionary and ecological patterns of selenium utilization in bacteria. ISME J 10:2048–2059
Pettine M, Gennari F, Campanella L (2013) The reaction of selenium (IV) with ascorbic acid: its relevance in aqueous and soil systems. Chemosphere 90:245–250
Pflugrath JW, Quiocho FA (1985) Sulphate sequestered in the sulphate-binding protein of *Salmonella typhimurium* is bound solely by hydrogen bonds. Nature 314:257–260
Ranjard L, Prigent-Combaret C, Nazaret S, Cournoyer B (2002) Methylation of inorganic and organic selenium by the bacterial thiopurine methyltransferase. J Bacteriol 184:3146–3149

Ranjard L, Nazaret S, Cournoyer B (2003) Freshwater bacteria can methylate selenium through the thiopurine methyltransferase pathway. Appl Environ Microbiol 69:3784–3790
Ranjard L, Prigent-Combaret C, Favre-Bonté S, Monnez C, Nazaret S, Cournoyer B (2004) Characterization of a novel selenium methyltransferase from freshwater bacteria showing strong similarities with the calicheamicin methyltransferase. Biochim Biophys Acta 1679:80–85
Rech SA, Macy JM (1992) The terminal reductases for selenate and nitrate respiration in *Thauera selenatis* are two distinct enzymes. J Bacteriol 174:7316–7320
Ridley H, Watts CA, Richardson DJ, Butler CS (2006) Resolution of distinct membrane-bound enzymes from *Enterobacter cloacae* SLD1a-1 that are responsible for selective reduction of nitrate and selenate oxyanions. Appl Environ Microbiol 72:5173–5180
Rother M, Wilting R, Commans S, Böck A (2000) Identification and characterisation of the selenocysteine-specific translation factor SelB from the archaeon *Methanococcus jannaschii*. J Mol Biol 299:351–358
Rother M, Resch A, Wilting R, Böck A (2001) Selenoprotein synthesis in archaea. Biofactors 14:75–83
Saier MH Jr, Tran CV, Barabote RD (2006) TCDB: the transporter classification database for membrane transport protein analyses and information. Nucleic Acids Res 34:D181–D186
Santos S, Ungureanu G, Boaventura R, Botelho C (2015) Selenium contaminated waters: an overview of analytical methods, treatment options and recent advances in sorption methods. Sci Total Environ 521-522C:246–260
Sawers RG, Blokesch M, Böck A (2004) Anaerobic formate and hydrogen metabolism. EcoSal Plus 1(1). doi:10.1128/ecosalplus.3.5.4
Schellenger AEP, Onnis-Hayden A, Jisi DP, Larese-Casanova P (2015) Oxygen kinetic isotope effects in selenate during microbial reduction. Appl Geochem 63:261–271
Schon A, Böck A, Ott G, Sprinzl M, Söll D (1989) The selenocysteine-inserting opal suppressor serine tRNA from *E. coli* is highly unusual in structure and modification. Nucleic Acids Res 17:7159–7165
Schräder T, Rienhöfer A, Andreesen JR (1999) Selenium-containing xanthine dehydrogenase from *Eubacterium barkeri*. Eur J Biochem 264:862–871
Schröder I, Rech S, Krafft T, Macy JM (1997) Purification and characterization of the selenate reductase from *Thauera selenatis*. J Biol Chem 272:23765–23768
Self WT, Stadtman TC (2000) Selenium-dependent metabolism of purines: a selenium-dependent purine hydroxylase and xanthine dehydrogenase were purified from *Clostridium purinolyticum* and characterized. Proc Natl Acad Sci U S A 97:7208–7213
Self WT, Wolfe MD, Stadtman TC (2003) Cofactor determination and spectroscopic characterization of the selenium-dependent purine hydroxylase from *Clostridium purinolyticum*. Biochemistry 42:11382–11390
Shirsat S, Kadam A, Naushad M, Mane RS (2015) Selenium nanostructures: microbial synthesis and applications. Roy Soc Chem Adv 5:92799–92811
Shrift A (1964) A selenium cycle in nature? Nature 201:1304–1305
Simmons DB, Wallschlaeger D (2005) A critical review of the biogeochemistry and ecotoxicology of selenium in lotic and lentic environments. Environ Toxicol Chem 24:1331–1343
Sirko A, Hryniewicz M, Hulanicka D, Böck A (1990) Sulfate and thiosulfate transport in *Escherichia coli* K-12: nucleotide sequence and expression of the cysTWAM gene cluster. J Bacteriol 172:3351–3357
Sohling B, Parther T, Rucknagel KP, Wagner MA, Andreesen JR (2001) A selenocysteine-containing peroxiredoxin from the strictly anaerobic organism *Eubacterium acidaminophilum*. Biol Chem 382:979–986
Sorgenfrei O, Klein A, Albracht SP (1993) Influence of illumination on the electronic interaction between ^{77}Se and nickel in active F420-non-reducing hydrogenase from *Methanococcus voltae*. FEBS Lett 332:291–297
Stadtman TC (1974) Selenium biochemistry. Science 183:915–922

Stadtman TC, Davis JN, Zehelein E, Böck A (1989) Biochemical and genetic analysis of *Salmonella typhimurium* and *Escherichia coli* mutants defective in specific incorporation of selenium into formate dehydrogenase and tRNAs. Biofactors 2:35–44

Staicu LC, van Hullebusch ED, Lens PNL (2015a) Production, recovery and reuse of biogenic elemental selenium. Environ Chem Lett 13:89–96

Staicu LC, Ackerson CJ, Cornelis P, Ye L, Berendsen RL, Hunter WJ, Noblitt SD, Henry CS, Cappa JJ, Montenieri RL, Wong AO, Musilova L, Sura-de Jong M, van Hullebusch ED, Lens PN, Reynolds RJ, Pilon-Smits EA et al (2015b) Pseudomonas moraviensis subsp. stanleyae: a bacterial endophyte capable of efficient selenite reduction to elemental selenium under aerobic conditions. J Appl Microbiol 119:400–410

Staicu LC, van Hullebusch ED, Oturan MA, Ackerson CJ, Lens PNL (2015c) Removal of colloidal biogenic selenium from wastewater. Chemosphere 125:130–138

Steinberg NA, Oremland RS (1990) Dissimilatory selenate reduction potentials in a diversity of sediment types. Appl Environ Microbiol 56:3550–3557

Stolz JF, Oremland RS (1999) Bacterial respiration of arsenic and selenium. FEMS Microbiol Rev 23:615–627

Stolz JF, Basu P, Santini JM, Oremland RS (2006) Arsenic and selenium in microbial metabolism. Annu Rev Microbiol 60:107–130

Sura-de Jong M, Reynolds RJ, Richterova MK, Musilova L, Staicu LC, Chocholata I, Cappa JJ, Taghavi S, van der Lelie D, Frantik T, Dolinova I, Strejcek M, Cochran AT, Lovecka P, Pilon-Smits EA et al (2015) Selenium hyperaccumulators harbor a diverse endophytic bacterial community characterized by high selenium tolerance and growth promoting properties. Front Plant Sci 6:113

Switzer Blum J, Burns Bindi A, Buzzelli J, Stolz JF, Oremland RS (1998) *Bacillus arsenicoselenatis*, sp. nov. and *Bacillus selenitireducens*, sp. nov.: two haloalkaliphiles from Mono Lake, California that respire oxyanions of selenium and arsenic. Arch Microbiol 171:19–30

Tamura T, Yamamoto S, Takahata M, Sakaguchi H, Tanaka H, Stadtman TC, Inagaki K (2004) Selenophosphate synthetase genes from lung adenocarcinoma cells: Sps1 for recycling L-selenocysteine and Sps2 for selenite assimilation. Proc Natl Acad Sci U S A 101:16162–16167

Tomei FA, Barton LL, Lemanski CL, Zocco TG (1992) Reduction of selenate and selenite to elemental selenium by *Wolinella succinogenes*. Can J Microbiol 38:1328–1333

Tomei FA, Barton LL, Lemanski CL, Zocco TG, Fink NH, Sillerud LO (1995) Transformation of selenate and selenite to elemental selenium by *Desulfovibrio desulfuricans*. J Ind Microbiol 14:329–336

Tormay P, Wilting R, Heider J, Böck A (1994) Genes coding for the selenocysteine-inserting tRNA species from *Desulfomicrobium baculatum* and *Clostridium thermoaceticum*: structural and evolutionary implications. J Bacteriol 176:1268–1274

Turner RJ, Weiner JH, Taylor DE (1998) Selenium metabolism in *Escherichia coli*. Biometals 11:223–227

Veres Z, Stadtman TC (1994) A purified selenophosphate-dependent enzyme from *Salmonella typhimurium* catalyzes the replacement of sulfur in 2-thiouridine residues in tRNAs with selenium. Proc Natl Acad Sci U S A 91:8092–8096

Veres Z, Tsai L, Scholz TD, Politino M, Balaban RS, Stadtman T (1992) Synthesis of 5-methylam inomethyl-2-selenouridine in tRNAs: ^{31}P NMR studies show the labile selenium donor synthesized by the *selD* gene product contains selenium bonded to phosphorus. Proc Natl Acad Sci U S A 89:2975–2979

Vorholt JA, Vaupel M, Thauer RK (1997) A selenium-dependent and a selenium-independent formylmethanofuran dehydrogenase and their transcriptional regulation in the hyperthermophilic *Methanopyrus kandleri*. Mol Microbiol 23:1033–1042

Wadhwani SA, Shedbalkar UU, Singh R, Chopade BA (2016) Biogenic selenium nanoparticles: current status and future prospects. Appl Microbiol Biotechnol 100:2555–2566

Wagner M, Sonntag D, Grimm R, Pich A, Eckerskorn C, Sohling B, Andreesen JR (1999) Substrate-specific selenoprotein B of glycine reductase from *Eubacterium acidaminophilum*. Biochemical and molecular analysis. Eur J Biochem 260:38–49
Wilting R, Schorling S, Persson BC, Böck A (1997) Selenoprotein synthesis in archaea: identification of an mRNA element of *Methanococcus jannaschii* probably directing selenocysteine insertion. J Mol Biol 266:637–641
Winkel LHE, Vriens B, Jones GD, Schneider LS, Pilon-Smits EAH, Banuelos GS (2015) Selenium cycling across soil-plant-atmosphere interfaces: a critical review. Forum Nutr 7:4199–4239
Wittwer AJ, Tsai L, Ching WM, Stadtman TC (1984) Identification and synthesis of a naturally occurring selenonucleoside in bacterial tRNAs: 5-[(methylamino)methyl]-2-selenouridine. Biochemistry 23:4650–4655
Wolfe MD, Ahmed F, Lacourciere GM, Lauhon CT, Stadtman TC, Larson TJ (2004) Functional diversity of the rhodanese homology domain: the *Escherichia coli ybbB* gene encodes a selenophosphate-dependent tRNA 2-selenouridine synthase. J Biol Chem 279:1801–1809
Wu Z, Bañuelos GS, Lin ZQ, Liu Y, Yuan L, Yin X, Li M (2015) Biofortification and phytoremediation of selenium in China. Front Plant Sci 6:136
Yamazaki S (1982) A selenium-containing hydrogenase from *Methanococcus vannielii*. Identification of the selenium moiety as a selenocysteine residue. J Biol Chem 257:7926–7929
Yasin M, El Mehdawi AF, Jahn CE, Turner MFS, Faisal M, Pilon-Smits EAH (2015a) Seleniferous soils as a source for production of selenium-enriched foods and potential of bacteria to enhance plant selenium uptake. Plant Soil 386:385–394
Yasin M, El-Mehdawi AF, Anwar A, Pilon-Smits EAH, Faisal M (2015b) Microbial-enhanced selenium and iron biofortification of wheat (*Triticum aestivum* L.) – applications in phytoremediation and biofortification. Int J Phytoremediation 17:341–347
Yuan J, Palioura S, Salazar JC, Su D, O'Donoghue P, Hohn MJ, Cardoso AM, Whitman WB, Söll D (2006) RNA-dependent conversion of phosphoserine forms selenocysteine in eukaryotes and archaea. Proc Natl Acad Sci U S A 103:18923–18927
Zannoni D, Borsetti F, Harrison JJ, Turner RJ (2008) The bacterial response to the chalcogen metalloids Se and Te. Adv Microb Physiol 53:1–72
Zawadzka AM, Crawford RL, Paszczynski AJ (2006) Pyridine-2,6-bis(thiocarboxylic acid) produced by *Pseudomonas stutzeri* KC reduces and precipitates selenium and tellurium oxyanions. Appl Environ Microbiol 72:3119–3129
Zhang Y, Romero H, Salinas G, Gladyshev VN (2006) Dynamic evolution of selenocysteine utilization in bacteria: a balance between selenoprotein loss and evolution of selenocysteine from redox active cysteine residues. Genome Biol 7:R94
Zhang Y, Turanov AA, Hatfield DL, Gladyshev VN (2008) *In silico* identification of genes involved in selenium metabolism: evidence for a third selenium utilization trait. BMC Genomics 9:251
Zhu J-M, Johnson TM, Clark SK, Wang X (2014) Selenium redox cycling during weathering of Se-rich shales: a selenium isotope study. Geochim Cosmochim Acta 126:228–249
Zinoni F, Birkmann A, Stadtman TC, Böck A (1986) Nucleotide sequence and expression of the selenocysteine-containing polypeptide of formate dehydrogenase (formate-hydrogen-lyase-linked) from *Escherichia coli*. Proc Natl Acad Sci U S A 83:4650–4654
Zinoni F, Birkmann A, Leinfelder W, Böck A (1987) Cotranslational insertion of selenocysteine into formate dehydrogenase from *Escherichia coli* directed by a UGA codon. Proc Natl Acad Sci U S A 84:3156–3160
Zolotarev AS, Unnikrishnan M, Shmukler BE, Clark JS, Vandorpe DH, Grigorieff N, Rubin EJ, Alper SL (2008) Increased sulfate uptake by *E. coli* overexpressing the SLC26-related SulP protein Rv1739c from *Mycobacterium tuberculosis*. A Mol Integr Physiol 149:255–266

Chapter 7
Selenium and the Plant Microbiome

Alyssa T. Cochran

Abstract Studies of plant microbiomes, which include all microorganisms that occur on and inside plants, are increasingly popular in multiple fields, particularly due to advances in next generation sequencing, a technique that has advantages over traditional culture-based methods. There are many advances yet to be made with regard to the interaction of selenium (Se) and the plant microbiome. This chapter will discuss aspects of the plant microbiome as well as the discoveries to date with regard to plant-associated microbes and Se, mostly explored through culture-dependent methods. Selenium hyperaccumulators appear to harbor equally diverse microbial communities as non-hyperaccumulators, although the microbial composition may vary. Investigations have isolated a variety of microbes from plants or soil in seleniferous areas including bacteria and fungi with enhanced Se tolerance. Inoculation of plants with individual strains or consortia of microbes was able to promote plant growth, Se uptake and/or Se volatilization, which shows promise for applications in phytoremediation or biofortification. Plant-derived microbes may also be applicable for cleanup of Se from wastewaters.

Keywords Microbiome • Rhizosphere • Endophytes • Hyperaccumulation

7.1 Introduction to the Plant Microbiome

7.1.1 General Overview

The plant microbiome, is becoming an increasingly popular area of study in plant sciences, but is still poorly understood. Microbiomes include all microorganisms of a particular environment, which include bacteria, archaea, fungi, and even some protists. Often, plant-associated microbes benefit their host via Plant Growth Promoting Properties (PGPP) while their host offers protection and nutrients to the microbial symbiont (Compant et al. 2010; Turner et al. 2013b). Microbiomes tend

A.T. Cochran (✉)
Department of Biology, Colorado State University, Fort Collins, CO 80525, USA
e-mail: cochranalyssa4@gmail.com

E.A.H. Pilon-Smits et al. (eds.), *Selenium in plants*, Plant Ecophysiology 11,
DOI 10.1007/978-3-319-56249-0_7

to be specific to plant species, geography, growth conditions and plant developmental stage (Redford et al. 2010; Chaparro et al. 2014; Mahnert et al. 2015). There is even evidence that the plant microbiome was responsible for the ability of early plants to colonize land (Knack et al. 2015).

Studying the composition of plant microbiomes has become increasingly popular with the onset of affordable next generation sequencing, offering a broader perspective on microbial diversity than culture-based methods. Studies have shown that microbial flora *in planta* is much more diverse and abundant than originally thought, with many samples containing hundreds more taxa via a 16S rRNA sequence analysis compared with culture-dependent methods (Kent and Triplett 2002; Visioli et al. 2015). Even though microbiomes are generally biogeography-specific, there tends to be taxonomic overlap between plant bacterial communities, with most samples containing *Actinobacteria, Proteobacteria*, and *Bacterioides* (Redford et al. 2010; Chaparro et al. 2014; Turner et al. 2013a; Panke-Buisse et al. 2014).

The plant microbiome is more easily studied when broken up into its three components, the rhizosphere, endosphere, and phyllosphere. Each of these spheres of the plant microbiome are unique and have their own intra- and intercommunity interactions relative to each other, which are dependent on biotic and abiotic conditions (Turner et al. 2013a). The phyllosphere includes the microbes that occur on the surface of plant shoots. This group of microbes, unlike the endosphere and rhizosphere, is exposed directly to the atmosphere and therefore must be resilient to many abiotic factors including high winds, UV, desiccation, and wet conditions (Turner et al. 2013a). Endophytes are microbes that live inside plant tissues (Alford et al. 2010) and can protect the plant from herbivores and pathogens was well as promote plant growth. The rhizosphere is the area underground that is within 5 mm of the roots. The microbes in the rhizosphere also include bacteria with PGPP, often referred to as PGP Rhizobacteria (PGPR). Like some endophytes, PGPR have been shown to produce or make available to plants compounds that promote plant growth including IAA, nutrients such as phosphates, nitrogen or iron (via iron-carrying siderophores), or compounds that inhibit pathogens or upregulate plant defenses (Jha et al. 2013). This chapter will focus on rhizosphere and endosphere microbes; the phyllosphere microbiome remains to be studied in relation to Se.

7.1.2 Introduction to Se in the Plant Microbiome System

Selenium is mainly taken up into plants as selenate and can leave plants in a volatile form, usually as dimethyl diselenide (DMDSe) or dimethyl selenide (DMSe) (Terry et al. 2000). Other ways in which Se can be deposited by plants is via litter, root turnover, or root exudation (Galeas et al. 2007; El Mehdawi et al. 2012). Depending on the species, plants also assimilate inorganic selenate to seleno-amino acids, and make this available to microbes (Terry et al. 2000). These seleno-amino acids are an attractive food source to microbes since they provide C, N and Se, all of which are essential nutrients for many bacteria. Inside the plant, several other forms of Se may

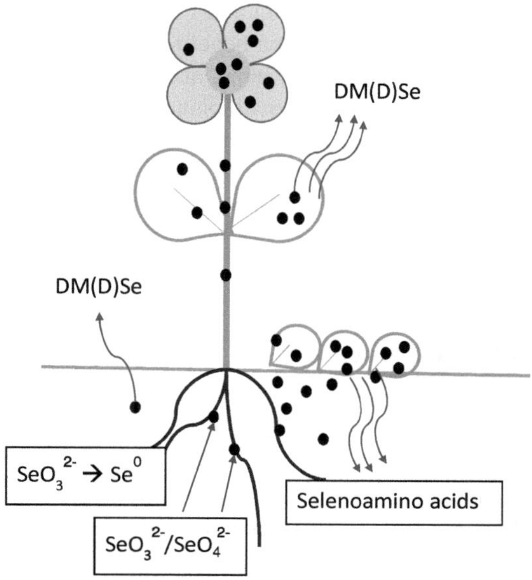

Fig. 7.1 Schematic depiction of plant and microbial processes in the plant-rhizosphere-soil system that affect the fate of selenium. Bacteria and fungi occur in rhizosphere, phyllosphere and endosphere of plants. Both plants and microbes can reduce selenate to selenite and produce organic forms of Se, including volatile DMSe/DMDSe. Microbes can also produce elemental Se (Se^0) and increase Se accumulation in plants from selenite and selenate (SeO_3^{2-} and SeO_4^{2-}). Hyperaccumulator plants produce selenoamino acids, which offer an additional source of (organic) Se to microbes when decomposed

be present, some of which are toxic to the plant. Se hyperaccumulator (HA) plant species have evolved ways to avoid this toxicity by converting selenate to methyl-selenocysteine, gamma-glutamyl-methylselenocysteine or selenocystathionine, which they can sequester in the vacuoles of epidermal tissues or transform to volatile DMDSe (Pilon-Smits and LeDuc 2009; Evans and Johnson 1967).

Bacterial Se metabolism shows similarities to that in plants, with capacity to assimilate inorganic Se to organic forms and to form organic volatile forms, DMSe or DMDSe (Frankenberger and Karlson 1994; Zayed and Terry 1994; Turner et al. 1998; Winkel et al. 2015). Bacteria are also capable of reducing selenite (and sometimes selenate) to elemental Se nanoparticles (Turner et al. 1998; Zayed et al. 1998; Husen and Siddiqi 2014; Staicu et al. 2015a, b; Winkel et al. 2015). These processes are shown in Fig. 7.1. Bacteria in general seem to be very tolerant to Se, some strains surviving and even benefiting from concentrations of selenate and selenite up to 200 mM; this capacity was not dependent on the Se concentration of the site or plant host they were isolated from (Sura-de Jong et al. 2015). In one study on Se-dependent litter decomposition by Quinn and coworkers it was found that litter from Se hyperaccumulator species harbored more culturable bacteria and decomposed faster than litter from related non-hyperaccumulator species (Quinn et al. 2011). Thus, while most other ecological partners associated with hyperaccumulators are by default sensitive to Se (El Mehdawi and Pilon-Smits 2012), bacteria appear to be by default Se-resistant and may even benefit from and seek out high-Se plant material to colonize. Fungi, on the other hand, are much more sensitive to Se than bacteria (Wangeline et al. 2011). Thus, not all plant-associated microbes are equally resistant to Se. It has been shown that systemic supply of Se was able to protect *Brassica juncea* from fungal pathogens *Fusarium sp.* and *Alternaria brassicicola* (Hanson et al. 2003). A protective effect of Se supply has not yet been investigated on bacterial pathogens.

7.2 Rhizosphere Microbes

7.2.1 Introduction to the Rhizosphere

The rhizosphere is a dynamic environment, constantly changing and influenced by multiple biotic and abiotic factors. Rhizosphere processes are a fascinating area of plant-microbe interaction research; the soil, host plants and microbial components of the system affect each other in a complex relationship triangle (Turner et al. 2013a). Among the different components of the plant microbiome, the rhizosphere has the highest abundance of microbes, about 1000 fold higher in microbial abundance than in bulk soil (Bergs and Smalla 2009). This phenomenon is often referred to as the rhizosphere effect. Plant exudates are rich in sugars and acids and maycontain specific secondary plant compounds that can induce bacterial pathways (Morgan and Whipps 2001; Bergs and Smalla 2009). The plant uses these strategies to build specific microbial communities in the soil to aid its survival and potentially that of its offspring (Lapsansky et al. 2016).

Some rhizosphere microbes, including strains of *Burkholderia*, *Ralstonia* and *Pseudomonas* are opportunistic pathogens, which can take advantage of a weakened immunity in the host (Berg et al. 2005; Mendes et al. 2013). Even though some rhizosphere microbes are pathogens or parasites, the majority of the bacteria found here is traditionally categorized as mutualistic with their hosts (Newton et al. 2010). There is a multitude of bacterial taxa that fall into the PGPR category, some of which can benefit a wide range of host plants and some of which are host-specific (Kloepper 1996). In order to identify PGPR, experiments are necessary showing that the host plant grows better after inoculation with the specific PGPR strain.

Rhizobacteria-legume interactions are one example of a widely studied host-specific interaction. These nitrogen (N_2)-fixing PGPR can enter into the roots and establish themselves inside root nodules, which gives the nodulated plant the ability to fix nitrogen. There are multiple genera capable of this symbiosis in the bacterial family Rhizobiaceae (Gray and Smith 2005). Among the most popular of these genera is *Rhizobium*, usually found in symbiosis with the plant family Fabaceae. The molecular cross-talk between the plant roots and the specific rhizobacteria often starts with plant root exudate signal compounds that induce bacterial signal compounds, which then leads to nodule formation (Gray and Smith 2005).

An example of a more promiscuous plant-microbe interaction is the large group of fungi that live in association with plant roots called mycorrhizae; in this mutualistic relationship the plant benefits from the fungus through increased water and nutrient uptake and the fungus benefits from the organic carbon compounds released by the plant (Marschner and Dell 1994). The most common mycorrhizae are the vesicular arbuscular mycorrhizae (VAM), defined by the colonization of the host root cortex by the fungal symbiont, which then uses its mycelium to reach into the soil to gather water and minerals (Barea et al. 2005; Wang and Qui 2006). The fungi responsible for these interactions are generally obligate in their symbioses, needing a host plant to colonize in order to survive and reproduce (Barea et al. 2005). Most plant families (92%) and even plant species (80%) are thought to have mycorrhizal partners (Wang and Qui 2006).

7.2.2 *Selenium and the Rhizosphere*

The plant family Fabaceae includes 25 species that hyperaccumulate Se, e.g. *Astragalus bisulcatus* (Beath et al. 1939). The enhanced nitrogen acquisition capacity associated with root nodulation is not only beneficial for plant growth, but also was found to enhance Se accumulation in the form of seleno-aminoacids in hyperaccumulators (HAs) including *A. bisculatus* (Alford et al. 2014). While it could be hypothesized that high plant concentrations of Se would inhibit root nodule formation in symbioses between *A. bisulcatus* and *Rhizobium*, there is no evidence for this (Alford et al. 2012). Increasing Se concentration in hyperaccumulators was associated with enhanced nodule formation, and tenfold higher levels of the N-rich compound gamma-glutamyl-MetSeCys (Alford et al. 2012, 2014). Thus, rhizobia in root nodules may play a role in Se hyperaccumulation in *A. bisculatus* by providing nitrogen for the selenoaminoacids that these plants accumulate up to 1% of their dry weight (Alford et al. 2012, 2014). Multiple species of *Rhizobium* have been shown to reduce selenite to elemental Se (Se^0), which may influence Se speciation in plants (Basaglia et al. 2007; Hunter and Kuykendall 2007; Valdez Barillas et al. 2012). While organic C-Se-C compounds make up close to 100% of Se in the roots of *A. bisulcatus*, it constituted only 75% of Se in root nodules, where the remaining substantial fraction (25%) was Se^0 (Valdez Barillas et al. 2012).

Like bacteria, many fungi have been shown to reduce selenite to Se^0, despite the generally lower Se tolerance of fungi to high concentrations of Se, as compared to bacteria (Gharieb et al. 1995; Wangeline et al. 2011; Lindblom et al. 2013). In a study by Wangeline and coworkers, hundreds of fungi were isolated from rhizosphere soil collected from seleniferous and non-seleniferous sites, identified, and characterized for their Se tolerance. The fungi isolated from seleniferous soils were more tolerant to Se than those isolated from non-seleniferous soils, indicating that

Table 7.1 Overview of plant inoculation studies that used fungi or bacteria from Se hyperaccumulators s and their effects on plant Se metabolism

	Promoted growth	Affected Se speciation	Can tolerate high Se	Increased Se accumulation
Fungi from HA	Lindblom et al. (2012b)	Lindblom et al. (2012a, b)	Wangeline et al. (2011)	Lindblom et al. (2013)
Bacteria from HA	Alford et al. (2014)	di Gregorio et al. (2005, 2006)	Di Gregorio et al. (2005)	de Souza et al. (1999a, b)
	Sura-de Jong et al. (2015)	Valdez Barillas et al. (2012)	Sura-de Jong et al. (2015)	di Gregorio et al. (2005)
	Yasin et al. (2015)	Alford et al. (2014)		Alford et al. (2014)
		Staicu et al. (2015b)		Yasin et al. (2015)
Microbe Consortium from HA	El Mehdawi et al. (2015)	***	***	Quinn et al. (2011)
				El Mehdawi et al. (2015)

Boxes with stars denote areas for future research

the fungi living in seleniferous soils have evolved to be more resilient to the high concentrations of Se in the soil (Wangeline et al. 2011).

In addition to reduction to Se^0, rhizobacteria and fungi isolated from rhizosphere soils have been shown to volatilize Se in the forms of DMSe or DMDSe from selenate or selenite (de Souza et al. 1999a). Because these volatile forms of Se are less toxic and remove Se from the site, Se volatilization has applications in bioremediation (Barkes and Fleming 1974; Azaizeh et al. 1997, 2003).

There have been multiple studies on the effects of rhizosphere microbes on growth and plant accumulation of Se and other elements. These studies are summarized in Table 7.1, showing that the presence of rhizosphere microbes can contribute to the growth and Se accumulation of HAs as well as non-HAs. In many instances, bacterial inoculation increased the biomass of the inoculated plant and enhanced Se accumulation (de Souza et al. 1999a, b; di Gregorio et al. 2005; Wenzel 2009; Durán et al. 2013; El Mehdawi et al. 2015; Sura-de Jong et al. 2015). In one study, rhizosphere soil slurry of HA *Symphyotrichum ericoides* stimulated growth and Se accumulation in the same species when grown from surface-sterilized seed on autoclaved naturally seleniferous or non-seleniferous soils (El Mehdawi et al. 2015). In another study, inoculation with a single environmental strain enabled wheat to take up more Se, as well as iron (Yasin et al. 2015). Furthermore, Se accumulation and volatilization could be enhanced in *Brassica juncea* and several aquatic species by inoculation with environmental bacteria isolated from a Se-rich sediment (de Souza et al. 1999a, b) or from the rhizosphere of Se hyperaccumulator *A. bisulcatus* (di Gregorio et al. 2005).

Effects on plant Se accumulation were also observed after inoculating plants with rhizosphere fungi. Some rhizosphere fungi isolated from Se HA were shown to increase Se accumulation in the roots of Se HA *Stanleya pinnata* (Lindblom et al. 2013). There have also been several studies on the effects of mycorrhizal fungi on Se accumulation and uptake. Most of these studies found that when a mycorrhizal relationship formed, the concentration of Se increased in the plant compared to a plant growing in seleniferous conditions without mycorrhizal inoculation (Wanek et al. 1999; Larsen et al. 2006; Yu et al. 2011). In some cases the effects of inoculation depended on the Se concentration supplied; even though the rhizosphere fungi were shown to able to increase Se concentrations in plants, at lower concentrations of Se it was actually shown to decrease Se accumulation (Munier-Lamy et al. 2007; Yu et al. 2011).

7.3 Endosphere Microbes

7.3.1 Introduction to Endophytes

Endophytes are bacteria and fungi that live inside plants, colonizing the roots, shoots, and reproductive portions (Jha et al. 2013). These microbes can either be inherited from the parent plant, introduced via a vector, or can colonize the plant during its life through sites of lateral root emergence or open areas in the plant epidermis (Reinhold-Hurek and Hurek 2011). As with rhizosphere microbes, endophytes can be pathogenic, parasitic or mutualistic. The PGPP endophytes are generally host-specific, and the mechanism(s) by which PGPP bacteria promote plant growth are similar to those of PGPR (Long et al. 2008). Endophytes have been shown to promote plant growth via production of IAA, nitrogen fixation, phosphate solubilization and production of metal chelating agents like siderophores (Hardoim et al. 2008; Long et al. 2008; Weyens et al. 2009a,b; Durán et al. 2014; Lins et al. 2014).

The ability of endophytes to escape the host immunity response is still poorly understood. It is known, however, that endophytes are able to modulate ethylene levels in plants, which could have some role in the plant immune response (Hardoim et al. 2008; Reinhold-Hurek and Hurek 2011). Some endophytes can induce attack against endophytic pathogens, increasing the host immunity to defend against these pathogens (Nejad and Johnson 2000; Arnold et al. 2003). It has been shown that some endophytes do this by triggering the host's systemic jasmonic acid or salicylic acid responses and can prime the plant immune response in preparation for future attacks (Van Wees et al. 2008; Reinhold-Hurek and Hurek 2011). In addition to growth promotion and immune regulation, endophytes are able to alleviate plant abiotic stresses and increase nutrient availability by regulation of host genes and increasing levels of abscisic acid (Hesse et al. 2003; Sziderics et al. 2007; Jha et al. 2013).

7.3.2 Selenium and the Endosphere

In a study done by Sura-de Jong et al. (2015), endophytic bacteria were isolated from Se hyperaccumulators *A. bisulcatus* and *S. pinnata* and tested for physiological properties as well as the ability to enhance growth and Se uptake in plants. When exposed to Se, the isolates were shown to be tolerant to high concentrations (up to 200 mM) of selenate and selenite, and to have the ability to reduce selenite to Se^0 (Sura-de Jong et al. 2015; Staicu et al. 2015a, b). A selection of endophytes from Se hyperaccumulators were inoculated to *Brassica juncea* and *Medicago sativa*, resulting in increased dry weight when compared to un-inoculated control plants; Se accumulation was not significantly affected (Sura-de Jong et al. 2015).

In addition, studies evaluated the potential use of bacterial endophytes in Se biofortification and phytoremediation. Durán et al. (2014) found that endophytic bacteria including *Acinetobacter, Bacillus* and *Klebsiella* tolerated high levels of Se and promoted plant growth (Durán et al. 2014). In addition to these properties, these endophytic bacteria were able to protect wheat crops from *Gaeumannomyces graminis*, a soil-borne pathogen that destroys many cereal crops (Durán et al. 2014). Since endophytes live in the plant and are generally host specific, they often possess abilities to degrade certain pollutants in the host plant environment (Doty 2008). Endophytic microbes have been used in a number of studies on other pollutants and have potential uses in cleaning up polluted areas (Doty 2008). For example, it was shown that an endophytic *Pseudomonas* strain isolated from Se hyperaccumulator *Stanleya pinnata* was able to completely remove up to 100 mM of selenite from water by precipitating it as Se^0 (Staicu et al. 2015a).

A study done by Lindblom et al. (2012a) showed that speciation in HAs may be affected by microbial endophytes that produce Se^0: the HA accumulated this form of Se up to 30% of their total Se in the field, but not in a laboratory setting. Elemental Se was particularly found in root nodules (Lindblom et al. 2012a). A selenophilic fungus known as *Alternaria astragali* which was isolated from the root of *A. bisulcatus* was used for further studies. Seeds of *A. bisulcatus* containing *A. astragali* had a significantly higher fraction of Se^0 (up to 30%) than those without this endophytic fungus (Valdez Barillas et al. 2012). A follow up study showed that *A. astragali* enhanced the growth of some *Astragalus* species but inhibited the growth of others (Lindblom et al. 2012b). This indicates that, like bacteria, endophytic fungi may also be capable of enhancing growth, changing Se speciation and affecting the Se accumulation of inoculated plants. These studies are also included in Table 7.1.

7.4 Future Directions

There have been many recent advances and discoveries in the area of plant microbiomes and Se. However, there is still much to be discovered and there are many research questions to be addressed. For instance, is there a core microbiome associated with Se hyperaccumulators that may contribute to Se accumulation and that can be used for bioremediation and phytoremediation? Do individual plants select their microbiomes or have plant species and their microbiomes coevolved? The advances in the understanding of plant microbiomes and Se could very well be useful to increase effectiveness of bioremediation, phytoremediation, and biofortification. Overall, the phyllosphere and endosphere need more attention, with almost no studies on the phyllosphere and Se to date. It is expected that the implementation of next generation sequencing will give additional insights into the plant microbiome, which will complement the limited existing studies.

Acknowledgements The author would like to acknowledge contributors to this project including the Earth Microbiome Project and a Nick Stavros graduate student fellowship from the Colorado State University Biology Department. I would also like to thank Erin Lapsansky and Michela Schiavon for providing insight and edits.

References

Alford ER, Pilon-Smits EAH, Paschke MW (2010) Metallophytes- a view from the rhizosphere. Plant Soil 337(1):33–50

Alford ER, Pilon-Smits EAH, Marcus MA, Fakra SC, Paschke MW (2012) No evidence for a cost of tolerance: selenium hyperaccumulation by *Astragalus* does not inhibit root nodule symbiosis. Am J Bot 99:1930–1941

Alford ER, Lindblom SD, Pittarello M, Freeman JL, Fakra SC, Marcus MA, Broeckling C, Pilon-Smits EAH, Paschke MW (2014) Roles of Rhizobial symbionts in *Astragalus* selenium hyperaccumulation. Am J Bot 101:1895–1905

Arnold AE, Mejía LC, Kyllo D, Rojas EI, Maynard Z, Robbins N, Herre EA (2003) Fungal endophytes limit pathogen damage in a tropical tree. Proc Natl Acad Sci U S A 100(26):15649–15654

Azaizeh AH, Gowthaman S, Terry N (1997) Microbial selenium volatilization in rhizosphere and bulk soils from a constructed wetland. J Environ Qual 26(3):666–672

Azaizeh AH, Salhani N, Sebesvari Z, Emons H (2003) The potential of rhizosphere microbes isolated from a constructed wetland to biomethylate selenium. J Environ Qual 325:55–62

Barea JM, Pozo MJ, Azcón R, Azcón-Aguilar C (2005) Microbial co-operation in the rhizosphere. J Exp Bot 56(417):1761–1778

Barkes L, Fleming RW (1974) Production of dimethylselenide gas from inorganic selenium by eleven soil fungi. Bull Environ Contam Toxicol 12:308–311

Basaglia M, Toffanin A, Baldan E, Bottegal M, Shapleigh JP, Casella S (2007) Selenite-reducing capacity of the copper-containing nitrite reductase of *Rhizobium sullae*. FEMS Microbiol Lett 269:124–130

Beath OA, Gilbert CS, Eppson HF (1939) The use of indicator plants in locating seleniferous areas in western United States. I. General. Am J Bot 26(4):257–269

Berg G, Eberl L, Hartmann A (2005) The rhizosphere as a reservoir for opportunistic human pathogenic bacteria. Environ Microbiol 7(11):1673–1685

Bergs G, Smalla K (2009) Plant species and soil type cooperatively shape the structure and function of microbial communities in the rhizosphere. FEMS Microbiol Ecol 68:1–13

Chaparro JM, Badri DV, Vivance JM (2014) Rhizosphere microbiome assemblage is affected by plant development. Int Soc Microb Ecol J 8:790–803

Compant S, Clemént C, Sessitsch A (2010) Plant growth-promoting bacteria in the rhizo- and endosphere of plants: their role, colonization, mechanisms involved and prospects for utilization. Soil Biol Biochem 42:669–678
de Souza MP, Chu D, Zhao M, Zayed AM, Ruzin SE, Schichnes D, Terry N (1999a) Rhizosphere bacteria enhance selenium accumulation and volatilization by Indian mustard. Plant Physiol 119:565–573
de Souza MP, Huang CPA, Chee N, Terry N (1999b) Rhizosphere bacteria enhance the accumulation of selenium and mercury in wetland plants. Planta 209:259–263
Di Gregorio S, Lampis S, Vallini G (2005) Selenite precipitation by a rhizospheric strain of *Stenotrophomonas sp.* isolated from the root system of *Astragalus bisulcatus*: a biotechnological perspective. Environ Int 31:233–241
Di Gregorio S, Lampis S, Malorgio F, Petruzzelli G, Pezzarossa B, Vallini G (2006) *Brassica juncea* can improve selenite and selenate abatement in contaminated soils through the aid of its rhizospheric bacterial population. Plant Soil 285:233–244
Doty SL (2008) Enhancing phytoremediation through the use of transgenics and endophytes. New Phytol 179:318–333
Durán P, Acuña JJ, Jorquera MA, Azcón R, Borie F, Cornejo P, Mora ML (2013) Enhanced selenium content in wheat grain by co-inoculation of selenobacteria and arbuscular mycorrhizal fungi: a preliminary study as a potential Se biofortification strategy. J Cereal Sci 57(3):275–280
Durán P, Acuña JJ, Jorquera MA, Azcón R, Paredes C, Rengel Z, de la Luz MM (2014) Endophytic bacteria from selenium-supplemented wheat plants could be useful for plant-growth promotion, biofortification and *Gaeumannomyces graminis* biocontrol in wheat production. Biol Fertil Soils 50(6):983–990
El Mehdawi AF, Pilon-Smits EAH (2012) Ecological aspects of plant selenium hyperaccumulation. Plant Biol 14(1):1–10
El Mehdawi AF, Paschke M, Pilon-Smits EAH (2015) Symphyotrichum ericoides populations from seleniferous and non- seleniferous soil display striking variation in selenium accumulation. New Phytol 206:231–242
El Mehdawi AF, Cappa JJ, Fakra SC, Self J, Pilon-Smits EAH (2012) Interactions of selenium hyperaccumulators and nonaccumulators during cocultivation on seleniferous or nonseleniferous soil- the importance of having good neighbors. New Phytol 194:264–277
Evans CJACS, Johnson CM (1967) Collection and partial characterization of volatile selenium compounds from *Medicago sativa* L. Aust J Biol Sci 20:737–748
Frankenberger WT, Karlson U (1994) Microbial volatilization of selenium from soils and sediments. In: Frankenberger WT Jr, Benson S (eds) Selenium in the environment. Marcek Dekker, New York, pp 369–387
Galeas ML, Zhang LH, Freeman JL, Wegner M, Pilon-Smits EAH (2007) Seasonal fluctuations of selenium and sulfur accumulation in selenium hyperaccumulators and related nonaccumulators. New Phytol 173:517–525
Gharieb MM, Wilkinson SC, Gadd GM (1995) Reduction of selenium oxyanions by unicellular, polymorphic and filamentous fungi: cellular location of reduced selenium and implications for tolerance. J Ind Microbiol 14:300–311
Gray EJ, Smith DL (2005) Intracellular and extracellular PGPR: commonalities and distinctions in the plant-bacterium signaling process. Soil Biol Biochem 37:395–412
Hanson B, Garifullina GF, Lindblom SD, Wangeline A, Ackley A, Kramer K, Norton AP, Lawrence CB, Pilon-Smits EAH (2003) Selenium accumulation protects Brassica juncea from invertebrate herbivory and fungal infection. New Phytol 159(2):461–469
Hardoim PR, van Overbeek LS, van Elsas JD (2008) Properties of bacterial endophytes and their proposed role in plant growth. Trends Microbiol 16(10):463–471
Hesse U, Schöberlein W, Wittenmayer L, Förster K, Warnstorff K, Diepenbrock W, Merbach W (2003) Effects of *Neotyphoidum* enophytes on growth, reproduction and drought-stress tolerance of three *Lolium perenne* L. genotypes. Grass Forage Sci 58(4):407–415

Hunter WJ, Kuykendall DL (2007) Reduction of selenite to elemental red selenium by *Rhizobium* sp. strain B1. Curr Microbiol 55(4):344–349
Husen A, Siddiqi KS (2014) Plants and microbes assisted selenium nanoparticles: characterization and application. J Nanobiotechnol 12:28
Jha PN, Gupta G, Jha P, Rajesh M (2013) Association of rhizospheric/endophytic bacteria with plants: a potential gateway to sustainable agriculture. GJAS 3(2):73–84
Kent AD, Triplett EW (2002) Microbial communities and their interactions in soil and rhizosphere ecosystems. Annu Rev Microbiol 56:211–236
Kloepper JW (1996) Host specificity in microbe-microbe interactions. Bioscience 46(6):406–409
Knack JJ, Wilcox LW, Delaux PM, Ané JM, Piotrowski MJ, Cook ME, Graham JM, Graham LE (2015) Microbiomes of streptophyte algae and bryophytes suggest that a functional suite of microbiota fostered plant colonization of land. Int J Plant Sci 176(5):405–420
Lapsansky ER, Milroy AM, Andales MJ, Vivanco JM (2016) Soil memory as a potential mechanism for encouraging sustainable plant health and productivity. Curr Opin Biotechnol 38:137–142
Larsen EH, Lobinski R, Burger-Meÿer K, Hansen M, Ruzik R, Mazurowska L, Ramussen PH, Sloth JJ, Scholten O, Kik C (2006) Uptake and speciation of selenium in garlic cultivated in soil amended with symbiotic fungi (mycorrhiza) and selenate. Anal Bioanal Chem 385:1098
Lindblom SD, Valdez-Barillas JR, Fakra SC, Marcus MA, Wangeline AL, Pilon-Smits EAH (2012a) Influence of microbial associations on Selenium localization and speciation in roots of *Astragalus* and *Stanleya* hyperaccumulators. Environ Exp Bot 88:33–42
Lindblom SD, Fakra SC, Landon J, Schulz P, Tracy B, Pilon-Smits EAH (2012b) Co-cultivation of *Astragalus racemosus* and *Astragalus convallarius* with selenium-hyperaccumulator rhizosphere fungi: effects on plant growth and accumulation of selenium and other elements. Planta 237:717–729
Lindblom SD, Fakra SC, Landon J, Schulz P, Tracy B, Pilon-Smits EAH (2013) Inoculation of selenium hyperaccumulator *Stanleya pinnata* and related non-accumulator *Stanleya elata* with hyperaccumulator rhizosphere fungi – effects on Se accumulation and speciation. Physiol Plant 150:107–118
Lins MRCR, Fontes JM, de Vasconcelos NM, da Silva Santos DM, Ferreira OE, de Azevedo JL, de Araújo JM, de Souza Lima GM (2014) Plant growth promoting potential of endophytic bacteria isolated from cashew leaves. Afr J Biotechnol 13(33):3360–3365
Long HH, Schmidt DD, Baldwin IT (2008) Native bacterial endophytes promote host growth in a species-specific manner; Phytohormone manipulations do not result in common growth responses. PLoS One 3(7):2702
Mahnert A, Moissl-Eichinger C, Berg G (2015) Microbiome interplay: plants alter microbial abundance and diversity within the built environment. Front Microbiol 6:887
Marschner H, Dell B (1994) Nutrient uptake in mycorrhizal symbiosis. Plant Soil 159:89–102
Mendes R, Garbeva P, Raaijmakers JM (2013) The rhizosphere microbiome: significance of plant beneficial, plant pathogenic, and human pathogenic microorganisms. FEMS Microbiol Rev 37(5):634–663
Morgan JAW, Whipps JM (2001) Methodological approaches to the study of rhizosphere carbon ow and microbial population dynamics. In: Pinton A, Varanini Z, Nannipieri P (eds) The rhizosphere. Biochemistry and organic substances at the soil-plant interface. Marcel Dekker, New York, pp 373–409
Munier-Lamy C, Deneux-Mustin S, Mustin C, Merlet D, Berthelin J, Leyval C (2007) Selenium bioavailability and uptake as affected by four different plants in a loamy clay soil with particular attention to mycorrhizae inoculated ryegrass. J Environ Radioact 97(2–3):148–158
Nejad P, Johnson PA (2000) Endophytic bacteria induce growth promotion and wilt disease suppression in oilseed rape and tomato. Biol Control 18(3):208–215
Newton AC, Fitt BDL, Atkins SD, Walters DR, Daniell TJ (2010) Pathogenesis, parasitism and mutualism in the trophic space of microbe-plant interactions. Trends Microbiol 18(8):365–373

Panke-Buisse K, Poole AC, Goodrich JK, Ley RE, Kao-Kniffin J (2014) Selection on soil microbiomes reveals reproducible impacts on plant function. ISME J 9:980–989

Pilon-Smits EAH, LeDuc DL (2009) Phytoremediation of selenium using transgenic plants. Curr Opin Biotechnol 20(2):207–212

Quinn CF, Wyant K, Wangeline AL, Shulman J, Galeas ML, Valdez JR, Paschke MW, Pilon-Smits EAH (2011) Selenium hyperaccumulation increases leaf decomposition rate in a seleniferous habitat. Plant Soil 341:51–61

Redford AJ, Bowers RM, Knight R, Linhart Y, Fierer N (2010) The ecology of the phyllosphere: geographic and phylogenetic variability in the distribution of bacteria on tree leaves. Environ Microbiol 12(11):2885–2893

Reinhold-Hurek B, Hurek T (2011) Living inside plants: bacterial endophytes. Curr Opin Plant Biol 14(4):435–443

Staicu LC, van Hullebusch ED, Lens PNL, Pilon-Smits EAH, Oturan MA (2015a) Electrocoagulation of colloidal biogenic selenium. Environ Sci Pollut Res 22:3127–3137

Staicu LC, Ackerson CJ, Cornelis P, Ye L, Berendsen RL, Hunter WJ, Noblitt SD, Henry CS, Cappa JJ, Montenieri RL, Wong AO, Musilova L, Sura-de Jong M, van Hullebusch ED, Lens PNL, Reynolds RJB, Pilon-Smits EAH (2015b) *Pseudomonas moraviensis* subsp. stanleyae, a bacterial endophyte of hyperaccumulator *Stanleya pinnata*, is capable of efficient selenite reduction to elemental selenium under aerobic conditions. J Appl Microbiol 119:400–410

Sura-de Jong M, Reynolds RJ, Richterova K, Musilova L, Hrochova I, Frantik T, Sakmaryova I, Strejcek M, Cochran A, Staicu L, Cappa JJ, van der Lelie D, Pilon-Smits EAH (2015) Selenium hyperaccumulators harbor a diverse endophytic bacterial community characterized by extreme selenium tolerance and plant growth promoting properties. Front Plant Sci 6:113

Sziderics AH, Rasche F, Trognitz F, Sessitsch A, Wilhelm E (2007) Bacterial endophytes contribute to abiotic stress adaptation in pepper plants (*Capsicum annuum* L.) Can J Microbiol 53:1195–1202

Terry N, Zayed AM, de Souza MP, Tarun AS (2000) Selenium in higher plants. Annu Rev Plant Physiol Plant Mol Biol 51:401–432

Turner RJ, Weiner JH, Taylor DE (1998) Selenium metabolism in *Escherichia coli.* Biometals 11:223–227

Turner TR, James EK, Poole PS (2013a) The plant microbiome. Genome Biol 14:209

Turner TR, Ramakrishnan K, Walshaw J, Heavens D, Alston M, Swarbreck D, Osbourn A, Grant A, Poole PS (2013b) Comparative metatranscriptomics reveals kingdom level changes in the rhizosphere microbiome of plants. ISME J 7:2248–2258

Valdez Barillas JR, Quinn CF, Freeman JL, Lindblom SD, Marcus MS, Fakra SC, Gilligan TM, Alford ER, Wangeline AL, Pilon-Smits EAH (2012) Selenium distribution and speciation in hyperaccumulator *Astragalus bisulcatus* and associated ecological partners. Plant Physiol 159:1834–1844

Van Wees SCM, Van der Ent S, Pieterse CMJ (2008) Plant immune responses triggered by beneficial microbes. Curr Opin Plant Biol 11:443–448

Visioli G, D'Egidio S, Sanangelantoni AM (2015) The bacterial *rhizobiome* of hyperaccumulators: future perspectives based on omics analysis and advanced microscopy. Front Plant Sci 5:752

Wanek PL, Vance GF, Stahi PD (1999) Selenium uptake by plants: effects of soil steaming, root addition, and selenium augmentation. Commun Soil Sci Plant Anal 30(1–2):265–278

Wang B, Qui Y-L (2006) Phylogenetic distribution and evolution of mycorrhizas in land plants. Mycorrhiza 16(5):229–363

Wangeline AL, Valdez JR, Lindblom SD, Bowling KL, Reeves FB, Pilon-Smits EAH (2011) Selenium tolerance in rhizosphere fungi from Se hyperaccumulator and non-hyperaccumulator plants. Am J Bot 98:1139–1147

Wenzel WW (2009) Rhizosphere processes and management in plant-assisted bioremediation (phytoremediation) of soils. Plant Soil 321:385–408

Weyens N, van der Lelie D, Taghavi S, Vangronsveld J (2009a) Phytoremediation: plant-endophyte partnerships take the challenge. Curr Opin Biotechnol 20(2):248–254

Weyens N, van der Lelie D, Taghavi S, Newman L, Vangronsveld J (2009b) Exploiting plant-microbe partnerships to improve biomass production and remediation. Trends Biotechnol 27(10):591–598

Winkel LHE, Vriens B, Jones GD, Schneider LS, Pilon-Smits EAH, Banuelos GS (2015) Selenium cycling across soil-plant-atmosphere interfaces: a critical review. Forum Nutr 7(1):4199–4239

Yasin M, El-Mehdawi AF, Anwar A, Pilon-Smits EAH, Faisal M (2015) Microbial-enhanced selenium and iron biofortification of wheat (*Triticum aestivum* L.) – applications in phytoremediation and biofortification. Int J Phytoremediation 17:341–347

Yu Y, Luo L, Yang K, Zhang S (2011) Influence of mycorrhizal inoculation on the accumulation and speciation of selenium in maize growing in selenite and selenate spiked soils. Pedobiologica-Int J Soil Biol 54:267–272

Zayed AM, Terry N (1994) Selenium volatilization in roots and shoots: effects of shoot removal and sulfate level. J Plant Physiol 143(1):8–14

Zayed A, Lytle CM, Terry N (1998) Accumulation and volatilization of different chemical species of selenium by pants. Planta 206:284–292

Chapter 8
Selenium Metabolism in Herbivores and Higher Trophic Levels Including Mammals

Lutz Schomburg and Elias S.J. Arnér

Abstract Plants provide dietary selenium (Se) for higher trophic levels including livestock and humans that depend on Se for their survival, but may also suffer Se toxicity upon excessive ingestion. Therefore, the fate of Se in plants and algae is relevant for mammals as well as for Se cycling in ecosystems. Conversely, the fate of Se in higher trophic levels is relevant for Se cycling back to soil and plants. This chapter focuses on the fate of dietary Se in humans and other animals, and health issues related to Se status. Selenium is an essential trace element for all mammals, implying that mammals develop disease upon Se deficiency. This is explained by the fact that certain selenoproteins, i.e. proteins containing the rare co-translationally inserted amino acid selenocysteine, have essential functions for physiology and health. Humans have 25 selenoprotein genes, about 0.1% of all protein-encoding genes. Among these, two thioredoxin reductases and one isoenzyme of glutathione peroxidase are essential for embryogenesis. For maintaining health and minimizing disease risks the human recommended daily intake is at least 50 μg Se per day. The majority of nutritional Se is provided in the form of two amino acids, selenocysteine or selenomethionine, and their derivatives, mainly in form of ingested proteins, as well as low molecular weight selenocompounds or rarely, in form of inorganic Se salts. If chronically ingested at high levels, above approximately 1 mg per day for an adult human, Se may become toxic. In this chapter, we give a brief overview of the main nutritional sources of Se for mammals, mammalian pathways of Se metabolism and excretion, and the biochemical and physiological functions that are sustained by mammalian selenoproteins. Established health risks of Se deficiency along with rare cases of inherited defects in selenoprotein biosynthesis will complement the picture on its essentiality.

L. Schomburg (✉)
Institute for Experimental Endocrinology, Charité – University Medical School, Suedring 10, CVK, D-13353 Berlin, Germany
e-mail: lutz.schomburg@charite.de

E.S.J. Arnér (✉)
Division of Biochemistry, Department of Medical Biochemistry and Biophysics, Karolinska Institutet, SE-171 77 Stockholm, Sweden
e-mail: elias.arner@ki.se

E.A.H. Pilon-Smits et al. (eds.), *Selenium in plants*, Plant Ecophysiology 11,
DOI 10.1007/978-3-319-56249-0_8

Keywords Selenium metabolism • Selenomethionine • Selenocysteine • Mammals

8.1 Introduction

Selenium (Se) was discovered about 200 years ago by the Swedish chemist Jöns Jacob Berzelius (Berzelius 1818), but it was not until the mid-1900s that Se intake levels were found to affect the well-being of mammals. First, it was recognized that livestock ingestion of plants containing high levels of Se could cause toxicity (Franke and Potter 1935). The symptoms of intoxicated mammals were later confirmed to include impaired vision, low appetite and wandering in circles (called "blind staggers disease") and/or emaciation, loss of hair, deformation and shedding of hooves, loss of vitality and erosion of the joints of long bones (called "alkali disease"). If acute and severe, Se intoxication can also yield paralysis and death by respiratory failure (Barceloux 1999). However, Se is not only toxic at high levels, but also an essential trace element for mammals, i.e. a severe Se deficiency predisposes to developmental abnormalities and increases disease risks. In the 1950s, Schwarz and Foltz observed that trace amounts of Se can protect against vitamin E-induced liver necrosis in rats (Schwarz and Foltz 1957). Indeed, Se deficiency can lead to wasting syndromes and white muscle disease in livestock, wherefore additional Se was later started to be given as supplement to cattle and other livestock and proved effective to prevent disease (Oldfield 1997; Whanger et al. 1977). Thus, Se intake levels are of major importance for maintaining health and reducing disease risks in mammals.

As herbivores feed on local plants, the Se content of the flora is a major factor controlling their Se status. The flora, in turn, reflects soil Se availability which depends on environmental deposition, pH, mineral pattern and other soil-specific characteristics (Winkel et al. 2015). While certain Se accumulating plants have been identified, the major factor controlling plant Se content is soil quality. This interaction in turn highlights that the nutritional intake of Se by herbivores may differ considerably, depending on their geographical area of residency. Soil differences in Se content and availability are thus affecting local flora, fauna as well as the human residents and are of high importance for outcome of clinical studies relating Se, selenoproteins and health in different populations (Rayman 2008). While it is generally accepted that the populations in mainland Europe and large parts of Africa and Asia are rather insufficiently supplied with Se through their regular nutritional intake, the soils in North America used for agricultural production of plants and animals feeding thereon are typically rich sources of Se, conferring an important difference in basal Se status between North Americans and many other human populations (Combs 2015). Here, we shall briefly discuss this topic, with a focus on the biochemistry of Se and the molecular mechanisms that underpin Se pathology in mammals.

8.2 Nutritional and Toxic Levels of Se for Mammals

As afore mentioned, Se intake of herbivores differs considerably, depending on food preferences and local Se availability in soils. Even within a given country, Se intake may vary grossly, e.g. in China, both areas with low average Se intake and exceptionally high average Se intake are described. Health effects of insufficient and excess Se intake can therefore be observed within the same population although otherwise being highly similar in way of living, genetics and other environmental factors. Indeed, symptoms of selenosis are well-described from residents in the Enshi region in Hubei, China, who chronically consumed >500 μg Se/day and developed nail and hair loss among other symptoms (Huang et al. 2013). Another source of information for defining the upper limit of safe Se intake is from intentional or unintentional oversupply, as observed e.g. in the course of homo- or suicide attempts (Hunsaker et al. 2005), as consequence of nutritional overconsumption (Senthilkumaran et al. 2012), or in response to the intake of wrongly formulated commercially available supplements (Sutter et al. 2008). A phase I clinical trial has intentionally tested the upper safe limit of selenite intake in cancer patients, the so-called maximum tolerated dose (MTD), which was determined as 10.2 mg Se per m^2 body surface per day, i.e. around 20 mg Se per day per individual (O Brodin et al. 2015 Nutrients 7:4978); however, selenite is not a major nutritional Se source.

Collectively, the examples indicate that fatigue, hair and nail loss besides a number of additional symptoms (skin, teeth, gastrointestinal, nervous) are among the most evident, early and indicative symptoms of Se intoxication, besides the generally described unpleasant smell of exhaled selenocompounds. As the average intake of Se in Enshi has declined in recent years, together with a disappearance of the overt selenosis symptoms, a chronic intake of >500 μg Se/day can be assumed to be close to the toxic limit. In the US, lack of adverse effects is reported from areas with an average intake of 700 μg/day (Longnecker et al. 1991), indicating that other dietary factors, body weight or form of the Se source contribute to the risk for selenosis upon chronic excessive intake. Especially the correction to body weight seems to be of physiological importance when considering health effects of Se intake. A recent study has convincingly demonstrated that daily Se intake is inversely associated to markers of obesity, while no statistically significant interaction between Se intake and biomarker data remained if body weight of the Se consuming subject was ignored during the biostatistical analysis (Wang et al. 2016).

The same dependence on total body mass is likely underlying our current assumptions on the minimally needed Se supply in humans. Clinical signs of Se deficiency in the poor Se areas of China include the appearance of Kashin-Beck and Keshan disease, which show some common molecular alterations (Wang et al. 2013). In both cases, low Se intake appears as a common risk factor for the development of these two endemic diseases, without being the sole cause. The intake determined in different studies of subjects living in the Se-poor area of China (the so-called Keshan disease belt, an area extending from the northeast to the southwest) averages at 20 μg Se/day or even less, hereby clearly defining a minimally needed essential amount of Se intake. It can safely be assumed that subjects with higher body weight are in need of higher daily intakes for avoiding deficiency symptoms. Collectively, these studies indicate that Se intakes of below 20 μg/day or above 500–800 μg/day can be considered as a health risk. Hence, there is only a factor of 30 between clear deficiency and risk of selenosis, which constitutes a relatively small margin.

Considering the available studies in more detail, and focusing on a single human disease which has been intensively studied in recent years, it becomes obvious that already small differences in Se intake can have a significant effect on disease risks. Thyroid disease prevalence has been determined and compared between two neighboring counties of China with considerably different soil Se concentrations; in this study, mainly farmers were included to obtaining a more reliable reflection of soil Se dependence. Indeed, average serum Se status differed two-fold (103.6 vs 57.4 μg/l) in the subjects residing in these two neighboring regions, while other anthropometric, life-style or likely also genetic parameters were highly similar. Notable, this moderate difference in Se intake and Se status already associated with a two-fold higher incidence of thyroid diseases in those subjects from the Se-deficient area (Wu et al. 2015a). Similar significant associations between Se intake, Se status and disease risk are clearly documented for cancer at various sites in different studies, among other wide-spread diseases (see further below). It can thus be summarized that there is a margin of about 30-fold difference between toxicity and deficiency in chronic daily Se intake for humans, likely depending on the form of the Se compound, individual genotype, other nutritional components and body weight. There are also a number of clinical studies reporting differences in disease risks upon very slight differences in average Se intake levels. From these studies, it can be extrapolated that an optimal intake should exceed 50 μg Se/day, with a likely optimum in the range of 2–3 μg Se/day/kg body weight.

8.3 Major Molecular Se Species in Mammalian Food Sources

As mentioned, Se is a micronutrient and essential trace element for animals. From a physico-chemical point of view, Se belongs to the group of chalcogens, group 16 (VIa), period 4, with chemical properties similar to sulphur (S), same group, period 3, in the table of elements. The chemical similarity also explains much of its distribution in food sources, as it can replace S in a number of molecules, and S-rich food sources typically also contain high Se levels. Selenium enters the food chain mainly from soil via uptake into plants, in the form of the inorganic ions selenite (SeIV) or selenate (SeVI) (White 2016), or even in the form of the Se-containing amino acids selenomethionine (SeMet) and selenocysteine (Sec) (White and Broadley 2009). However, Se is not an essential trace element for plants, rather a growth promoting

Table 8.1 Major molecular Se species in mammalian food sources

Selenium compound	Major food sources
selenocysteine (Sec)	Animals having selenoproteins, to some extent also plants
selenomethionine (SeMet)	Plants, especially high-Se accumulators, and to some extent animals
Methylated inorganic and organic Se species	Plants, especially high-Se accumulators
Selenite and selenite	Plants grown in selenium-rich soil

Levels and species of Se compounds can vary much between different mammalian food sources and is also highly variable between regions. This table only serves as a short illustration of the major Se compounds found in different food sources

factor within a certain concentration range, depending on the plant species and genotype.

Plants can be divided into Se-sensitive, Se-indicating and Se-accumulating species, depending on their response to different Se supply. The Se concentrations reached in accumulating plants can reach up to >1 g/kg dry mass (Cappa et al. 2014), and reflect soil Se phytoavailability, which is a function of soil type, soil pH, mineral composition, humidity and many more biogeochemical parameters. Upon uptake, selenite becomes fast metabolized to organic selenocompounds while selenate can enter the xylem for systemic transport and uptake into shoot plastids, from where it can be activated and reduced to selenite in order to enter biochemical pathways (Pilon-Smits et al. 2009). Further enzymatic steps convert the inorganic selenite mainly to SeMet, Sec, Se-cystine and methylated derivatives of these selenoaminoacids, predominantly methyl-selenocysteine (MSec). Besides these amino acids, selenocystathionine and selenohomocysteine are formed as Se-containing intermediate metabolites. Several additional Se-containing molecules have been detected in plants, including the pungent but potentially healthy selenoglucosinolates and a number of sweet selenosugars. The relative amount of these organic selenocompounds differs considerably between different plant species, plant compartments and may be modulated by plant genotypes (Thavarajah et al. 2011). Elemental Se is found in some plants, especially when carrying symbiotic bacteria or fungi, which are capable of reducing selenocompounds to elemental Se and forming Se-rich granules. Several Se-accumulating plants deposit Se in their vacuoles, potentially as a detoxification strategy, as also shown in yeast (Gharieb and Gadd 1998). Food sources of Se for omnivores naturally encompass also animals. Table 8.1 summarizes the major food sources of Se for mammals.

Different strategies have been followed in order to increase Se concentrations in plants, including regular breeding and selection, transgenic approaches or supplementing the soil or the plants directly via increasing Se in the fertilizers or via foliar spray (Wu et al. 2015b). In combination with the different atmospheric and geological selenocompounds, the environmental influences on the Se supply to and uptake in different species of plants are subject to variation, as are the absolute quantities of the different plant selenocompounds (Winkel et al. 2015). For all of these reasons, it is impossible to predict the form and quantity in which Se is present in a given plant harvested from a natural environment, e.g. even the famous Se-rich Brazil nuts differ in Se concentration depending on their regional source within Brazil from 0.03 to 512.0 mg/kg (Chang et al. 1995). Quantitative analysis and speciation are thus essentially needed for determining Se content and composition of selenocompounds in plants in order to obtain a reliable picture. Importantly, this information should be provided with commercial plant-derived products in case these plants are known to be effective accumulators of Se, in order to inform the customer and enable an adequate Se intake via these products. But despite the complexity of Se metabolism in plants and the heterogeneity of plant selenocompounds, the nutritional and quantitative most relevant selenocompounds in plants are the organic Se-containing amino acids SeMet and Sec and their derivatives, which constitute the largest fraction of selenocompounds in both naturally grown and Se supplemented plants (Winkel et al. 2015; Rayman et al. 2008).

The situation is fundamentally different in herbivores and higher trophic levels including mammals, as Se constitutes an essential micronutrient, indispensable for the biosynthesis of the essential selenoproteins. Hence, the metabolism of Se is regulated in a way that tries to ensure a sufficiently high biosynthetic rate of selenoproteins, while at the same time avoiding Se concentrations exceeding the threshold towards toxicity. To this end, a complex and tightly regulated molecular machinery has been developed for controlling biosynthesis of Sec-containing selenoproteins according to Se availability, tissue requirements, and other inputs from the environment and feedback regulatory signals (Gladyshev and Hatfield 1999). In parallel, a relaxed and promiscuous pathway is in operation for generating a pool of reserve Se in form of SeMet-containing selenoproteins (Lyons et al. 2007). Both systems are connected by the methionine gamma-lyase, capable of releasing methylselenol from SeMet as substrate for the anabolic reactions towards selenoproteins (Esaki et al. 1979), or via the transsulphuration/transselenation pathway converting SeMet into Sec which can then be degraded further by the beta-lyase for selenoprotein biosynthesis (Beilstein and Whanger 1992).

Accordingly, the major molecular Se species from plant food sources are SeMet, either free or in proteins, followed by Sec derivatives including Se-methylselenocysteine and γ-glutamyl-Se-methyl-selenocysteine, along with some selenite, selenate and a set of selenosugars and intermediary selenocompounds. The spectrum of selenocompounds from animal food sources is more restricted, mainly to the two organic selenoamino acids SeMet and Sec as part of selenoproteins and SeMet-containing proteins, respectively (1).

Depending on the Se status and plant-derived SeMet consumption rate of the animal, the relative concentrations of SeMet versus Sec may vary considerably. From a quantitative perspective, the intermediary selenocompounds generated and metabolized during SeMet and Sec turnover and selenoprotein biosynthesis are only minor selenocompounds in mammals. As soil Se concentrations and bioavailability vary drastically between different regions used for plant and animal production, the Se concentrations in plants may vary as drastically as the soil contents; however, as Se is essential for animals and disease symptoms along with poor fertility result

Table 8.2 Major molecular Se species in mammalian excretion of Se

Selenium species	Route of excretion
Dimethyl-selenide	Breath
Selenosugars	Urine
Selenonium	Urine, possible sweat
Selenoproteins	Shedded skin and intestinal cells
SELENOP, GPx3	Blood (plasma)
SELENOP, GPx4	Ejaculate

This table shortly summarizes routes for mammalian excretion of Se

from Se deficiency, the range of Se concentrations found in mammals is much more restricted than in plants, e.g. Brazil nut Se concentrations may differ up to 10.000-fold (Chang et al. 1995), while human serum Se concentrations may be as low as 21 μg/l in the Se-poor areas of China (Xia et al. 2005), and reach up to 600 μg/l in subjects on high dosage Se supplementation (Schrauzer and White 1978), hereby showing a maximal difference of 30-fold only.

8.4 Major Molecular Se Species in Mammalian Excretion of Se

The Se status of mammals is controlled by both intake and excretion, and feedback regulatory systems are in place trying to maintain an optimal level. Among the routes via which Se becomes excreted, one has to take the Se status into account, as the relative contributions of different excretory products are being adapted to the supply, the needs and other metabolic and health-related parameters. Under normal conditions, the selenosugars, especially methyl-2-acetamido-2-deoxy-1-seleno-beta-D-galactopyranoside, dominate Se excretion via urine (Kobayashi et al. 2002). High Se exposure progressively leads to the formation of methylated Se compounds, such as trimethylselenonium ions that can be secreted via urine (Nahapetian et al. 1983), and volatile dimethylselenide that can be exhaled contributing to the garlic odor as a characteristic sign of selenosis (McConnell and Portman 1952).

Besides these quantitatively adaptable and regulated routes, some Se is regularly lost in fertile women in the course of blood loss during menstruation and in men during ejaculation (Michaelis et al. 2014). In addition, a constant loss can be expected in the form of cellular selenoproteins in faeces from intestinal epithelial cell shedding and as consequence of apoptosis (Suzuki et al. 2013). Interestingly, some sex-specific differences in Se excretion in humans have been observed in a large supplementation trial (Combs et al. 2012). The major forms of Se species in mammalian excretion are summarized in Table 8.2.

8.5 Mammalian Selenoproteins

Selenoproteins are a unique family of proteins defined by containing one, or rarely several, selenocysteine (Sec) residues in their amino acid sequence. Selenocysteine, today recognized as the 21st proteinogenic amino acid, is in many aspects truly unique. It is co-translationally inserted into a growing polypeptide chain by a redefinition of a predefined UGA amber codon that usually leads to termination (stop codon) of translation. In the context of selenoprotein synthesis, UGA is interpreted as a sense codon for Sec insertion through the use of Sec-specific synthesis and translation machineries that interact with a secondary structure in the selenoprotein-encoding mRNA that, furthermore, are species specific and different when comparing different subsets of bacteria with archaea or mammals (Allmang et al. 2009; Castellano et al. 2008; Zhang et al. 2006; Schomburg et al. 2004; Mehta et al. 2004; Kryukov and Gladyshev 2004; Lescure et al. 2002; Nasim et al. 2000). It is also quite extraordinary that many organisms completely lack selenoproteins, such as higher plants, yeasts and many microorganisms, while others have only one (e.g. *C. elegans*), three (e.g. *E. coli*) or up to more than 35 selenoproteins (e.g. many fish species) (Castellano et al. 2008, 2009; Schomburg et al. 2004; Kryukov and Gladyshev 2004; Lobanov et al. 2009; Kim et al. 2009; Lobanov et al. 2007, 2008; Zhang et al. 2005; Taskov et al. 2005; Kryukov et al. 2003). Among mammals, humans have 25 selenoproteins and mouse has 24 (Kryukov et al. 2003), and some of these are essential for embryonic survival, as shown with knockout studies in mice for the two genes for mitochondrial or cytosolic thioredoxin reductase (Bondareva et al. 2007; Jakupoglu et al. 2005; Conrad et al. 2004), phospholipid hydroperoxide glutathione peroxidase (Yant et al. 2003) and also the tRNA for carrying Sec to the ribosome (Bosl et al. 1997).

Apart from selenoprotein P (SePP), which is the only form with several Sec residues, is synthesized in the liver and secreted into plasma to have a Se transport function (Renko et al. 2008; Hill et al. 2007), the mammalian selenoproteins are all believed to be enzymes with oxidoreductase activities, utilizing Sec as a catalytic residue in their active sites (Schomburg et al. 2004; Fomenko et al. 2008; Nauser et al. 2006; Johansson et al. 2005; Reich and Hondal 2016; Arnér 2010; Wessjohann et al. 2007; Papp et al. 2007; Birringer et al. 2002; Köhrle et al. 2000; Flohe et al. 2000). Indeed, Sec has rather specific biochemical and chemical features as a catalytic residue compared with its sulfur-containing Cys analog, including higher nucleophilicity, lower pKa, stronger resistance to overoxidation and generally higher reactivity (Schomburg et al. 2004; Reich and Hondal 2016; Arnér 2010; Wessjohann et al. 2007; Fomenko et al. 2008; Nauser et al. 2006; Johansson et al. 2005; Papp et al. 2007; Birringer et al. 2002; Köhrle et al. 2000; Flohe et al. 2000; Metanis and Hilvert 2014; Castellano 2009). Such features of Sec are believed to be evolutionary drivers explaining its inclusion in selenoproteins, although major questions regarding the necessity of Sec remain because of the fact that many organisms, such as plants, can thrive without selenoproteins (Castellano et al. 2009; Reich and Hondal 2016; Arnér 2010; Lothrop et al. 2009, 2014). Still, it is clear that

Table 8.3 Human selenoproteins

Human selenoprotein-encoding genes, in alphabetical order	Identities of the corresponding selenoproteins
DIO1	Type 1 thyroid hormone deiodinase, Dio1
DIO2	Type 2 thyroid hormone deiodinase, Dio2
DIO3	Type 3 thyroid hormone deiodinase, Dio3
GPX1	Cytosolic glutathione peroxidase, Gpx1
GPX2	Gastrointestinal glutathione peroxidase, GPx2/GI-GPx
GPX3	Plasma glutathione peroxidase, pGPx/GPx3
GPX4	Phospholipid hydroperoxide glutathione peroxidase, GPx4
GPX6	GPx6, less studied GPx in embryos and olfactory epithelium
MSRB1	Methionine sulfoxide reductase B1, MsrB1
SELENOF	Selenoprotein of 15 kDa in endoplasmic reticulum, Sel15/Sep15
SELENOH	Chromosome 11 Open Reading Frame 31, Selenoprotein H/SelH
SELENOI	Ethanolaminephosphotransferase 1, Selenoprotein I/SelI
SELENOK	Selenoprotein K in the endoplasmic reticulum, SelK
SELENOM	Perinuclear selenoprotein M, SelM
SELENON	Selenoprotein N in endoplasmic reticulum, SelN
SELENOO	Selenoprotein O with yet unknown function, SelO
SELENOP	Selenoprotein P in plasma, SEPP1/SelP
SELENOS	Selenoprotein S in the endoplasmic reticulum, SelS/VIMP
SELENOT	Selenoprotein T with yet unknown function, SelT
SELENOV	Paralog of Selenoprotein W of yet unknown function, SelV
SELENOW	Cytosolic Selenoprotein W expressed in muscle tissue, SelW
SEPHS2	Selenophosphate synthetase, SPS2
TXNRD1	Cytosolic thioredoxin reductase, TrxR1/TR1
TXNRD2	Mitochondrial thioredoxin reductase, TrxR2/TR3
TXNRD3	Testis-specific thioredoxin glutathione reductase, TGR

The human genome encompasses 25 genes encoding selenoproteins, several of which giving rise to alternative forms of protein products by splicing or posttranslational modification. This table summarizes the 25 human genes encoding selenoproteins, using the newly approved HUGO gene names and selected keywords regarding function

selenoproteins carry out a number of important functions in mammals, with the human genes encoding selenoproteins being summarized in Table 8.3. The interested reader is referred to the literature cited in this chapter for further in-depth information on specific selenoproteins, as that topic is beyond the scope of this contribution.

8.6 Mammalian Se Metabolism – Putting It All Together

With different Se species derived from food ingestion being metabolized and secreted or excreted in other forms it is important to note that Se, of course, is absolutely required for the synthesis of selenoproteins. For selenoprotein synthesis, Se metabolites must be reduced to selenide, which is the precursor of selenophosphate that needs to be formed for conversion of the phosphoseryl moiety of the tRNA for selenocysteine into a final selenized variant that can be utilized for selenoprotein biosynthesis (Xu et al. 2007). In addition, selenoproteins can themselves catalyze the metabolism of certain Se-containing metabolites, such as reduction of methylseleninate or selenite by thioredoxin reductase (Gromer and Gross 2002; Björnstedt et al. 1996; Kumar et al. 1992), and regulate tissue-specific uptake of Se, such as uptake of SePP as a Se source from plasma (Renko et al. 2008; Hill et al. 2007; Valentine et al. 2008). Selenium can also be reutilized from the breakdown of Sec derived from selenoproteins through the actions of selenocysteine lyase, the lack of which leads to metabolic syndrome in mice (Seale et al. 2012). Thus, Se metabolism in mammals, and their health, is intricately linked to the status of selenoproteins, their translation and their activities. The importance of Se metabolism (Fig. 8.1) and selenoproteins (Table 8.3) is illustrated by the mammalian phenotypes and diseases linked to Se deficiency, as discussed next (Table 8.4).

8.7 Established Health Risks of Se Deficiency and Lessons from Inherited Diseases in Selenoprotein Biosynthesis

By the mid-1950s it was discovered that supplementation with Se to livestock supports good health and since then it has remained common practice in agricultural production to provide additional Se to cattle and other domestic animals, with significant value for this industry (Oldfield 1997). In humans it has long been known that people living in Se-deprived areas and only eating local food, i.e. establishing nutritional Se deficiency, have high risks of developing Keshan or Kashin-Beck disease with cardiomyopathy, immune dysfunction, impaired bone development and joint inflammation, male infertility and possibly increased incidence of cancer (Köhrle et al. 2000; Rayman 2000). The exact underlying molecular mechanisms explaining such disease spectra upon Se deficiency have, however, remained uncertain. As genetic traits with links between aberrations in selenoprotein synthesis and disease are discovered, specific phenotypes may increasingly be understood and causally explained (Schomburg 2010).

Of interest in this context are patients with defects in the SECISBP2 protein, also known as SBP2, participating in co-translational Sec insertion, which present with a syndrome encompassing azoospermia, muscular dystrophy, immune dysfunction, thyroid hormone insufficiency and enhanced insulin sensitivity, with these complex symptoms likely reflecting a mixture of effects from deficiency of several different selenoproteins (Schoenmakers et al. 2010). For example, muscular dystrophy may be

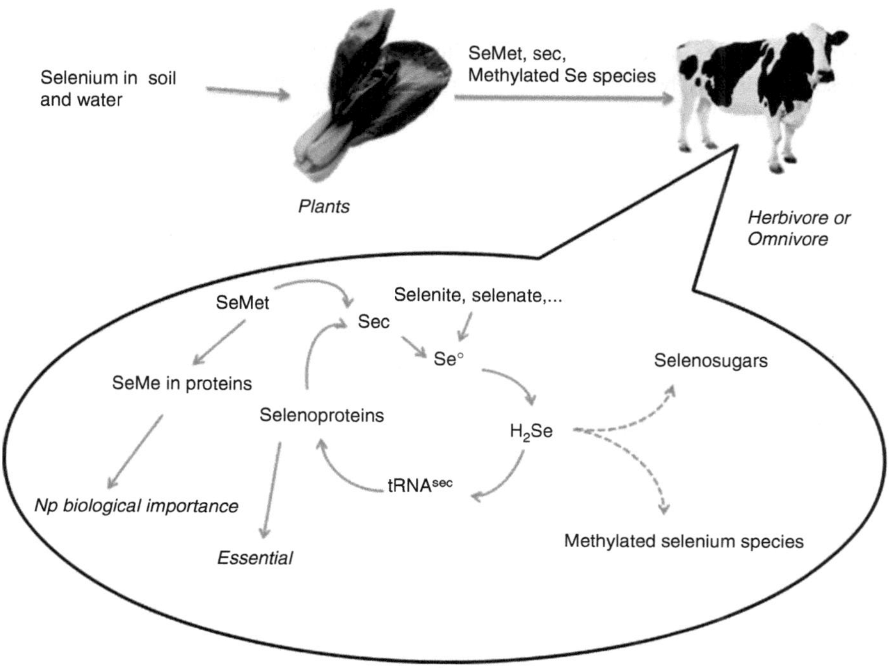

Fig. 8.1 Schematic overview of mammalian Se metabolism. This figure schematically summarizes the major forms of Se found in herbivores or omnivores, their metabolism (*solid arrows*) and forms in excretion (*dashed arrows*)

Table 8.4 Clinical symptoms related to defects in selenoprotein-related genes

Gene	Year	Symptoms/Disease	References
SELENON	2001	Muscular dystrophy with spinal rigidity	(94) Moghadaszadeh et al. (2001)
SECISBP2	2005	Resistance to thyroid hormones, delayed development	(95) Dumitrescu et al. (2005)
SEPSECS	2010	Progressive cerebello-cerebral atrophy	(96) Agamy et al. (2010)
tRNA[Ser]Sec	2016	Abdominal pain, fatigue, muscle weakness	(97) Schoenmakers et al. (2016)
TXN2	2016	Cerebellar atrophy, epilepsy, dystonia, neuropathy	(98) Holzerova et al. (2016)
TXNRD2	2015	Glucocorticoid deficiency syndrome	(87) Prasad et al. (2014)

This table summarizes the findings of clinical symptoms or diseases yet linked to specific human selenoprotein-linked genes together with their years of discovery

explained by deficiency of selenoprotein N (SEPN), where a single nucleotide polymorphism decreasing binding of the SEPN-encoding mRNA to SECISBP2 has been linked to disease (Allamand et al. 2006). In another study it was found that deficiency in the mitochondrial thioredoxin reductase 2 (TXNRD2) leads to a glucocorticoid deficiency syndrome in humans (Prasad et al. 2014). Not many monogenic selenoprotein-related deficiencies have yet been discovered in humans (4), but quite a number of knockout mouse models have been made that further help to unravel the functions of mammalian selenoproteins. Among these it was shown that, in mouse, both the cytosolic and mitochondrial thioredoxin reductases are essential for healthy embryogenesis (Bondareva et al. 2007; Jakupoglu et al. 2005; Conrad et al. 2004) as is glutathione peroxidase 4 (Yant et al. 2003). Several additional mouse models have revealed important roles of Se and selenoproteins in signaling or specific tissue-related diseases, with the interested reader being referred to more extensive review articles for details (Allmang et al. 2009; Schomburg et al. 2004; Lobanov et al. 2009; Papp et al. 2007; Lei et al. 2016; Kim and Gladyshev 2007; Hatfield et al. 2006). Here we shall solely note that it is well established that the symptoms of Se deficiency syndromes are related to insufficient functions of one or several selenoproteins, while the exact mechanisms that link selenoprotein function to disease in many cases awaits to be characterized at the molecular level.

8.8 Conclusions

Higher herbivores including mammals metabolize Se compounds present in ingested plants and utilize the trace element by well-controlled and tightly regulated pathways for biosynthesis of selenoproteins, many of which have crucial functions for maintaining health. Excessive Se can be excreted by different routes in the form of distinct metabolites. Conversely, too low daily intake of Se can result in endemic deficiency syndromes and increases the risk for a number of common human diseases including cancer at various sites, infections, autoimmune diseases and others. With the identification of distinct mutations in human selenoprotein genes or components of the selenoprotein biosynthesis machinery, complemented with detailed studies of targeted genetic mouse models, the understanding of molecular links between Se metabolism, selenoproteins and mammalian pathology and health is rapidly increasing.

Acknowledgements Work of the authors has support from Swedish Cancer Society, Swedish Research Council (Medicine and Natural Sciences sections), Karolinska Institutet, DFG and Charité Berlin.

References

Agamy O, Ben Zeev B, Lev D, Marcus B, Fine D, Su D et al (2010) Mutations disrupting selenocysteine formation cause progressive cerebello-cerebral atrophy. Am J Hum Genet 87(4):538–544

Allamand V, Richard P, Lescure A, Ledeuil C, Desjardin D, Petit N et al (2006) A single homozygous point mutation in a 3′untranslated region motif of selenoprotein N mRNA causes SEPN1-related myopathy. EMBO Rep 7(4):450–454

Allmang C, Wurth L, Krol A (2009) The selenium to selenoprotein pathway in eukaryotes: more molecular partners than anticipated. Biochim Biophys Acta 1790(11):1415–1423
Arnér ESJ (2010) Selenoproteins-What unique properties can arise with selenocysteine in place of cysteine? Exp Cell Res 316(8):1296–1303
Barceloux DG (1999) Selenium. J Toxicol Clin Toxicol 37(2):145–172
Beilstein MA, Whanger PD (1992) Selenium metabolism and glutathione peroxidase activity in cultured human lymphoblasts. Effects of transsulfuration defects and pyridoxal phosphate. Biol Trace Elem Res 35(2):105–118
Berzelius JJ (1818) Undersökning af en ny Mineral-kropp, funnen i de orenare sorterna af det i Falun tillverkade svaflet. Afhandlingar i fysik, kemi och mineralogi 6:42–144
Birringer M, Pilawa S, Flohe L (2002) Trends in selenium biochemistry. Nat Prod Rep 19(6):693–718
Björnstedt M, Odlander B, Kuprin S, Claesson HE, Holmgren A (1996) Selenite incubated with NADPH and mammalian thioredoxin reductase yields selenide, which inhibits lipoxygenase and changes the electron spin resonance spectrum of the active site iron. Biochemistry 35:8511–8516
Bondareva AA, Capecchi MR, Iverson SV, Li Y, Lopez NI, Lucas O et al (2007) Effects of thioredoxin reductase-1 deletion on embryogenesis and transcriptome. Free Radic Biol Med 43(6):911–923
Bosl MR, Takaku K, Oshima M, Nishimura S, Taketo MM (1997) Early embryonic lethality caused by targeted disruption of the mouse selenocysteine tRNA gene (Trsp). Proc Natl Acad Sci U S A 94:5531–5534
Brodin O, Eksborg S, Wallenberg M, Asker-Hagelberg C, Larsen EH, Mohlkert D et al (2015) Pharmacokinetics and toxicity of sodium selenite in the treatment of patients with carcinoma in a phase I clinical trial: the SECAR study. Forum Nutr 7(6):4978–4994
Cappa JJ, Cappa PJ, El Mehdawi AF, McAleer JM, Simmons MP, Pilon-Smits EA (2014) Characterization of selenium and sulfur accumulation across the genus Stanleya (Brassicaceae): a field survey and common-garden experiment. Am J Bot 101(5):830–839
Castellano S (2009) On the unique function of selenocysteine – insights from the evolution of selenoproteins. Biochim Biophys Acta 1790(11):1463–1470
Castellano S, Gladyshev VN, Guigo R, Berry MJ (2008) SelenoDB 1.0: a database of selenoprotein genes, proteins and SECIS elements. Nucleic Acids Res 36(Database issue):D332–D338
Castellano S, Andres AM, Bosch E, Bayes M, Guigo R, Clark AG (2009) Low exchangeability of selenocysteine, the 21st amino acid, in vertebrate proteins. Mol Biol Evol 26(9):2031–2040
Chang JC, Gutenmann WH, Reid CM, Lisk DJ (1995) Selenium content of Brazil nuts from two geographic locations in Brazil. Chemosphere 30(4):801–802
Combs GF Jr (2015) Biomarkers of selenium status. Forum Nutr 7(4):2209–2236
Combs GF Jr, Jackson MI, Watts JC, Johnson LK, Zeng H, Idso J et al (2012) Differential responses to selenomethionine supplementation by sex and genotype in healthy adults. Br J Nutr 107(10):1514–1525
Conrad M, Jakupoglu C, Moreno SG, Lippl S, Banjac A, Schneider M et al (2004) Essential role for mitochondrial thioredoxin reductase in hematopoiesis, heart development, and heart function. Mol Cell Biol 24(21):9414–9423
Dumitrescu AM, Liao XH, Abdullah MS, Lado-Abeal J, Majed FA, Moeller LC et al (2005) Mutations in SECISBP2 result in abnormal thyroid hormone metabolism. Nat Genet 37(11):1247–1252
Esaki N, Tanaka H, Uemura S, Suzuki T, Soda K (1979) Catalytic action of L-methionine gamma-lyase on selenomethionine and selenols. Biochemistry 18(3):407–410
Flohe L, Andreesen JR, Brigelius-Flohe R, Maiorino M, Ursini F (2000) Selenium, the element of the moon, in life on earth. IUBMB Life 49(5):411–420
Fomenko DE, Marino SM, Gladyshev VN (2008) Functional diversity of cysteine residues in proteins and unique features of catalytic redox-active cysteines in thiol oxidoreductases. Mol Cells 26(3):228–235

Franke KW, Potter VR (1935) A new toxicant occurring naturally in certain samples of plant foodstuffs. IX Toxic effects of orally ingested selenium. J Nutr 10(2):213–221

Gharieb MM, Gadd GM (1998) Evidence for the involvement of vacuolar activity in metal(loid) tolerance: vacuolar-lacking and -defective mutants of *Saccharomyces cerevisiae* display higher sensitivity to chromate, tellurite and selenite. Biometals 11(2):101–106

Gladyshev VN, Hatfield DL (1999) Selenocysteine-containing proteins in mammals. J Biomed Sci 6(3):151–160

Gromer S, Gross JH (2002) Methylseleninate is a substrate rather than an inhibitor of mammalian thioredoxin reductase. Implications for the antitumor effects of selenium. J Biol Chem 277(12):9701–9706

Hatfield DL, Carlson BA, Xu XM, Mix H, Gladyshev VN (2006) Selenocysteine incorporation machinery and the role of selenoproteins in development and health. Prog Nucleic Acid Res Mol Biol 81:97–142

Hill KE, Zhou J, Austin LM, Motley AK, Ham AJ, Olson GE et al (2007) The selenium-rich C-terminal domain of mouse selenoprotein P is necessary for the supply of selenium to brain and testis but not for the maintenance of whole body selenium. J Biol Chem 282(15):10972–10980

Holzerova E, Danhauser K, Haack TB, Kremer LS, Melcher M, Ingold I et al (2016) Human thioredoxin 2 deficiency impairs mitochondrial redox homeostasis and causes early-onset neurodegeneration. Brain 139:346–354

Huang Y, Wang Q, Gao J, Lin Z, Banuelos GS, Yuan L et al (2013) Daily dietary selenium intake in a high selenium area of Enshi, China. Forum Nutr 5(3):700–710

Hunsaker DM, Spiller HA, Williams D (2005) Acute selenium poisoning: suicide by ingestion. J Forensic Sci 50(4):942–946

Jakupoglu C, Przemeck GK, Schneider M, Moreno SG, Mayr N, Hatzopoulos AK et al (2005) Cytoplasmic thioredoxin reductase is essential for embryogenesis but dispensable for cardiac development. Mol Cell Biol 25(5):1980–1988

Johansson L, Gafvelin G, Arnér ESJ (2005) Selenocysteine in proteins – properties and biotechnological use. Biochim Biophys Acta 1726(1):1–13

Kim HY, Gladyshev VN (2007) Methionine sulfoxide reductases: selenoprotein forms and roles in antioxidant protein repair in mammals. Biochem J 407(3):321–329

Kim HY, Zhang Y, Lee BC, Kim JR, Gladyshev VN (2009) The selenoproteome of Clostridium sp. OhILAs: characterization of anaerobic bacterial selenoprotein methionine sulfoxide reductase A. Proteins 74(4):1008–1017

Kobayashi Y, Ogra Y, Ishiwata K, Takayama H, Aimi N, Suzuki KT (2002) Selenosugars are key and urinary metabolites for selenium excretion within the required to low-toxic range. Proc Natl Acad Sci U S A 99(25):15932–15936

Köhrle J, Brigelius-Flohe R, Böck A, Gartner R, Meyer O, Flohe L (2000) Selenium in biology: facts and medical perspectives. Biol Chem 381(9–10):849–864

Kryukov GV, Gladyshev VN (2004) The prokaryotic selenoproteome. EMBO Rep 5(5):538–543

Kryukov GV, Castellano S, Novoselov SV, Lobanov AV, Zehtab O, Guigo R et al (2003) Characterization of mammalian selenoproteomes. Science 300(5624):1439–1443

Kumar S, Björnstedt M, Holmgren A (1992) Selenite is a substrate for calf thymus thioredoxin reductase and thioredoxin and elicits a large non-stoichiometric oxidation of NADPH in the presence of oxygen. Eur J Biochem 207(2):435–439

Lei XG, Zhu JH, Cheng WH, Bao Y, Ho YS, Reddi AR et al (2016) Paradoxical roles of antioxidant enzymes: basic mechanisms and health implications. Physiol Rev 96(1):307–364

Lescure A, Fagegaltier D, Carbon P, Krol A (2002) Protein factors mediating selenoprotein synthesis. Curr Protein Pept Sci 3(1):143–151

Lobanov AV, Fomenko DE, Zhang Y, Sengupta A, Hatfield DL, Gladyshev VN (2007) Evolutionary dynamics of eukaryotic selenoproteomes: large selenoproteomes may associate with aquatic life and small with terrestrial life. Genome Biol 8(9):R198

Lobanov AV, Hatfield DL, Gladyshev VN (2008) Selenoproteinless animals: selenophosphate synthetase SPS1 functions in a pathway unrelated to selenocysteine biosynthesis. Protein Sci 17(1):176–182

Lobanov AV, Hatfield DL, Gladyshev VN (2009) Eukaryotic selenoproteins and selenoproteomes. Biochim Biophys Acta 1790(11):1424–1428

Longnecker MP, Taylor PR, Levander OA, Howe M, Veillon C, McAdam PA et al (1991) Selenium in diet, blood, and toenails in relation to human health in a seleniferous area. Am J Clin Nutr 53(5):1288–1294

Lothrop AP, Ruggles EL, Hondal RJ (2009) No selenium required: reactions catalyzed by mammalian thioredoxin reductase that are independent of a selenocysteine residue. Biochemistry 48(26):6213–6223

Lothrop AP, Snider GW, Ruggles EL, Hondal RJ (2014) Why is mammalian thioredoxin reductase 1 so dependent upon the use of selenium? Biochemistry 53(3):554–565

Lyons MP, Papazyan TT, Surai PF (2007) Selenium in food chain and animal nutrition: lessons from nature. Asian Austral J Anim 20(7):1135–1155

McConnell CK, Portman OW (1952) Excretion of dimethyl selenide by the rat. J Biol Chem 195(1):277–282

Mehta A, Rebsch CM, Kinzy SA, Fletcher JE, Copeland PR (2004) Efficiency of mammalian selenocysteine incorporation. J Biol Chem 279(36):37852–37859

Metanis N, Hilvert D (2014) Natural and synthetic selenoproteins. Curr Opin Chem Biol 22:27–34

Michaelis M, Gralla O, Behrends T, Scharpf M, Endermann T, Rijntjes E et al (2014) Selenoprotein P in seminal fluid is a novel biomarker of sperm quality. Biochem Biophys Res Commun 443(3):905–910

Moghadaszadeh B, Petit N, Jaillard C, Brockington M, Quijano Roy S, Merlini L et al (2001) Mutations in SEPN1 cause congenital muscular dystrophy with spinal rigidity and restrictive respiratory syndrome. Nat Genet 29(1):17–18

Nahapetian AT, Janghorbani M, Young VR (1983) Urinary trimethylselenonium excretion by the rat: effect of level and source of selenium-75. J Nutr 113(2):401–411

Nasim MT, Jaenecke S, Belduz A, Kollmus H, Flohé L, McCarthy JE (2000) Eukaryotic selenocysteine incorporation follows a nonprocessive mechanism that competes with translational termination. J Biol Chem 275:14846–14852

Nauser T, Dockheer S, Kissner R, Koppenol WH (2006) Catalysis of electron transfer by selenocysteine. Biochemistry 45(19):6038–6043

Oldfield JE (1997) Observations on the efficacy of various forms of selenium for livestock: a review. Biomed Environ Sci 10(2–3):280–291

Papp LV, Lu J, Holmgren A, Khanna KK (2007) From selenium to selenoproteins: synthesis, identity, and their role in human health. Antioxid Redox Signal 9(7):775–806

Pilon-Smits EAH, Quinn CF, Tapken W, Malagoli M, Schiavon M (2009) Physiological functions of beneficial elements. Curr Opin Plant Biol 12(3):267–274

Prasad R, Chan LF, Hughes CR, Kaski JP, Kowalczyk JC, Savage MO et al (2014) Thioredoxin Reductase 2 (TXNRD2) mutation associated with familial glucocorticoid deficiency (FGD). J Clin Endocrinol Metab 99(8):E1556–E1563

Rayman MP (2000) The importance of selenium to human health. Lancet 356:233–241

Rayman MP (2008) Food-chain selenium and human health: emphasis on intake. Br J Nutr 100(2):254–268

Rayman MP, Infante HG, Sargent M (2008) Food-chain selenium and human health: spotlight on speciation. Br J Nutr 100(2):238–253

Reich HJ, Hondal RJ (2016) Why nature chose selenium. ACS Chem Biol 11(4):821–841

Renko K, Werner M, Renner-Muller I, Cooper TG, Yeung CH, Hollenbach B et al (2008) Hepatic selenoprotein P (SePP) expression restores selenium transport and prevents infertility and motor-incoordination in Sepp-knockout mice. Biochem J 409(3):741–749

Schoenmakers E, Agostini M, Mitchell C, Schoenmakers N, Papp L, Rajanayagam O et al (2010) Mutations in the selenocysteine insertion sequence-binding protein 2 gene lead to a multisystem selenoprotein deficiency disorder in humans. J Clin Invest 120(12):4220–4235
Schoenmakers E, Carlson B, Agostini M, Moran C, Rajanayagam O, Bochukova E et al (2016) Mutation in human selenocysteine transfer RNA selectively disrupts selenoprotein synthesis. J Clin Invest 126(3):992–996
Schomburg L (2010) Genetics and phenomics of selenoenzymes – how to identify an impaired biosynthesis? Mol Cell Endocrinol 322(1–2):114–124
Schomburg L, Schweizer U, Köhrle J (2004) Selenium and selenoproteins in mammals: extraordinary, essential, enigmatic. Cell Mol Life Sci 61(16):1988–1995
Schrauzer GN, White DA (1978) Selenium in human nutrition: dietary intakes and effects of supplementation. Bioinorg Chem 8(4):303–318
Schwarz K, Foltz CM (1957) Selenium as an integral part of factor 3 against dietary necrotic liver degeneration. J Am Chem Soc 79(12):3292–3293
Seale LA, Hashimoto AC, Kurokawa S, Gilman CL, Seyedali A, Bellinger FP et al (2012) Disruption of the selenocysteine lyase-mediated selenium recycling pathway leads to metabolic syndrome in mice. Mol Cell Biol 32(20):4141–4154
Senthilkumaran S, Balamurugan N, Vohra R (2012) Thirumalaikolundusubramanian P. Paradise nut paradox: alopecia due to selenosis from a nutritional therapy. Int J Trichol 4(4):283–284
Sutter ME, Thomas JD, Brown J, Morgan B (2008) Selenium toxicity: a case of selenosis caused by a nutritional supplement. Ann Intern Med 148(12):970–971
Suzuki Y, Hashiura Y, Sakai T, Yamamoto T, Matsukawa T, Shinohara A et al (2013) Selenium metabolism and excretion in mice after injection of (82)Se-enriched selenomethionine. Metallomics 5(5):445–452
Taskov K, Chapple C, Kryukov GV, Castellano S, Lobanov AV, Korotkov KV et al (2005) Nematode selenoproteome: the use of the selenocysteine insertion system to decode one codon in an animal genome? Nucleic Acids Res 33(7):2227–2238
Thavarajah D, Thavarajah P, Sarker A, Materne M, Vandemark G, Shrestha R et al (2011) A global survey of effects of genotype and environment on selenium concentration in lentils (*Lens culinaris* L.): implications for nutritional fortification strategies. Food Chem 125(1):72–76
Valentine WM, Abel TW, Hill KE, Austin LM, Burk RF (2008) Neurodegeneration in mice resulting from loss of functional selenoprotein P or its receptor apolipoprotein E receptor 2. J Neuropathol Exp Neurol 67(1):68–77
Wang X, Wang S, He S, Zhang F, Tan W, Lei Y et al (2013) Comparing gene expression profiles of Kashin-Beck and Keshan diseases occurring within the same endemic areas of China. Sci China Life Sci 56(9):797–803
Wang Y, Gao X, Pedram P, Shahidi M, Du J, Yi Y et al (2016) Significant beneficial association of high dietary selenium intake with reduced body fat in the CODING study. Forum Nutr 8(1):24
Wessjohann LA, Schneider A, Abbas M, Brandt W (2007) Selenium in chemistry and biochemistry in comparison to sulfur. Biol Chem 388(10):997–1006
Whanger PD, Weswig PH, Schmitz JA, Oldfield JE (1977) Effects of selenium and vitamin E on blood selenium levels, tissue glutathione peroxidase activities and white muscle disease in sheep fed purified or hay diets. J Nutr 107(7):1298–1307
White PJ (2016) Selenium accumulation by plants. Ann Bot 117(2):217–235
White PJ, Broadley MR (2009) Biofortification of crops with seven mineral elements often lacking in human diets – iron, zinc, copper, calcium, magnesium, selenium and iodine. New Phytol 182(1):49–84
Winkel LH, Vriens B, Jones GD, Schneider LS, Pilon-Smits E, Banuelos GS (2015) Selenium cycling across soil-plant-atmosphere interfaces: a critical review. Forum Nutr 7(6):4199–4239
Wu Q, Rayman MP, Lv H, Schomburg L, Cui B, Gao C et al (2015a) Low population selenium status is associated with increased prevalence of thyroid disease. J Clin Endocrinol Metab 100(11):4037–4047

Wu Z, Banuelos GS, Lin ZQ, Liu Y, Yuan L, Yin X et al (2015b) Biofortification and phytoremediation of selenium in China. Front Plant Sci 6:136

Xia Y, Hill KE, Byrne DW, Xu J, Burk RF (2005) Effectiveness of selenium supplements in a low-selenium area of China. Am J Clin Nutr 81(4):829–834

Xu XM, Carlson BA, Mix H, Zhang Y, Saira K, Glass RS et al (2007) Biosynthesis of selenocysteine on its tRNA in eukaryotes. PLoS Biol 5(1):e4

Yant LJ, Ran Q, Rao L, Van Remmen H, Shibatani T, Belter JG et al (2003) The selenoprotein GPX4 is essential for mouse development and protects from radiation and oxidative damage insults. Free Radic Biol Med 34(4):496–502

Zhang Y, Fomenko DE, Gladyshev VN (2005) The microbial selenoproteome of the Sargasso Sea. Genome Biol 6(4):R37

Zhang Y, Romero H, Salinas G, Gladyshev VN (2006) Dynamic evolution of selenocysteine utilization in bacteria: a balance between selenoprotein loss and evolution of selenocysteine from redox active cysteine residues. Genome Biol 7(10):R94

Part III
Genetic, Evolutionary and Ecological Aspects of Plant Se Accumulation

Chapter 9
The Genetics of Selenium Accumulation by Plants

Philip J. White

Abstract Genetics is broadly defined as the study of how genes control the characteristics of organisms. In this chapter, emphasis has been placed on differences in metabolic pathways, and their associated genes, that could account for variation in the ability of angiosperm species to tolerate large tissue selenium (Se) concentrations. The current view of the molecular biology of Se uptake and assimilation by plants is presented and differences between plant species likely to affect their ability to tolerate large tissue Se concentrations are identified. In particular, it is noted that plants that hyperaccumulate Se generally exhibit constitutive expression of genes encoding Se-transporters and enzymes involved in primary Se assimilation, biosynthesis of non-toxic Se metabolites and Se volatilisation. A plausible scheme for the evolution of differences in Se accumulation between angiosperm species is described. Since Se is an essential mineral element for animals, and the diets of many humans lack sufficient Se, the possibility of breeding crops with greater Se concentrations in their edible tissues is discussed. It is observed that, although Se concentrations in plants are largely determined by the phytoavailability of Se in the environment, there is significant intraspecific genetic variation in the Se concentrations of most edible crops that might be utilised to improve human diets. However, although molecular markers might be developed to known chromosomal quantitative trait loci (QTL) impacting Se concentration in edible tissues to assist breeding programmes, the actual genes underpinning this variation are largely unknown.

Keywords Genetic variation • Breeding • Nutrition • Evolution • Hyperaccumulation

---

P.J. White (✉)
Ecological Science Group, The James Hutton Institute, Invergowrie, Dundee DD2 5DA, UK
e-mail: philip.white@hutton.ac.uk

E.A.H. Pilon-Smits et al. (eds.), *Selenium in plants*, Plant Ecophysiology 11,
DOI 10.1007/978-3-319-56249-0_9

9.1 Introduction: An Ecological Perspective on Selenium Accumulation by Plants

Selenium (Se) is not an essential element for flowering plants (angiosperms), although there is evidence that it might benefit their growth and survival under certain circumstances (Pilon-Smits and LeDuc 2009; White and Brown 2010). In particular, adequate Se nutrition can mitigate the oxidative stresses caused by various environmental factors and, if present in sufficient amounts in plant tissues, Se can protect against herbivores and pathogens (Pilon-Smits et al. 2009; El Mehdawi and Pilon-Smits 2012; Feng et al. 2013). By contrast, excess Se accumulation is toxic to most plants, presumably because the indiscriminate incorporation of the Se-amino acids, selenocysteine (SeCys) and selenomethionine (SeMet) into proteins impairs their activity and results in biochemical and physiological malfunction (Brown and Shrift 1982; White et al. 2004; Van Hoewyk 2013).

The accumulation of Se by plants is largely influenced by Se phytoavailability in the soil, although there are differences between angiosperm species in their ability to acquire Se under identical environmental conditions (Fig. 9.1) (White et al. 2004, 2007a; White 2016). Soil Se concentrations differ greatly. Although most soils have Se concentrations between 0.01 and 2.0 mg/kg, some soils associated with particular geological formations can reach concentrations of 1200 mg/kg (Dhillon and Dhillon 2003; Fordyce 2013; Pilbeam et al. 2015). The phytoavailability of Se in soils is influenced by a number of factors including pH, redox potential and organic matter content (Mikkelsen et al. 1989; White et al. 2007b; Fordyce 2013; Pilbeam et al. 2015). Plant roots can acquire Se from the soil solution as selenate (SeO_4^{2-}), selenite (SeO_3^{2-}, $HSeO_3^-$, H_2SeO_3) or organoselenium compounds, such as SeCys and SeMet, but are unable to take up selenide (Se^{2-}) species or colloidal elemental Se (White and Broadley 2009; White 2016). In oxic soils ($pH + pE > 15$), selenate is the main water-soluble form of Se, whereas in anaerobic soils with a neutral to acidic pH ($pH + pE = 7.5–15$) selenite (SeO_3^{2-}) is the main water-soluble form. Selenides occur only in severely anaerobic and often acidic soils ($pH + pE < 7.5$). Selenate is relatively mobile in the soil solution, but selenite is strongly absorbed by iron and aluminum oxides/hydroxides and, to a lesser extent, by clays and organic matter (Fordyce 2013; Pilbeam et al. 2015).

Distinct plant communities exist on soils with high Se phytoavailability (Rosenfeld and Beath 1964; Brown and Shrift 1982). Since the accumulation of Se is potentially toxic, plant species growing on seleniferous soils must either exclude, or actively remove, Se from their tissues or tolerate the Se they accumulate. There are marked differences between plant species in their ability to tolerate Se in their tissues and, since all plants growing on seleniferous soils have elevated tissue Se concentrations, it is hypothesised that the colonisation of seleniferous soils required a minimal ability to tolerate elevated tissue Se concentrations (Rosenfeld and Beath 1964; Brown and Shrift 1982; El Mehdawi and Pilon-Smits 2012).

Angiosperm species have been divided into three ecological types according to their ability to accumulate Se in their tissues (Table 9.1) (Rosenfeld and Beath 1964;

Fig. 9.1 Relationship between shoot Se and S concentrations in 39 angiosperm species grown hydroponically with 0.91 mM sulphate and 0.63 μM selenate. The *line* indicates a shoot Se/S quotient of 1.725 mg Se per g S. *Closed symbols* represent species from the Brassicales (Data from White et al. 2007)

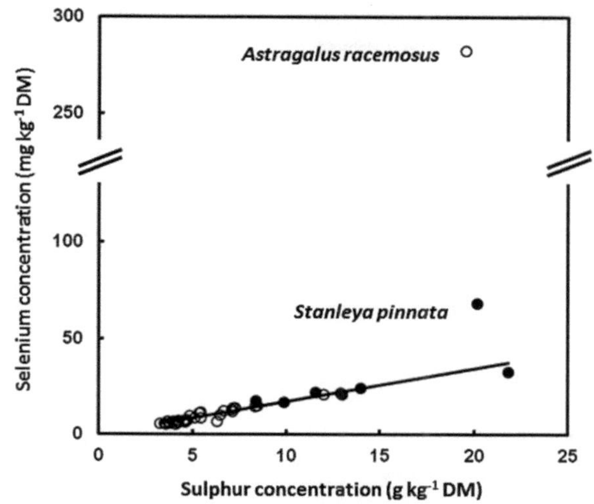

Brown and Shrift 1982; White et al. 2004, 2007a; El-Mehdawi and Pilon-Smits 2012; White 2016). Most angiosperm species are designated "non-accumulator" species. These species cannot tolerate tissue Se concentrations greater than 10–100 μg/g dry matter (DM) and cannot colonize seleniferous soils (Rosenfeld and Beath 1964; White et al. 2004; Fordyce 2013; White 2016). Species that can tolerate tissue Se concentrations approaching 1 mg/g DM are designated "Se-indicator" plants. These species can colonize both non-seleniferous and seleniferous soils (Rosenfeld and Beath 1964; Moreno Rodriguez et al. 2005). Their tissue Se concentrations are directly related to the Se phytoavailability in the soil and, thereby, 'indicate' soil Se phytoavailability (cf. Baker 1981). Species tolerating tissue Se concentrations greater than 1 mg/g DM are designated "Se-accumulator" species. These species are generally restricted to seleniferous soils (Rosenfeld and Beath 1964; Brown and Shrift 1982; El-Mehdawi and Pilon-Smits 2012). The appellation "Se-hyperaccumulator" is given to a plant whose leaves contain more than 1 mg/g DM when sampled from its natural environment (Reeves and Baker 2000; White 2016), although some experts argue that this threshold should be lowered to 100 μg/g DM (Reeves and Baker 2000; van der Ent et al. 2013).

Although there are notable differences between species in their ability to accumulate Se, which are manifested, for example, in their ecological strategies towards soils with high Se phytoavailability, there appear to be no systematic phylogenetic effects on shoot Se concentration within the angiosperms (White et al. 2004, 2007a; Watanabe et al. 2007). The majority of the variation in shoot Se concentrations among angiosperms appears to occur between species within orders and between ecotypes within species (White et al. 2004, 2007a; Watanabe et al. 2007; Feist and Parker 2001; Cappa et al. 2014; El Mehdawi et al. 2015; White 2016). Thus, variation in shoot Se concentration is most apparent within plant families that contain

Table 9.1 Classification of angiosperm species into ecological types according to their ability to accumulate Se in their tissues

Classification	Leaf Se concentration (μg/g DM)	Soil ecology	Biochemical and physiological characteristics
non-accumulator	< 10–100	non-seleniferous	Se uptake induced by S-starvation
			High S/Se selectivity for uptake
Se-indicator	< 1000	seleniferous & non-seleniferous	Se transport to shoot induced by S-starvation
			Se volatilisation as dimethylselenide (DMSe)
			Sequestration highest in leaves
Se-accumulator	> 100	seleniferous	Main Se form selenate (sometimes SeMet)
Se-hyperaccumulator	> 1000	seleniferous	Constitutive, S-independent Se uptake
			Low S/Se selectivity for uptake
			Constitutively large Se transport to shoot
			Constitutively large Se metabolic flux
			Large Se volatilisation as dimethyldiselenide (DMDSe)
			Sequestration highest in reproductive organs
			Main Se form MSeCys

The broad biochemical and physiological characteristics thought to differentiate Se-hyperaccumulator and non-hyperaccumulator species are also indicated. *DM* dry matter

Se-accumulator or Se-indicator plants (White et al. 2004; White 2016). Nevertheless, it is noteworthy that, when plants are sampled from the same environment, there is a stoichiometric relationship between shoot Se and sulphur (S) concentrations across most angiosperm species and plants. For example, alliums and brassicas, which have greater S concentrations also have greater Se concentrations (Fig. 9.1) (White et al. 2007a). This can, in part, be attributed to the fact that S and Se share common pathways for uptake by roots, translocation between organs, and metabolism within the plant (Fig. 9.2) (White et al. 2007b; White 2016).

9.2 Molecular Biology of Selenium Uptake and Assimilation by Plants

Selenate is taken up by plant roots through high-affinity sulphate transporters (HASTs) homologous to the arabidopsis (*Arabidopsis thaliana* [L.] Heynh.) AtSULTR1;1 and AtSULTR1;2 transporters (Sors et al. 2005b; White et al. 2007b;

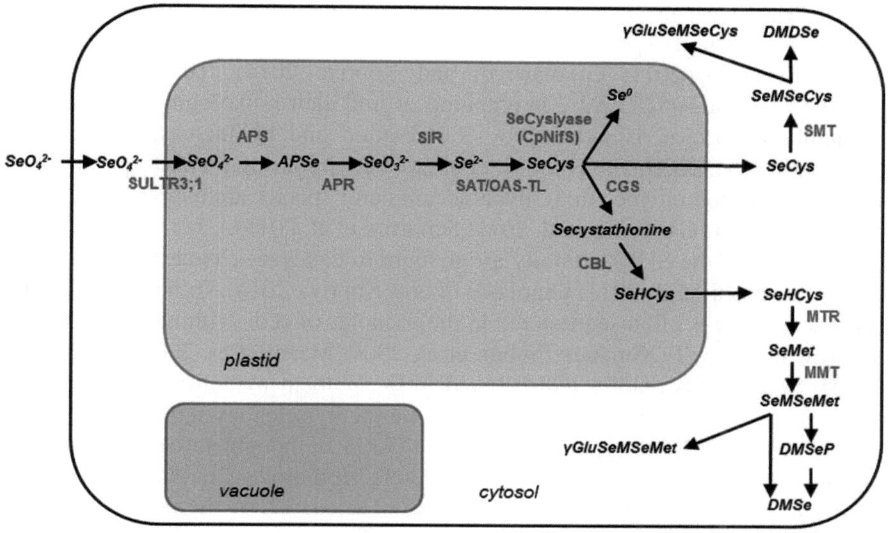

Fig. 9.2 Generalised scheme of Se metabolism in the leaves of higher plants. A detailed explanation can be found in the text and in papers by White et al. (2007b), Pilon-Smits and LeDuc (2009), Pilon-Smits (2012) and White (2016). Abbreviations: SeO_4^{2-} selenate, *SULTR3;1* transporter on plastid membrane, *APS* adenosine triphosphate sulphurylase, *APSe* adenosine 5′-phosphoselenate, *APR* adenosine 5′-phosphosulphate reductase, SeO_3^{2-} selenite, *SiR* sulphite reductase, Se^{2-} selenide, *SAT/OAS-TL* cysteine synthase, which contains both serine acetyl transferase (SAT) and O-acetylserine (thiol) lyase (OAS-TL) subunits, *SeCys* selenocysteine, *CGS* cystathionine γ-synthase, *CBL* cystathionine β-lyase, *SeHCys* selenohomocysteine, *MTR* methionine synthase, *SeMet* selenomethionine, *cpNifS* chloroplast SeCyslyase, Se^0 elemental Se, *SMT* selenocysteine methyltransferase, *SeMSeCys* Se-methylselenocysteine, *γ-GluSeMSeCys* γ-glutamyl-SeMSeCys, *DMDSe* dimethyldiselenide, *SeMSeMet* Se-methylselenomethionine, *MMT* S-adenosyl-methionine: methionine methyl transferase, *γ-GluSeMSeMet* γ-glutamyl-SeMSeMet, *DMSeP* dimethylselenonium propionate, *DMSe* dimethylselenide

Shinmachi et al. 2010; Gigolashvili and Kopriva 2014; White 2016). AtSULTR1;2 catalyses most Se uptake by roots of S-replete plants (Shibagaki et al. 2002; El Kassis et al. 2007; Barberon et al. 2008). AtSULTR1;1 generally contributes little to Se uptake in S-replete plants, but its relative contribution increases in S-deficient plants or when Se supply is increased (El Kassis et al. 2007; Rouached et al. 2008; Shinmachi et al. 2010). Selenite can be taken up by roots as $HSeO_3^-$ through phosphate transporters, such as rice OsPT2 (Zhang et al. 2014), and as H_2SeO_3 through aquaporins, such as OsNIP2;1 (Zhao et al. 2010). It is thought that roots might take up SeCys and SeMet through transporters for Cys and Met (Kikkert and Berkelaar 2013; White 2016). After uptake, selenate traverses the root cylinder and is loaded into the xylem for transport to the shoot, whereas selenite is rapidly converted to organoselenium compounds (White et al. 2004; Li et al. 2008; Wang et al. 2015). It is thought that a member of the ALMT family of transporters (organic acid transporters) loads selenate into the xylem, although AtSULTR2;1, AtSULTR2;2 and

AtSULTR3;5 have also been implicated in modulating the process by catalysing the uptake of selenate into pericycle and xylem parenchyma cells (Kataoka et al. 2004a; Takahashi et al. 2011; Gigolashvili and Kopriva 2014). The expression of *AtSULTR2;1* and *AtSULTR2;2* in arabidopsis, and their homologues in other non-accumulator species, is induced by S starvation and by increasing Se supply (Takahashi et al. 2000; Van Hoewyk et al. 2008b; Gigolashvili and Kopriva 2014), whereas their homologues in Se-hyperaccumulator species are expressed constitutively (Table 9.1; Cabannes et al. 2011; Schiavon et al. 2015).

Members of the SULTR family are thought to catalyse selenate uptake by leaf cells (Takahashi et al. 2011; Gigolashvili and Kopriva 2014). In non-accumulator species selenate is often sequestered in the vacuoles of cells within the vasculature and leaf mesophyll (Ximénez-Embún et al. 2004; Mazej et al. 2008; Wang et al. 2015). Selenium is readily redistributed in the phloem as both selenate and the organoselenium compounds SeMet and Se-methylselenocysteine (SeMSeCys) (Carey et al. 2012). In arabidopsis, AtSULTR1;3 is thought to catalyse selenate uptake into the phloem (Yoshimoto et al. 2003). Homologues of AtSULTR4;1 and AtSULTR4;2 are thought to catalyse the efflux of selenate from the vacuole (Kataoka et al. 2004b; Gigolashvili and Kopriva 2014). The expression of genes encoding these transporters is greater in shoots of Se-hyperaccumulator species, such as *Stanleya pinnata* (Pursch) Britton, than in congeneric Se-indicator or non-accumulator species (Freeman et al. 2010). It has also been suggested that genes encoding ABC transporters whose expression is increased upon exposure to Se might be involved in Se transport within the plant (Byrne et al. 2010), although there is no direct evidence of this.

Selenate is assimilated into organoselenium compounds in plastids (Fig. 9.2). Selenium and S share the same primary metabolic pathway (White et al. 2007b; Pilon-Smits and LeDuc 2009; Pilon-Smits 2012; White 2016). It is thought that AtSULTR3;1, which is located on the chloroplast membrane (Cao et al. 2013), catalyses selenate transport into plastids and is expressed constitutively in Se-hyperaccumulator species (Table 9.1). Selenate is first activated by adenosine triphosphate sulphurylase (APS) to form adenosine 5′-phosphoselenate (APSe), which is then reduced to selenite (SeO_3^{2-}) by adenosine 5′-phosphosulphate reductase (APR) using reduced glutathione (GSH) as the electron donor. The conversion of selenate to selenite appears to be the rate-limiting step in the assimilation of Se into organic compounds (Pilon-Smits et al. 1999b). Selenite is reduced to selenide enzymatically by sulphite reductase (SiR) or non-enzymatically by GSH. Selenide is then converted to SeCys by the enzyme complex cysteine synthase, which contains both serine acetyl transferase (SAT) and O-acetylserine (thiol) lyase (OAS-TL) subunits. The conversion of SeCys to SeMet proceeds via selenocystathionine and selenohomocysteine (SeHCys) and is catalysed by cystathionine γ-synthase (CGS), cystathionine β-lyase (CBL) and methionine synthase (MTR). Tissues of several Se-hyperaccumulator species accumulate high concentrations of selenocystathionine (Birringer et al. 2002; Freeman et al. 2006, 2010) and the predominant form of Se in the Se-hyperaccumulator species *Cardamine hupingshanensis* is selenocystine ($SeCys_2$) (Yuan et al. 2013). The enzymes catalysing Se metabolism are encoded by small multi-gene families. For example, the arabidopsis genome contains four genes encoding APS, three genes encoding APR, five genes encoding SAT, nine genes encoding OAS-TL, two genes encoding CGS, one gene encoding CBL, and three genes encoding MTR (Hesse et al. 2004; Bermúdez et al. 2013; Schiavon et al. 2015). In most angiosperm species, genes encoding enzymes involved in the primary S/Se assimilation pathway are upregulated when they become S-deficient or when Se supply is increased, whilst in Se-hyperaccumulator species genes encoding APS, APR, SAT/OAS-TL and MTP exhibit constitutively high expression (Table 9.1) Van Hoewyk et al. 2005, 2008b; Freeman et al. 2010; White 2016).

Many plants accumulate Se as SeCys or SeMet (Birringer et al. 2002; White et al. 2007b; Fairweather-Tait et al. 2011; Drahoňovský et al. 2016; White 2016). These Se-amino acids can replace Cys and Met in proteins, which impairs protein activities and ultimately results in physiological malfunction. The conversion of SeCys and SeMet to non-toxic or volatile Se metabolites can increase the tolerance of plants to Se in their tissues and in the environment (Sors et al. 2005b; White et al. 2007b; Pilon-Smits and LeDuc 2009; Van Hoewyk 2013). In addition, plant genomes contain genes encoding putative Se-binding proteins (SBPs) that might contribute to Se tolerance in plant tissues (Agalou et al. 2005; Dutilleul et al. 2008). Selenocysteine can be converted to elemental Se by a SeCys lyase (cpNifS) that is located in the chloroplast (Fig. 9.2) (van Hoewyk et al. 2008a; Pilon-Smits and LeDuc 2009). In addition, SeCys and SeMet can be methylated to form SeMSeCys and Se-methylselenomethionine (SeMSeMet) through the activities of SeCys methyltransferase (SMT) and S-adenosyl-methionine: methionine methyl transferase (MMT), respectively (Sors et al. 2005b; White et al. 2007b; Pilon-Smits and LeDuc 2009; Van Hoewyk 2013; White 2016). However, genes encoding functional SMT are not thought to exist in non-accumulator plants such as arabidopsis (Lyi et al. 2005; Van Hoewyk 2013; Zhao et al. 2015), and arabidopsis possesses only one gene encoding MMT (Tagmount et al. 2002). In species that hyperaccumulate Se, genes encoding functional SMT appear to be expressed constitutively (Table 9.1) (Pickering et al. 2003) and differences between species within the genera *Astragalus* and *Stanleya*, which both contain Se-hyperaccumulator species, in their ability to accumulate Se appear to be directly related to their SMT activity (Sors et al. 2005a, 2009; Freeman et al. 2010). The most abundant form of Se in Se-hyperaccumulator species, such as *Astragalus bisulcatus* (Hook.) A. Gray and *Stanleya pinnata*, is SeMSeCys (Birringer et al. 2002; Pickering et al. 2003; Sors et al. 2005a; Freeman et al. 2006, 2010; Lindblom et al. 2013; Alford et al. 2014). Large concentrations of SeMSeCys are also found in alliums (chive, garlic, leek, onion) and brassicas (broccoli, Brussels sprouts, cabbage, cauliflower, Chinese cabbage, kale) fertilised with selenate or selenite (Birringer et al. 2002; Fairweather-Tait et al. 2011; White 2016), although it has been speculated that the accumulation of SeMSeCys might be a characteristic of the Poales rather than other angiosperm species, which preferentially accumulate $SeCys_2$ (Drahoňovský et al. 2016). Both SeMSeCys and SeMSeMet can be conjugated with glutamate to form γ-glutamyl-SeMSeCys (γ-GluSeMSeCys) or γ-glutamyl-SeMSeMet (γ-GluSeMSeMet), or converted to dimethylselenide (DMSe) or dimethyldiselenide (DMDSe) and volatilized (Fig. 9.2)

(Sors et al. 2005b; White et al. 2007b; Pilon-Smits and LeDuc 2009; Van Hoewyk 2013). The production of DMSe appears to be limited by the conversion of SeCys to SeMet (Van Huysen et al. 2003; Pilon-Smits and LeDuc 2009). SeMSeMet can also be converted to dimethylselenonium propionate (DMSeP) and thence to DMSe (Grant et al. 2004). Many Se-hyperaccumulator species, such as *A. bisulcatus* (Freeman et al. 2006; Alford et al. 2014), and allium crops grown on Se-rich soils can contain large concentrations of γ-GluSeMeSeCys (White et al. 2007b; Fairweather-Tait et al. 2011; White 2016). In general, Se is volatilized as DMSe in non-hyperaccumulator species and as DMDSe in Se-hyperaccumulator species (Table 9.1) (Pilon-Smits and LeDuc 2009), and angiosperm species differ greatly in their ability to volatilize Se (Terry et al. 1992; Pilon-Smits et al. 1999a; de Souza et al. 2000).

Differences in Se uptake and metabolism between genotypes underlie differences in their ability to tolerate Se in the rhizosphere and within their tissues. These differences can be exploited for practical purposes, such as the phytoremediation of soils contaminated with excess Se or the delivery of Se to livestock and humans, for whom Se is an essential mineral nutrient that is often in insufficient supply in their diets (White 2016).

9.3 Evolution of Differences in Selenium Accumulation Between Angiosperm Species

Natural genetic variation in Se uptake and metabolism has determined the ecological strategies of angiosperm species towards soils with high Se phytoavailability, and the evolutionary progression towards Se-hyperaccumulation has fascinated plant scientists for many years. The core eudicot families Amaranthaceae (Caryophyllales), Asteraceae (Asterales), Brassicaceae (Brassicales), Fabaceae (Fabales), Orobanchaceae (Lamiales) and Rubiaceae (Gentianales) all contain Se-hyperaccumulator species (Fig. 9.3). It is thought that tissue Se tolerance and Se-hyperaccumulation arose independently in these plant families by convergent evolution of appropriate biochemical pathways (Brown and Shrift 1982; White et al. 2004; Cappa and Pilon-Smits 2014; White 2016). However, the phylogeny of Se-hyperaccumulation has been studied in detail in only a few genera, namely *Astragalus* (Fabaceae) and *Stanleya* (Brassicaceae). The trait of Se-hyperaccumulation appears to have evolved several times among the North American *Astragalus*: in the Homaloboid Phalanx within the seleniferous Homalobi, for which it can be used as a taxonomic character (Barneby 1964), and the Preussiani (Fig. 9.4), and in the Piptoloboid and Ceridothrix Phalanxes (White 2016). By contrast, the trait of Se-hyperaccumulation seems to have evolved only once among the North American *Stanleya*, on the eastern side of the Rocky Mountains in the *bipinnata*/*pinnata* clade, and was subsequently lost in various ecotypes (Fig. 9.5) (Cappa et al. 2014, 2015). The trait of Se-tolerance appears to have evolved before, and was possibly a

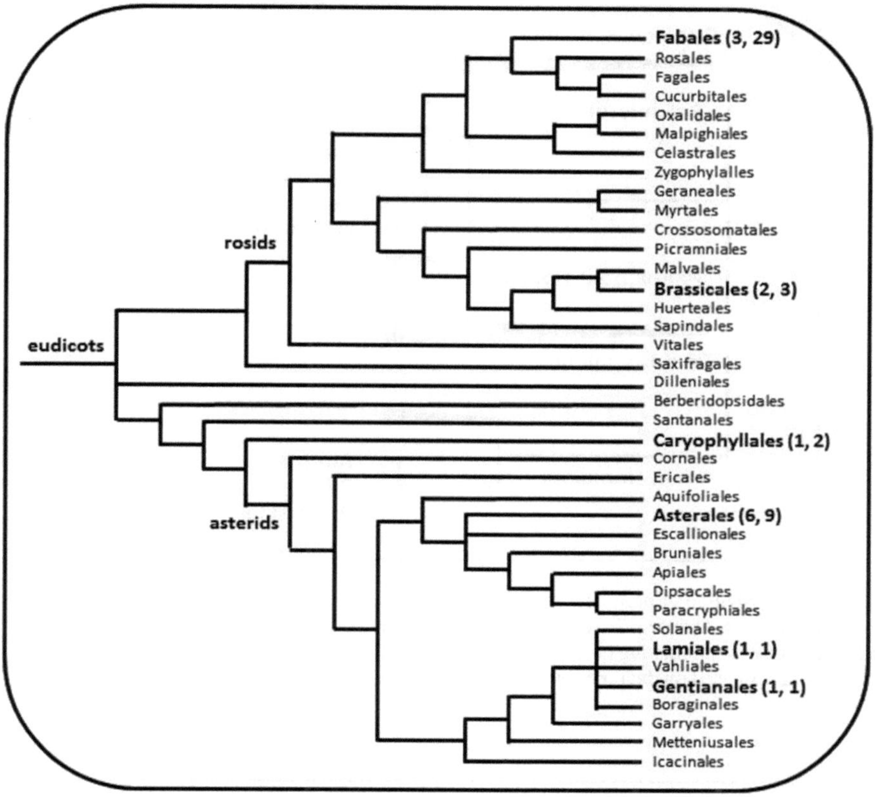

Fig. 9.3 The distribution of Se-hyperaccumulator species among the core eudicot orders. Phylogenetic relationships between orders are reproduced from the Angiosperm Phylogeny Group (2016). The number of Se-hyperaccumulating genera and Se-hyperaccumulating species in each order are given in *parentheses* based on data presented by White (2016)

prerequisite for, the evolution of Se-hyperaccumulation in *Stanleya* (Fig. 9.5) (Cappa et al. 2015). These two genera provide model species for studies of the physiology and genetics of Se-hyperaccumulation.

El-Mehdawi and Pilon-Smits (2012) have suggested a plausible sequence of events leading to the evolution of Se-hyperaccumulator species. First, since Se is a beneficial element, variation in the ability of plants to acquire Se led to the selection of individuals with greater tissue Se concentrations in which Se mitigated the oxidative stresses caused by various environmental factors and improved growth. There is considerable genetic variation in the ability to accumulate Se, both within and between plant species, that would facilitate the evolution of species with greater tissue Se concentrations (White et al. 2004, 2007a; Watanabe et al. 2007; Sasmaz et al. 2015; Szakova et al. 2015; Drahoňovský et al. 2016, White 2016). Second, further genetic variation in non-accumulator species allowed the selection of plants

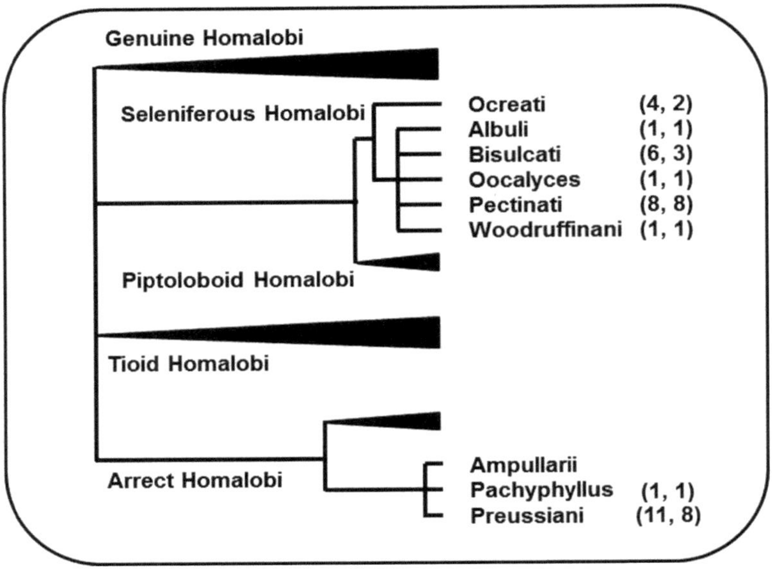

Fig. 9.4 Distribution of proposed Se-hyperaccumulating taxa among sections of the Homaloboid astragali of North America. Taxonomic relationships are derived from Barneby (1964). The number of Se-hyperaccumulating taxa and Se-hyperaccumulating species in each section and are given in *parentheses* based on data presented by White (2016)

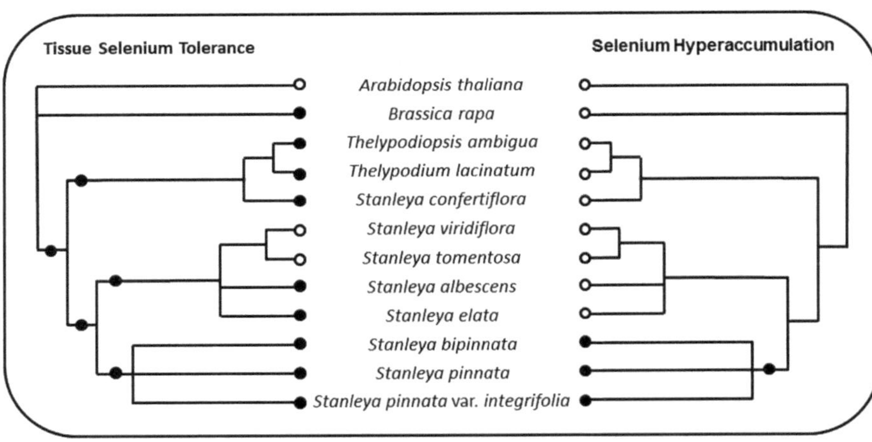

Fig. 9.5 Phylogenetic relationships within the Brassicaceae suggesting that the evolution of tolerance to large tissue Se concentrations (*Left*: biomass production exceeding half-maximum at tissue Se concentrations >1000 μg/g dry matter) occurred before the evolution of Se hyperaccumulation (*Right*: maximum shoot Se concentration > 1000 μg/g dry matter when sampled from the natural environment). Taxa exhibiting tolerance of large tissue Se concentrations or exhibiting Se hyperaccumulation are indicated by filled circles (Data from Cappa et al. 2015)

with greater tissue Se tolerance that were able to colonize seleniferous soils and evolve into Se-indicator species. This could have been achieved by the evolution of metabolic pathways that allow more Se to be accumulated in tissues in non-toxic forms and Se removal from tissues by volatilization. These characteristics are shared by many Se-indicator and Se-accumulator species and, again, there is considerable genetic variation in their expression both within and between plant species (White et al. 2007b; Pilon-Smits and LeDuc 2009; White 2016). Third, Se-hyperaccumulation evolved because large Se concentrations in tissues protect them against pathogens and herbivores and, thereby, confers an evolutionary advantage (Quinn et al. 2007). This is consistent with the greatest Se concentrations occurring in the reproductive tissues of Se-hyperaccumulator species (El-Mehdawi and Pilon-Smits 2012) and the accumulation of large Se concentrations leaf trichomes and epidermal cells (Freeman et al. 2006, 2010). Large tissue Se concentrations in leaf litter might also prevent competition by plant species with less tolerance of Se in the rhizosphere (El Mehdawi et al. 2011). Ecotypes of Se-hyperaccumulator species growing in the same environment can differ significantly in their shoot Se concentrations (Feist and Parker 2001; Cappa et al. 2014; El Mehdawi et al. 2015), which suggests that the evolution of Se-hyperaccumulation required not only the ability to tolerate high tissue Se concentrations but also the efficient acquisition of Se by roots and transport of Se within the plant. Thus, Se-hyperaccumulator species are characterised by constitutive expression of (1) genes encoding Se transporters, which promote Se uptake and movement to the shoot, (2) genes involved in primary Se assimilation, which allows a large metabolic flux to SeCys and SeMet, (3) genes involved in the conversion of SeCys and SeMet to non-toxic or volatile compounds, such as SeMSeCys, γ-GluSeMeSeCys and DMDSe (Table 9.1). Species that hyperaccumulate Se also differ from other angiosperms by exhibiting (1) elevated leaf Se/S quotients and (2) a reduction in their Mo accumulation with increasing rhizosphere sulphate or selenate concentrations (White et al. 2007a; Harris et al. 2014). This is likely to reflect differences in the selectivity and regulation of their complement of Se transporters.

Insights into the molecular mechanisms of tissue Se tolerance and Se-hyperaccumulation have not only been gained by comparing the biochemistry and physiology of Se-hyperaccumulator species with other angiosperms but have also been provided by the manipulation of gene expression in non-accumulator species, such as arabidopsis, and Se-indicator species, such as Indian mustard (*Brassica juncea* [L.] Czern.). Arabidopsis mutants lacking AtSULTR1;2 and rice mutants lacking OsPT2 take up less Se than wild-type plants and tolerate greater rhizosphere concentrations of selenate and selenite, respectively (Shibagaki et al. 2002; El Kassis et al. 2007; Barberon et al. 2008; Zhang et al. 2014). The overexpression of genes encoding SMT, CpNifS or cytosolic SeCys lyase enables arabidopsis to accumulate greater Se concentrations in shoot tissues and confers tolerance to greater concentrations of selenate and selenite in the rhizosphere (Ellis et al. 2004; LeDuc et al. 2004; Pilon et al. 2003; Van Hoewyk et al. 2005). The overexpression of *AbSMT1* in arabidopsis also results in increased production of volatile Se-compounds (LeDuc et al. 2004). The overexpression in arabidopsis of genes encoding APS, APR or SAT result in greater concentrations of organic Se shoot tissues, but lower

Se concentrations, and the overexpression of genes encoding APR and SBP confer tolerance to greater concentrations of selenate and selenite in the rhizosphere, respectively (Agalou et al. 2005; Sors et al. 2005a). Consistent with these observations, arabidopsis mutants lacking *AtAPR2* and accessions with less APR activity have larger Se concentrations in their shoots (Chao et al. 2014). An arabidopsis mutant lacking CSM1, which exhibits elevated activities or glutathione peroxidase and other peroxidases, shows greater Se tolerance (Jiang et al. 2015). Similar results were obtained in Indian mustard following the overexpression of genes encoding SMT, but the overexpression of genes encoding APS not only conferred tolerance to greater concentrations of selenate in the rhizosphere, but also resulted in increased concentrations of both total and organic Se in the shoot (Pilon-Smits et al. 1999b; Van Huysen et al. 2004; Bañuelos et al. 2005, 2007; LeDuc et al. 2004, 2006; Kubachka et al. 2007). The overexpression of CGS in Indian mustard resulted in greater tolerance of selenite in the rhizosphere, reduced shoot Se concentrations and greater Se volatilization (Van Huysen et al. 2003, 2004) and overexpression of genes involved in glutathione synthesis, such as glutathione synthetase and γ-glutamyl-cysteine synthetase, resulted in greater shoot Se concentrations and improved growth on seleniferous soils (Bañuelos et al. 2005).

9.4 Natural Genetic Variation in Selenium Accumulation Within Crop Species

Selenium is an essential mineral nutrient for humans and other animals (White and Broadley 2009; Fairweather-Tait et al. 20,011; Fordyce 2013). Unfortunately, up to 15% of the world's human population might lack sufficient Se in their diets (Combs 2001). One strategy to rectify this, termed "biofortification", is to produce crops with greater Se concentrations in their edible portions (Broadley et al. 2006; White and Broadley 2009). However, since Se deficiency in humans and livestock often occurs when crops are grown on soils with little phytoavailable Se, Se-biofortification must be achieved through the application of Se-fertilisers (Broadley et al. 2006; White and Broadley 2009; Alfthan et al. 2015; Joy et al. 2015; White 2016). The efficiency of Se-fertilisers in Se-biofortification of crops can be improved by the development of crop genotypes that can acquire more of the Se applied and accumulate it in their edible portions. There has, therefore, been considerable work to determine the genetic basis of variation in Se accumulation in edible portions of numerous crops (Table 9.2) (White and Broadley 2009; Pilbeam et al. 2015; White 2016).

Significant genetic variation in grain Se concentration has been reported for several cereals including bread wheat (*Triticum aestivum* L.) (Garvin et al. 2006; Murphy et al. 2008; Rodríguez et al. 2011; Pu et al. 2014), durum wheat (*Triticum turgidum* L.; Rodríguez et al. 2011; Yang et al. 2013), barley (*Hordeum vulgare* L.) (Ilbas et al. 2012; Mangan et al. 2015), wild barley (*Hordeum spontaneum* K. Koch) (Yan et al. 2011), oat (*Avena sativa* L.) (Eurola et al. 2004) and rice (*Orzya sativa* L.)

Table 9.2 Variation in the Se concentration of edible tissues among genotypes of common crops grown in the same environment

Crop	Plant Species	Tissue	Studies	ns	P<0.05	P<0.01	P<0.001
Wheat	*Triticum aestivum* L.	Grain	12	9	1		2
Durum Wheat	*Triticum turgidum* L.	Grain	2	1	1		
Einkorn Wheat	*Triticum monococcum* L.	Grain	1	1			
Emmer Wheat	*Triticum dicoccon* (Schrank) Schübl.	Grain	2	2			
Spelt Wheat	*Triticum spelta* L.	Grain	2	2			
Barley	*Hordeum vulgare* L.	Grain	3	2	1		
Wild Barley	*Hordeum spontaneum* K.Koch	Grain	1	1			
Oat	*Avena sativa* L.	Grain	2	1			1
Rice	*Oryza sativa* L.	Brown grain	1		1		
Common Bean	*Phaseolus vulgaris* L.	Seed	3	3			
Field Pea	*Pisum sativum* L.	Seed	1	1			
Chickpea	*Cicer arietinum* L.	Seed	2		1	1	
Lentil	*Lens culinaris* Medik.	Seed	13	4	8	1	
Mung bean	*Vigna radiata* (L.) R.Wilczek	Seed	1			1	
Soybean	*Glycine max* (L.) Merr.	Seed	1			1	
Onion	*Allium cepa* L.	Bulb	1		1		
Broccoli	*Brassica oleracea* L.	Leaves/Floret	6	3	2	1	
Cauliflower	*Brassica oleracea* L.	Sprouts	1		1		
Kale	*Brassica oleracea* L.	Sprouts	1		1		
Chinese Cabbage	*Brassica rapa* L.	Sprouts	1		1		
Indian Mustard	*Brassica juncea* (L.) Czern.	Leaves	1		1		
Kale	*Brassica oleracea* L.	Sprouts	1		1		
Chinese Cabbage	*Brassica rapa* L.	Sprouts	1		1		
Indian Mustard	*Brassica juncea* (L.) Czern.	Leaves	1		1		
Chicory	*Cichorium intybus* L.	Leaves	1		1		
Tomato	*Solanum lycopersicum* L.	Fruit	1		1		
Pepper	*Capsicum annuum* L.	Fruit	1		1		
Potato	*Solanum tuberosum* L.	Tubers	1		1		

Data indicate the total number of studies and, of these, the number of studies reporting no significant differences between genotypes (ns) or differences between genotypes with $P < 0.05$, $P < 0.01$, and $P < 0.001$. Original data are presented by White (2016)

(Zhang et al. 2006; Norton et al. 2010, 2012; Huang et al. 2015). In addition, chromosomal loci (QTLs) influencing grain Se concentration have been identified using populations derived from (1) crosses between bread wheat genotypes (Pu et al. 2014), (2) a cross between wild emmer wheat (*Triticum dicoccoides* [Körn. ex Asch. and Graebn.] Schweinf.) and tetraploid durum wheat (Yang et al. 2013), (3) a cross between an indica and a japonica rice variety (Norton et al. 2010, 2012), and (4) an association mapping panel of rice accessions (Huang et al. 2015). However, no genes affecting grain Se concentrations have yet been identified in any cereal species. Similarly, significant genetic variation in seed Se concentration has been reported for several legumes including chickpea (*Cicer arietinum* L.) (Thavarajah and Thavarajah 2012; Ray et al. 2014), lentil (*Lens culinaris* Medik.) (Thavarajah et al. 2011; Ray et al. 2014; Rahman et al. 2015), mung bean (*Vigna radiata* [L.] R.Wilczek) (Nair et al. 2015) and soybean (*Glycine max* [L.] Merr.) (Yang et al. 2003; Ramamurthy et al. 2014). Two QTLs affecting seed Se concentrations were identified using a population derived from a cross between two soybean cultivars, one of which includes a gene encoding GmSULTR2;1 that might facilitate Se translocation from the root to the shoot (Ramamurthy et al. 2014).

Significant genetic variation has also been reported for Se concentrations in various leafy vegetables (Table 9.2) including onion bulbs (*Allium cepa* L.) Kopsell and Randle 1997), leaves of rapid-cycling *Brassica oleracea* (Kopsell and Randle 2001), broccoli florets (*B. oleracea L. Italica* Group) (Bañuelos et al. 2003; Farnham et al. 2007; Ramos et al. 2011b), sprouts of cauliflower (*B. oleracea* L. Botrytis Group), kale (*B. oleracea* L. Acephala Group) and Chinese cabbage (*Brassica rapa* L.) (Ávila et al. 2014), leaves of Indian mustard (Bañuelos et al. 1997), leaves of chicory (*Cichorium intybus* L.) (Mazej et al. 2008) and leaves of lettuce (*Lactuca sativa* L.) (Ramos et al. 2011a). The ability of lettuce genotypes to accumulate Se supplied as selenate was positively correlated with the expression of *LsSULTR1;1*, *LsAPS1* and *LsAPR1* (Ramos et al. 2011a). Significant genetic variation in Se concentration has also been reported for leaves of tea (*Camelia sinensis* [L.] Kuntze) (Zhao et al. 2016) and fruit of tomato (*Solanum lycopersicum* L.) (Guil-Guerrero and Rebolloso-Fuentes 2009) and pepper (*Capsicum annuum* L.) (Guil-Guerrero et al. 2006) and for tubers of potato (*Solanum tuberosum* L.) (Perla et al. 2012).

9.5 Conclusions: The Genetics of Selenium Accumulation by Plants

Selenium is an essential element for animals, but is not essential for plants (White and Brown 2010). Nevertheless, small concentrations of Se in plant tissues can mitigate the oxidative stresses caused by various environmental factors and greater concentrations can protect plants against herbivores and pathogens (El Mehdawi and Pilon-Smits 2012). However, Se is toxic to most plants when it is present at excessive concentrations in their tissues and only plant species that can tolerate elevated tissue Se concentrations are able to colonise seleniferous soils (Rosenfeld and

Beath 1964; Brown and Shrift 1982). The ability of Se-hyperaccumulator species to tolerate large (>1 mg/g DW) tissue Se concentrations appears to have evolved by convergent evolution in several angiosperm clades by appropriate modification of their Se metabolism (Brown and Shrift 1982; White et al. 2004; Cappa and Pilon-Smits 2014; White 2016).

Genetics is broadly defined as the scientific study of how genes control the characteristics of organisms. It encompasses both knowledge of the genetic variation in a characteristic and the genes, and their alleles, affecting the characteristic. In this chapter, emphasis has been placed on differences in metabolic pathways, and their associated genes, that could account for variation among angiosperm species in their ability to tolerate large tissue Se concentrations (Table 9.1). First, the current view of the molecular biology of Se uptake and assimilation by plants is presented (Fig. 9.2) and differences between plant species likely to affect their ability to tolerate large tissue Se concentrations are identified. In particular, it is noted that plants that hyperaccumulate Se generally exhibit constitutive expression of genes encoding Se-transporters and enzymes involved in primary Se assimilation, biosynthesis of non-toxic Se metabolites and Se volatilisation (Table 9.1). A plausible scheme for the evolution of differences in Se accumulation between angiosperm species is described. Since Se is an essential mineral element for animals (White and Broadley 2009), and the diets of many humans lack sufficient Se (Combs 2001), the possibility of breeding crops with greater Se concentrations in their edible tissues is also discussed. It is observed that, although Se concentrations in plants are largely determined by the phytoavailability of Se in the environment, there is significant genetic variation in the Se concentrations of most edible crops (Table 9.2) that might be utilised to improve human diets. However, although molecular markers might be developed to QTL impacting Se concentration in edible tissues to assist breeding programmes, the actual genes underpinning this variation are still largely unknown. Nevertheless, our knowledge of the genetics of Se acquisition, metabolism and accumulation in plants is increasing rapidly and the combined application of multi-omics technologies is likely to reveal more genes affecting Se accumulation in the immediate future.

Acknowledgements This work was supported by the Rural and Environment Science and Analytical Services Division (RESAS) of the Scottish Government and by a Fellowship funded by The National Council for Scientific and Technological Development (CNPq) of Brazil (Grant #402868/2012-9). I thank Dr. Paula Pongrac for reading my original manuscript.

References

Agalou A, Roussis A, Spaink HP (2005) The *Arabidopsis* selenium-binding protein confers tolerance to toxic levels of selenium. Funct Plant Biol 32:881–890

Alford ÉR, Lindblom SD, Pittarello M et al (2014) Roles of rhizobial symbionts in selenium hyperaccumulation in *Astragalus* (Fabaceae). Am J Bot 101:1895–1905

Alfthan G, Eurola M, Ekholm P et al (2015) Effects of nationwide addition of selenium to fertilizers on foods, and animal and human health in Finland: from deficiency to optimal selenium status of the population. J Trace Elem Med Biol 31:142–147

Angiosperm Phylogeny Group (2016) An update of the Angiosperm Phylogeny Group classification for the orders and families of flowering plants: APG IV. Bot J Linn Soc 181:1–20

Ávila FW, Yang Y, Faquin V et al (2014) Impact of selenium supply on Se-methylselenocysteine and glucosinolate accumulation in selenium-biofortified Brassica sprouts. Food Chem 165:578–586

Baker AJM (1981) Accumulators and excluders – strategies in the response of plants to heavy metals. J Plant Nutr 3:643–654

Bañuelos GS, Ajwa HA, Wu L et al (1997) Selenium-induced growth reduction in Brassica land races considered for phytoremediation. Ecotoxicol Environ Saf 36:282–287

Bañuelos GS, Pasakdee S, Finley JW (2003) Growth response and selenium and boron distribution in broccoli varieties irrigated with poor quality water. J Plant Nutr 26:2537–2549

Bañuelos G, Terry N, LeDuc DL et al (2005) Field trial of transgenic Indian mustard plants shows enhanced phytoremediation of selenium-contaminated sediment. Environ Sci Technol 39:1771–1777

Bañuelos G, LeDuc DL, Pilon-Smits EAH et al (2007) Transgenic Indian mustard overexpressing selenocysteine lyase or selenocysteine methyltransferase exhibit enhanced potential for selenium phytoremediation under field conditions. Environ Sci Technol 41:599–605

Barberon M, Berthomieu P, Clairotte M et al (2008) Unequal functional redundancy between the two *Arabidopsis thaliana* high-affinity sulphate transporters *SULTR1;1* and *SULTR1;2*. New Phytol 180:608–619

Barneby RC (1964) Atlas of North American *Astragalus*. Volumes I and II. The New York Botanical Garden, New York

Bermúdez MA, Páez-Ochoa MA, Gotor C et al (2013) *Arabidopsis* S-sulfocysteine synthase activity is essential for chloroplast function and long-day light-dependent redox control. Plant Cell 22:403–416

Birringer M, Pilawa S, Flohé L (2002) Trends in selenium biochemistry. Nat Prod Rep 19:693–718

Broadley MR, White PJ, Bryson RJ et al (2006) Biofortification of UK food crops with selenium. Proc Nutr Soc 65:169–181

Brown TA, Shrift A (1982) Selenium: toxicity and tolerance in higher plants. Biol Rev 57:59–84

Byrne SL, Durandeau K, Nagy I et al (2010) Identification of ABC transporters from *Lolium perenne* L. that are regulated by toxic levels of selenium. Planta 231:901–911

Cabannes E, Buchner P, Broadley MR et al (2011) A comparison of sulfate and selenium accumulation in relation to the expression of sulfate transporter genes in *Astragalus* species. Plant Physiol 157:2227–2239

Cao MJ, Wang Z, Wirtz M et al (2013) SULTR3;1 is a chloroplast-localized sulphate transporter in *Arabidopsis thaliana*. Plant J 73:607–616

Cappa JJ, Pilon-Smits EAH (2014) Evolutionary aspects of elemental hyperaccumulation. Planta 239:267–275

Cappa JJ, Cappa PJ, El Mehdawi AF et al (2014) Characterization of selenium and sulfur accumulation across the genus *Stanleya* (Brassicaceae): a field survey and common-garden experiment. Am J Bot 101:830–839

Cappa JJ, Yetter C, Fakra S et al (2015) Evolution of selenium hyperaccumulation in *Stanleya* (Brassicaceae) as inferred from phylogeny, physiology and X-ray microprobe analysis. New Phytol 205:583–595

Carey A-M, Scheckel KG, Lombi E et al (2012) Grain accumulation of selenium species in rice (*Oryza sativa* L.) Environ Sci Technol 46:5557–5564

Chao D-Y, Baraniecka P, Danku J et al (2014) Variation in sulfur and selenium accumulation is controlled by naturally occurring isoforms of the key sulfur assimilation enzyme ADENOSINE 5′-PHOSPHOSULFATE REDUCTASE2 across the Arabidopsis species range. Plant Physiol 166:1593–1608

Combs GF (2001) Selenium in global food systems. Br J Nutr 85:517–547 de Souza MP, Pilon-Smits EAH, Terry N (2000) The physiology and biochemistry of selenium volatilization by plants. In: Raskin I, Ensley BD (eds) Phytoremediation of toxic metals: using plants to clean-up the environment. Wiley, New York, pp 171–190

Dhillon KS, Dhillon SK (2003) Distribution and management of seleniferous soils. Adv Agron 79:119–184

Drahoňovský J, Száková J, Mestek O et al (2016) Selenium uptake, transformation and inter-element interactions by selected wildlife plant species after foliar selenate application. Environ Exp Bot 125:12–19

Dutilleul C, Jourdain A, Bourguignon J et al (2008) The Arabidopsis putative selenium-binding protein family: expression study and characterization of SBP1 as a potential new player in cadmium detoxification processes. Plant Physiol 147:239–251

El Kassis E, Cathala N, Rouached H et al (2007) Characterization of a selenate-resistant Arabidopsis mutant. Root growth as a potential target for selenite toxicity. Plant Physiol 143:1231–1241

El Mehdawi AF, Pilon-Smits EAH (2012) Ecological aspects of plant selenium hyperaccumulation. Plant Biol 14:1–10

El Mehdawi AF, Quinn CF, Pilon-Smits EAH (2011) Effects of selenium hyperaccumulation on plant–plant interactions: evidence for elemental allelopathy? New Phytol 191:120–131

El Mehdawi AF, Paschke MW, Pilon-Smits EAH (2015) *Symphyotrichum ericoides* populations from seleniferous and nonseleniferous soil display striking variation in selenium accumulation. New Phytol 206:231–242

Ellis DR, Sors TG, Brunk DG et al (2004) Production of Se-methylselenocysteine in transgenic plants expressing selenocysteine methyltransferase. BMC Plant Biol 4:1. doi:10.1186/1471-2229-4-1

van der Ent A, Baker AJM, Reeves RD et al (2013) Hyperaccumulators of metal and metalloid trace elements: facts and fiction. Plant Soil 362:319–334

Eurola M, Hietaniemi V, Kontturi M et al (2004) Selenium content of Finnish oats in 1997–1999: effect of cultivars and cultivation techniques. Agric Food Sci 13:46–53

Fairweather-Tait SJ, Bao Y, Broadley MR et al (2011) Selenium in human health and disease. Antioxid Redox Signal 14:1337–1383

Farnham MW, Hale AJ, Grusak MA et al (2007) Genotypic and environmental effects on selenium concentration of broccoli heads grown without supplemental selenium fertilizer. Plant Breed 126:195–200

Feist LJ, Parker DR (2001) Ecotypic variation in selenium accumulation among populations of *Stanleya pinnata*. New Phytol 149:61–69

Feng R, Wei C, Tud S (2013) The roles of selenium in protecting plants against abiotic stresses. Environ Exp Bot 87:58–68

Fordyce FM (2013) Selenium deficiency and toxicity in the environment. In: Selinus O, Alloway B, Centeno JA et al (eds) Essentials of medical geology, revised edn. Springer, Dordracht, pp 375–416

Freeman JL, Zhang L, Marcus MA et al (2006) Spatial imaging, speciation and quantification of Se in the hyperaccumulator plants *Astragalus bisulcatus* and *Stanleya pinnata*. Plant Physiol 142:124–134

Freeman JL, Tamaoki M, Stushnoff C et al (2010) Molecular mechanisms of selenium tolerance and hyperaccumulation in *Stanleya pinnata*. Plant Physiol 153:1630–1652

Garvin DF, Welch RM, Finley JW (2006) Historical shifts in the seed mineral micronutrient concentration of US hard red winter wheat germplasm. J Sci Food Agric 86:2213–2220

Gigolashvili T, Kopriva S (2014) Transporters in plant sulfur metabolism. Front Plant Sci 5:422. doi:10.3389/fpls.2014.00442

Grant TD, Montes-Bayón M, LeDuc D et al (2004) Identification and characterization of Se-methyl selenomethionine in *Brassica juncea* roots. J Chromatogr A 1026:159–166

Guil-Guerrero JL, Martínez-Guirado C, del Mar R-FM et al (2006) Nutrient composition and antioxidant activity of 10 pepper (*Capsicum annuum*) varieties. Eur Food Res Technol 224:1–9

Guil-Guerrero JL, Rebolloso-Fuentes MM (2009) Nutrient composition and antioxidant activity of eight tomato (*Lycopersicon esculentum*) varieties. J Food Comp Anal 22:123–129
Harris J, Schneberg KA, Pilon-Smits EAH (2014) Sulfur–selenium–molybdenum interactions distinguish selenium hyperaccumulator *Stanleya pinnata* from non-hyperaccumulator *Brassica juncea* (Brassicaceae). Planta 239:479–491
Hesse H, Kreft O, Maimann S et al (2004) Current understanding of the regulation of methionine biosynthesis in plants. J Exp Bot 55:1799–1808
Huang Y, Sun C, Min J et al (2015) Association mapping of quantitative trait loci for mineral element contents in whole grain rice (*Oryza sativa* L.) J Agric Food Chem 63:10885–10892
Ilbas AI, Yılmaz S, Akbulut M, Bogdevich O (2012) Uptake and distribution of selenium, nitrogen and sulfur in three barley cultivars subjected to selenium applications. J Plant Nutr 35:442–452
Jiang L, Gao QC, Chen ZP et al (2015) Selenium tolerance of an *Arabidopsis* drought-resistant mutant *csm1-1*. Russ J Plant Physiol 62:625–631
Joy EJM, Broadley MR, Young SD et al (2015) Soil type influences crop mineral composition in Malawi. Sci Total Environ 505:587–595
Kataoka T, Hayashi N, Yamaya T et al (2004a) Root-to-shoot transport of sulfate in Arabidopsis. Evidence for the role of SULTR3;5 as a component of low-affinity sulfate transport system in the root vasculature. Plant Physiol 136:4198–4204
Kataoka T, Watanabe-Takahashi A, Hayashi N et al (2004b) Vacuolar sulphate transporters are essential determinants controlling internal distribution of sulfate in Arabidopsis. Plant Cell 16:2693–2704
Kikkert J, Berkelaar E (2013) Plant uptake and translocation of inorganic and organic forms of selenium. Arch Environ Contam Toxicol 65:458–465
Kopsell DA, Randle WM (1997) Short-day onion cultivars differ in bulb selenium and sulphur accumulation which can affect bulb pungency. Euphytica 96:385–390
Kopsell DA, Randle WM (2001) Genetic variances and selection potential for selenium accumulation in a rapid-cycling *Brassica oleracea* population. J Am Soc Hortic Sci 126:329–335
Kubachka KM, Meija J, LeDuc DL et al (2007) Selenium volatiles as proxy to the metabolic pathways of selenium in genetically modified *Brassica juncea*. Environ Sci Technol 41:1863–1869
LeDuc DL, Tarun AS, Montes-Bayon M et al (2004) Overexpression of selenocysteine methyltransferase in Arabidopsis and Indian mustard increases selenium tolerance and accumulation. Plant Physiol 135:377–383
LeDuc DL, AbdelSamie M, Montes-Bayon M et al (2006) Overexpressing both ATP sulfurylase and selenocysteine methyltransferase enhances selenium phytoremediation traits in Indian mustard. Environ Pollut 144:70–76
Li H-F, McGrath SP, Zhao F-J (2008) Selenium uptake, translocation and speciation in wheat supplied with selenate or selenite. New Phytol 178:92–102
Lindblom SD, Valdez-Barillas JR, Fakra SC et al (2013) Influence of microbial associations on selenium localization and speciation in roots of *Astragalus* and *Stanleya* hyperaccumulators. Environ Exp Bot 88:33–42
Lyi SM, Heller LI, Rutzke M et al (2005) Molecular and biochemical characterization of the selenocysteine Se-methyltransferase gene and Se-methylselenocysteine synthesis in broccoli. Plant Physiol 138:409–420
Mangan B-u-N, Hui L, Lashari MS et al (2015) Nutritional characteristics and starch properties of Tibetan barley. Int J Agric Policy Res 3:293–299
Mazej D, Osvald J, Stibilj V (2008) Selenium species in leaves of chicory, dandelion, lamb's lettuce and parsley. Food Chem 107:75–83
Mikkelsen RL, Page AL, Bingham FT (1989) Factors affecting selenium accumulation by agricultural crops. In: Jacobs LW, Chang AC, Dowdy RH et al (eds) Selenium in agriculture and the environment. Soil Science Society of America, Special Publication 23, pp 65–94

Moreno Rodriguez MJ, Cala Rivero V, Jiménez Ballesta R (2005) Selenium distribution in topsoils and plants of a semi-arid Mediterranean environment. Environ Geochem Health 27:513–519

Murphy KM, Reeves PG, Jones SS (2008) Relationship between yield and mineral nutrient concentrations in historical and modern spring wheat cultivars. Euphytica 163:381–390

Nair RM, Thavarajah P, Giri RR et al (2015) Mineral and phenolic concentrations of mungbean [*Vigna radiata* (L.) R. Wilczek var. *radiata*] grown in semi-arid tropical India. J Food Comp Anal 39:23–32

Norton GJ, Deacon CM, Xiong L et al (2010) Genetic mapping of the rice ionome in leaves and grain: identification of QTLs for 17 elements including arsenic, cadmium, iron and selenium. Plant Soil 329:139–153

Norton GJ, Duan GL, Lei M et al (2012) Identification of quantitative trait loci for rice grain element composition on an arsenic impacted soil: influence of flowering time on genetic loci. Ann Appl Biol 161:46–56

Perla V, Holm DG, Jayanty SS (2012) Selenium and sulfur content and activity of associated enzymes in selected potato germplasm. Am J Potato Res 89:111–120

Pickering IJ, Wright C, Bubner B et al (2003) Chemical form and distribution of selenium and sulfur in the selenium hyperaccumulator *Astragalus bisulcatus*. Plant Physiol 131:1460–1467

Pilbeam DJ, Greathead HMR, Drihem K (2015) Selenium. In: Barker AV, Pilbeam DJ (eds) A handbook of plant nutrition, 2nd edn. CRC Press, Boca Raton, pp 165–198

Pilon M, Owen JD, Garifullina GF et al (2003) Enhanced selenium tolerance and accumulation in transgenic Arabidopsis expressing a mouse selenocysteine lyase. Plant Physiol 131:1250–1257

Pilon-Smits EAH (2012) Plant selenium metabolism – genetic manipulation, phytotechnological applications, and ecological implications. In: Wong MH (ed) Environmental contamination: health risks and ecological restoration. CRC Press, Boca Raton, pp 293–311

Pilon-Smits EAH, LeDuc DL (2009) Phytoremediation of selenium using transgenic plants. Curr Opin Biotechnol 20:207–212

Pilon-Smits EAH, de Souza MP, Hong G et al (1999a) Selenium volatalization and accumulation by twenty aquatic plant species. J Environ Qual 28:1011–1018

Pilon-Smits EAH, Hwang SB, Lytle CM et al (1999b) Overexpression of ATPsulphurylase in *Brassica juncea* leads to increased selenite uptake, reduction and tolerance. Plant Physiol 119:123–132

Pilon-Smits EAH, Quinn CF, Tapken W et al (2009) Physiological functions of beneficial elements. Curr Opin Plant Biol 12:267–274

Pu ZE, Yu M, He QY et al (2014) Quantitative trait loci associated with micronutrient concentrations in two recombinant inbred wheat lines. J Integr Agric 13:2322–2329

Quinn CF, Galeas ML, Freeman JL et al (2007) Selenium: deterrence, toxicity, and adaptation. Integr Environ Assess Manag 3:460–462

Rahman MM, Erskine W, Materne MA et al (2015) Enhancing selenium concentration in lentil (*Lens culinaris subsp. culinaris*) through foliar application. J Agric Sci 153:656–665

Ramamurthy RK, Jedlicka J, Graef GL et al (2014) Identification of new QTLs for seed mineral, cysteine, and methionine concentrations in soybean [*Glycine max* (L.) Merr.] Mol Breed 34:431–445

Ramos SJ, Rutzke MA, Hayes RJ et al (2011a) Selenium accumulation in lettuce germplasm. Planta 233:649–660

Ramos SJ, Yuan Y, Faquin V et al (2011b) Evaluation of genotypic variation of broccoli (*Brassica oleracea var. italica*) in response to selenium treatment. J Agric Food Chem 59:3657–3665

Ray H, Bett K, Tar'an B et al (2014) Mineral micronutrient content of cultivars of field pea, chickpea, common bean, and lentil grown in Saskatchewan, Canada. Crop Sci 54:1698–1708

Reeves RD, Baker AJM (2000) Metal-accumulating plants. In: Raskin I, Ensley BD (eds) Phytoremediation of toxic metals: using plants to clean up the environment. Wiley, New York, pp 193–229

Rodríguez LH, Morales DA, Rodríguez ER et al (2011) Minerals and trace elements in a collection of wheat landraces from the Canary Islands. J Food Comp Anal 24:1081–1090
Rosenfeld I, Beath OA (1964) Selenium: geobotany, biochemistry, toxicity, and nutrition. Academic Press, New York
Rouached H, Wirtz M, Alary R et al (2008) Differential regulation of the expression of two high-affinity sulfate transporters, SULTR1.1 and SULTR1.2, in Arabidopsis. Plant Physiol 147:897–911
Sasmaz M, Akgül B, Sasmaz A (2015) Distribution and accumulation of selenium in wild plants growing naturally in the Gumuskoy (Kutahya) mining area Turkey. Bull Environ Contam Toxicol 94:598–603
Schiavon M, Pilon M, Malagoli M et al (2015) Exploring the importance of sulphate transporters and ATPsulphurylases for selenium hyperaccumulation – comparison of *Stanleya pinnata* and *Brassica juncea* (Brassicaceae). Front Plant Sci 6:2
Shibagaki N, Rose A, McDermott JP et al (2002) Selenate-resistant mutants of *Arabidopsis thaliana* identify *Sultr1;2*, a sulfate transporter required for efficient transport of sulfate into roots. Plant J 29:475–486
Shinmachi F, Buchner P, Stroud JL et al (2010) Influence of sulfur deficiency on the expression of specific sulfate transporters and the distribution of sulfur, selenium, and molybdenum in wheat. Plant Physiol 153:327–336
Sors TG, Ellis DR, Na GN et al (2005a) Analysis of sulfur and selenium assimilation in *Astragalus* plants with varying capacities to accumulate selenium. Plant J 42:785–797
Sors TG, Ellis DR, Salt DE (2005b) Selenium uptake, translocation, assimilation and metabolic fate in plants. Photosynth Res 86:373–389
Sors TG, Martin CP, Salt DE (2009) Characterization of selenocysteine methyltransferases from *Astragalus* species with contrasting selenium accumulation capacity. Plant J 59:110–122
Szakova J, Tremlova J, Pegova K et al (2015) Soil-to-plant transfer of native selenium for wild vegetation cover at selected locations of the Czech Republic. Environ Monit Assess 187:358. doi:10.1007/s10661-015-4588-1
Tagmount A, Berken A, Terry N (2002) An essential role of S-adenosyl-Lmethionine: L-methionine S-methyltransferase in selenium volatilization by plants. Methylation of selenomethionine to selenium-methyl-L-seleniummethionine, the precursor of volatile selenium. Plant Physiol 130:847–856
Takahashi H, Watanabe-Takahashi A, Smith FW et al (2000) The roles of three functional sulphate transporters involved in uptake and translocation of sulphate in *Arabidopsis thaliana*. Plant J 23:171–182
Takahashi H, Kopriva S, Giordano M et al (2011) Sulfur assimilation in photosynthetic organisms: molecular functions and regulations of transporters and assimilatory enzymes. Annu Rev Plant Biol 62:157–184
Terry N, Carlson C, Raab TK et al (1992) Rates of selenium volatilization among crop species. J Environ Qual 21:341–344
Thavarajah D, Thavarajah P (2012) Evaluation of chickpea (*Cicer arietinum* L.) micronutrient composition: biofortification opportunities to combat global micronutrient malnutrition. Food Res Int 49:99–104
Thavarajah D, Thavarajah P, Sarker A et al (2011) A global survey of effects of genotype and environment on selenium concentration in lentils (*Lens culinaris* L.): implications for nutritional fortification strategies. Food Chem 125:72–76
Van Hoewyk D (2013) A tale of two toxicities: malformed selenoproteins and oxidative stress both contribute to selenium stress in plants. Ann Bot 112:965–972
Van Hoewyk D, Garifullina GF, Ackley AR et al (2005) Overexpression of AtCpNifS enhances selenium tolerance and accumulation in Arabidopsis. Plant Physiol 139:1518–1528
Van Hoewyk D, Pilon M, Pilon-Smits EAH (2008a) The functions of NifS-like proteins in plant sulfur and selenium metabolism. Plant Sci 174:117–123

Van Hoewyk D, Takahashi H, Hess A et al (2008b) Transcriptome and biochemical analyses give insights into selenium-stress responses and selenium tolerance mechanisms in Arabidopsis. Physiol Plant 132:236–253

Van Huysen T, Abdel-Ghany S, Hale KL et al (2003) Overexpression of cystathionine-γ-synthase enhances selenium volatilisation in *Brassica juncea*. Planta 218:71–78

Van Huysen T, Terry N, Pilon-Smits EAH (2004) Exploring the selenium phytoremediation potential of transgenic *Brassica juncea* overexpressing ATP sulfurylase or cystathionine-c-synthase. Int J Phytorem 6:111–118

Wang P, Menzies NW, Lombi E et al (2015) Synchrotron-based X-ray absorption near-edge spectroscopy imaging for laterally resolved speciation of selenium in fresh roots and leaves of wheat and rice. J Exp Bot 66:4795–4806

Watanabe T, Broadley MR, Jansen S et al (2007) Evolutionary control of leaf element composition in plants. New Phytol 174:516–523

White PJ (2016) Selenium accumulation by plants. Ann Bot 117:217–235

White PJ, Broadley MR (2009) Biofortification of crops with seven mineral elements often lacking in human diets – iron, zinc, copper, calcium, magnesium, selenium and iodine. New Phytol 182:49–84

White PJ, Brown PH (2010) Plant nutrition for sustainable development and global health. Ann Bot 105:1073–1080

White PJ, Bowen HC, Parmaguru P et al (2004) Interactions between selenium and sulphur nutrition in *Arabidopsis thaliana*. J Exp Bot 55:1927–1937

White PJ, Bowen HC, Marshall B et al (2007a) Extraordinarily high leaf selenium to sulphur ratios define 'Se-accumulator' plants. Ann Bot 100:111–118

White PJ, Broadley MR, Bowen HC et al (2007b) Selenium and its relationship with sufur. In: Hawkesford MJ, de Kok LJ (eds) Sulfur in plants – an ecological perspective. Springer, Dordrecht, pp 225–252

Ximénez-Embún P, Alonso I, Madrid-Albarráan Y et al (2004) Establishment of selenium uptake and species distribution in lupine, Indian mustard, and sunflower plants. J Agric Food Chem 52:832–838

Yan J, Wang F, Qin H et al (2011) Natural variation in grain selenium concentration of wild barley, *Hordeum spontaneum*, populations from Israel. Biol Trace Elem Res 142:773–786

Yang F, Chen L, Hu Q et al (2003) Effect of the application of selenium on selenium content of soybean and its products. Biol Trace Elem Res 93:249–256

Yang R, Wang R, Xue W et al (2013) QTL location and analysis of selenium content in tetraploid wheat grain. Guizhou Agric Sci 10:1–4 .[In Chinese]

Yoshimoto N, Inoue E, Saito K et al (2003) Phloem localizing sulfate transporter, Sultr1;3, mediates re-distribution of sulphur from source to sink organs in Arabidopsis. Plant Physiol 131:1511–1517

Yuan LX, Zhu YY, Lin ZQ et al (2013) A novel selenocystine-accumulating plant in selenium-mine drainage area in Enshi, China. PLoS One 8:e65615

Zhang L, Shi W, Wang X et al (2006) Genotypic differences in selenium accumulation in rice seedlings at early growth stage and analysis of dominant factors influencing selenium content in rice seeds. J Plant Nutr 29:1601–1618

Zhang L, Hu B, Li W et al (2014) OsPT2, a phosphate transporter, is involved in the active uptake of selenite in rice. New Phytol 201:1183–1191

Zhao XQ, Mitani N, Yamaji N et al (2010) Involvement of silicon influx transporter OsNIP2;1 in selenite uptake in rice. Plant Physiol 153:1871–1877

Zhao D-Y, Sun F-L, Zhang B et al (2015) Systematic comparisons of orthologous selenocysteine methyltransferase and homocysteine methyltransferase genes from seven monocots species. Not Sci Biol 7:210–216

Zhao H, Huang J, Li Y et al (2016) Natural variation of selenium concentration in diverse tea plant (*Camellia sinensis*) accessions at seedling stage. Sci Hort 198:163–169

Chapter 10
Manipulating Selenium Metabolism in Plants: A Simple Twist of Metabolic Fate Can Alter Selenium Tolerance and Accumulation

Doug Van Hoewyk and Ozgur Çakir

Abstract Selenium (Se) is a micronutrient for many organisms including humans. But like many trace elements, Se can be toxic at high concentrations and become a public health concern if it accumulates in soils or groundwater. Although higher plants don't require Se, plants can still accumulate and metabolize Se *via* the sulfur assimilatory pathway. Genetic manipulation of plant selenium metabolism primarily stems from two areas of interest: it has the potential to improve the phytoremediation of Se in contaminated areas, and it may aid the development of Se-containing phytochemical compounds that possess health benefits. This review highlights studies that have successfully altered Se metabolism in plants, and concludes by focusing on novel genes and pathways that might be targeted to manipulate Se metabolic processes.

Keywords Selenium • Metabolism • Oxidative stress • Selenoprotein • miRNA

10.1 Introduction

Selenium (Se) is an essential trace element for mammals, bacteria, and some green algae (Stadtman 1996). However, it is unlikely to be required by higher plants even though it can be beneficial (El Mehdawi and Pilon-Smits EAH 2012; Feng et al. 2013). As a nutrient in humans, Se is an essential component of the 21st amino acid selenocysteine, which is used to make 25 selenoproteins (Papp et al. 2007). A daily intake of 55 micograms of Se is recommended (Institute of Medicine 2000).

D. Van Hoewyk (✉)
Department of Biology, Coastal Carolina University, Conway, SC 29526, USA
e-mail: dougvh@coastal.edu

O. Çakir
Department of Molecular Biology and Genetics, Faculty of Science, Istanbul University, 34134 Vezneciler, Istanbul, Turkey
e-mail: ozgurckr@istanbul.edu.tr

E.A.H. Pilon-Smits et al. (eds.), *Selenium in plants*, Plant Ecophysiology 11,
DOI 10.1007/978-3-319-56249-0_10

A deficiency in dietary Se decreases the abundance of selenoproteins, and can lead to Kashin-Beck and Keshan disease, which alters bone and cardiac function, respectively. Additionally, numerous *in vitro* studies have reported the protective properties of Se compounds, particularly against cancer (Davis 2012). Due to its anticarcinogenic properties, Se supplementation or Se-fortified crops may be beneficial, but this is still controversial.

While Se levels in most soils are between 0.01 and 2 Se/kg, Se naturally accumulates in certain Cretaceous shale sediments. Such seleniferous soils can contain up to 100 mg/kg Se (Pilbeam et al. 2015). Anthropogenic activities, such as irrigation, can also result in Se accumulation in soil and potentially crops (Zhu et al. 2009). Selenium is chemically very similar to S, and its inadvertent accumulation in plants occurs primarily when selenate is transported into roots via sulfate transporters (White 2015). Selenate is readily translocated into shoot tissue, where is can be metabolized by chloroplastic enzymes involved in S assimilation.

Se decreases growth of most plants at concentrations exceeding 10–25 μM (Zhang et al. 2006). In non-hyperaccumulating plants- including most crops- Se toxicity occurs if its foliar accumulation exceeds 10–100 μg, as recently reviewed (White 2015). Selenium toxicity stems from two separate processes (Van Hoewyk 2013). Inorganic Se, particularly selenite, can redox cycle with thiols and generate reactive oxygen species (Spallholz 1994), including hydrogen peroxide (Tamaoki et al. 2008) and mitochondrial superoxide that alters respiration (Dimkovikj and Van Hoewyk 2014). Additionally, Se is likely toxic when it replaces S in protein. This hypothesis originated by the discovery that Se-tolerant *Astragalus* species were found to have nearly tenfold lower concentration of Se in protein compared with non-tolerant *Astragalus* species (Brown and Shrift 1981). In particular, the substitution of cysteine with selenocysteine (Sec) is believed to cause protein misfolding (Stadtman 1990). Several lines of evidence bolster the hypothesis that Se causes protein misfolding. Selenocysteine causes severe toxicity in Arabidopsis plants with a mutation in Bip2 (Sabbagh and Van Hoewyk 2012), an endoplasmic reticulum protein that participates in the unfolded protein response and renders mutant plants sensitive to agents that cause protein misfolding. Arabidopsis plants with mutations in ER protein quality control are also sensitive when treated with selenate (Van Hoewyk 2016). Additional support for the malformed selenoprotein hypothesis comes from the observation that selenoproteins can be removed by the ubiquitin-proteasome pathway in a variety of plants, including the Se-hyperaccumulator *Stanleya pinnata* (Sabbagh and Van Hoewyk 2012), canola (Dimkovikj et al. 2015), and the green algae *Chlamydomonas* (Vallentine et al. 2014).

Averting Se toxicity may potentially improve efforts to clean polluted soils and water *via* phytoremediation (Pilon-Smits 2005). Additionally, the development of crops with fortified levels of Se is appealing, as a source of both nutrition and Se-based therapeutics. Therefore, in some circumstances it may be desirable to use plants more efficiently for phytoremediation or as Se-fortified foods. To meet this aim, several different plant genetic engineering strategies have been designed and used successfully to further enhance plant Se metabolism, including its uptake and

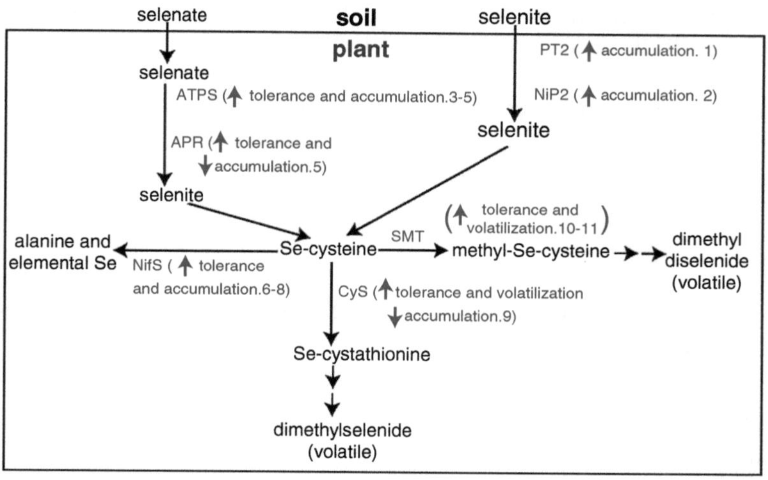

Fig. 10.1 Schematic diagram highlighting transgenic approaches that have altered selenium metabolism in plants. *Black*: Se metabolites; *Red*: manipulated enzymes and their impacts on selenium metabolism

accumulation, volatilization, and tolerance (Pilon-Smits and LeDuc 2009). Most genetic engineering approaches have targeted enzymes participating in S uptake or assimilation. However, a few studies have manipulated Se metabolism by focusing on genes unrelated to S metabolism, and there is compelling evidence that other unidentified pathways may also control plant Se tolerance and accumulation. These strategies are discussed below and summarized in Fig. 10.1.

10.2 Targeting Sulfur Transporters Alters Selenium Accumulation in Plants

In terrestrial ecosystems, selenate is the most bioavailable form of Se in soil (Terry et al. 2000). Although a selenate-specific transporter in plants remains elusive, it can be transported into roots *via* sulfate transporters. Mutation of sulfate transporter SULTR1;2 in Arabidopsis improved selenate tolerance by restricting selenate entry into the plant, and therefore decreased Se accumulation; mutations in ten other sulfate transporters did not affect selenate tolerance (El Kassis et al. 2007). Although it remains to be confirmed, overexpression of SULTR1;2 would likely increase Se accumulation in crops, but comes with the caveat that that increased Se content may also potentially decrease Se tolerance.

In aquatic ecosystems or flooding conditions that promote anaerobia, selenite is likely to be the predominant Se metabolite available for plant uptake. Similar to selenate, a selenite-specific transporter has yet to be identified. However, overexpression of phosphate transporter (OsPT2) in rice increased selenite uptake and Se accumulation in rice grains (Zhang et al. 2014). Additionally, a silicon transporter in rice (OsNiP2;1) appears to assist in selenite transport under acidic conditions (Zhao et al. 2010). Lastly, in addition to transporters, exogenous glutathione (GSH) can increase selenite transport in rice (Zhang 2015), indicating that perhaps GSH content in roots can control selenite uptake. Whether or not genetic engineering of GSH content in roots can augment selenite transport and accumulation remains to be determined.

10.3 Manipulation of the Sulfate Reduction Pathway Alters Metabolism

The sulfate reduction pathway controls the flux of the assimilation of sulfate into cysteine, as extensively reviewed (Çakır et al. 2012; Hawkesford and De Kok 2006; Pilon-Smits 2015; White 2015). Given that Se and S behave similarly, initial attempts aimed at manipulating Se metabolism have targeted enzymes involved in sulfate assimilation. The reduction of sulfate to sulfide occurs in plastids and involves the concerted actions of ATP sulfurylase (ATPS), adenosine 5-phosphoreductase (APR), and sulfite reductase (SiR). The reduction of selenate to selenite is likely a rate-limiting step for the assimilation of selenate into organic Se. This conclusion is based on studies reporting that plants treated with selenate accumulated mainly selenate, while plants that were fed selenite accumulated mainly organic Se (de Souza et al. 1998; Zayed et al. 1998). To overcome this apparent rate limitation in Se metabolism, Arabidopsis APTS- which activates sulfate- was overexpressed in *Brassica juncea* (Pilon-Smits et al. 1999). When treated with selenate, these transgenic plants accumulated an organic form of Se, in contrast to wild-type plants that accumulated selenate. Although Se volatilization was unaltered, the ATPS transgenics were more tolerant to selenate and accumulated threefold to fivefold more Se than wild type in both laboratory and in the field (Bañuelos et al. 2005); this phenotype was explained by their ability to quickly metabolize inorganic Se into organic forms. However, an alternative explanation to their improved Se tolerance may also be envisioned. When ATPS was overexpressed in Arabidopsis, it also resulted in increased Se accumulation and assimilation of organic Se, but was also accompanied by increased levels of cysteine and GSH (Sors et al. 2005). Elevated levels of GSH can maintain redox poise during oxidative stress (Noctor et al. 2012), and is associated with improved Se tolerance (Grant et al. 2011). Therefore, it is possible that improved Se tolerance in ATPS transgenics could have at least partially stemmed from an elevated GSH status.

APR catalyzes the reaction of activated sulfate to sulfite. When APR from *Pseudomonas aeruginosa* was overexpressed in Arabidopsis, it also increased the proportion of organic Se and improved tolerance when treated with selenate (Sors et al. 2005). Although an Arabidopsis APR isoform has not been overexpressed, knockout of APR in Arabidopsis was associated with decreased Se accumulation and tolerance, which was explained by the observed decrease in glutathione and superoxide accumulation (Grant et al. 2011). Taken together, these data indicate that APR also controls the flux of selenate into organic forms, similar to ATPS.

Sulfite is converted into sulfide *via* sulfite reductase (SiR), but it is doubtful that the enzyme also has selenite reductase activity (Ng and Anderson 1979). Rather, GSH likely non-enzymatically reduces selenite to selenide, and in doing so generates superoxide (Seko et al. 1989; Kessi and Hanselmann 2004). Additionally, Arabidopsis plants with decreased levels of SiR do not display altered tolerance when stressed with selenite (Fisher et al. 2016), suggesting that knockdown of SiR does not play an important role in determining Se tolerance or accumulation.

10.4 Minimizing Se-Cysteine Incorporation in Protein Improves Se Tolerance in Plants

Astragalus bisulcatus' tolerance to Se is attributable to the presence of a chloroplastic enzyme with selenocysteine methyltransferase (SMT) activity (Neuhierl and Bock 1996). This enzyme methylates Sec and prevents its incorporation into protein; therefore, the formation of malformed selenoproteins is avoided. Methyl-Sec is the predominant Se-containing metabolite in Se hyperaccumulators (Whanger 2002). *SMT* has been cloned and characterized from different plant species (Cakir and Ari 2013; Lyi et al. 2005; Neuhierl and Bock 1996; Sors et al. 2009; Zhu et al. 2008), and it is widely believed that this enzyme confers Se tolerance in Se-hyperaccumulating plants. Methyl-Sec can be further metabolized to non-toxic dimethyl-diselenide, a volatile molecule that is emitted into the atmosphere (de Souza et al. 1998). The *A. bisulcatus* SMT enzyme has been overexpressed in *A. thaliana* and *B. juncea* (Ellis et al. 2004; LeDuc et al. 2004). In both species, selenite-treated SMT-transgenic plants converted Sec to methyl-Sec. The ability to convert Sec to methyl-Sec was associated with increased total Se accumulation, improved Se tolerance, and enhanced volatilization of dimethyl-diselenide. The non-hyperaccumulator *Astragalus drummondii* also possesses an *SMT*-like gene (Sors et al. 2009). Despite its homology to the gene from *A. bisulcatus*, biochemical studies revealed that the enzyme from *A. drummondii* lacks SMT activity, thus likely rendering the plant intolerant to Se. Mutagenesis of the *A. drummondii* gene to make it more similar to the one from *A. bisulcatus* provided some SMT activity, but still the mutated enzyme was not as active as its homologue in *A. bisulcatus* (Sors et al. 2009). Additionaly, *B. juncea* over-expressing both APS and SMT increased Se accumulation up to ninefold compated to WT plants (LeDuc et al. 2006). Collectively, these experimental studies reveal that SMT activity plays a vital role in Se hyperaccumulation, and *A. bisulcatus* SMT provides both increased Se tolerance and accumulation when genetically engineered in non-hyperaccumulators. This may ultimately prove useful for the environmental cleanup of seleniferous soils or to fulfill the human dietary needs of Se.

Cystathionine gamma synthase (CgS) can also prevent the formation of non-specific selenoproteins by catalyzing the reaction of Sec to seleno-cystathionine, a precursor metabolite of Se-methionine. Overexpression of Arabidopsis *CgS* in *B. juncea* improved Se tolerance, which was explained by a twofold to threefold increase in Se volatilization (Van Huysen et al. 2003). As a result of enhanced volatilization, the CgS transgenics accumulated 40% less Se compared to wild-type plants. These results indicate that CgS is involved and rate limiting in Se volatilization.

In another approach to divert Sec from being incorporated into proteins, genetic engineering approaches have also targeted Sec-lyases, which catabolize Sec into alanine and elemental Se. Initially, a mouse Sec-lyase was over-expressed in Arabidopsis, which decreased the amount of Se in protein, yet increased Se accumulation (Pilon et al. 2003). Overexpression of Sec-lyase in the cytosol improved Se tolerance, but intriguingly, targeting of this enzyme to the chloroplast increased sensitivity to Se. This could potentially be explained by the ability of elemental Se to replace S in chloroplastic Fe-S proteins. Fe-Se clusters are known to be unstable and their incorporation into proteins can decrease activity (Hallenbeck et al. 2009). Sequencing of the Arabidopsis genome revealed a chloroplastic Sec-lyase called CpNifS. Overexpression of CpNifS in Arabidopsis increased Se accumulation and selenate tolerance almost twofold, and this phenotype was associated with a 33% decrease of Se in protein and increased S levels (Van Hoewyk et al. 2005). Additionally, *B. juncea* over-expressing a Sec-lyase also accumulated Se twofold when grown in soil polluted with Se (Bañuelos et al. 2007). In summary, these data indicate that overexpression of CpNifS prevents the formation of selenoproteins in plants, which likely explains their improved tolerance to selenate.

10.5 Manipulation of Oxidative Stress Response Genes Alters Se Metabolism

As mentioned above, Se is known to induce oxidative stress in plants. Thus, antioxidant systems may contribute to plant Se tolerance. Indeed, several studies have indicated that overexpression of genes associated with an oxidative stress response improve Se tolerance and alter plants' ability to accumulate Se.

Arabidopsis selenium-binding protein (*SBP1*) was the first gene unrelated to sulfur metabolism whose overexpression improved Se (Agalou et al. 2005). Expression of this gene is tightly linked to oxidative stress, and is also induced during sulfur starvation. Although its biological function remains unknown, SBP1 has been speculated to have antioxidant properties (Hugouvieux et al. 2009), as its overexpression in Arabidopsis also improves tolerance to cadmium and hydrogen. However, recently it was discovered that SPB1 can bind to a variety of heavy metals; additionally, it can bind to and reduce selenite, but not selenate (Schild et al. 2014). Therefore, increased tolerance in SBP1 transgenics may also be attributed to its capacity to prevent selenite-induced oxidative stress that can impair mitochondrial function

(Dimkovikj and Van Hoewyk 2014). In agreement with this conclusion, human cells with mutant SBP1 are sensitive to selenite and suffer from mitochondrial damage (Ying et al. 2015).

The story of a broccoli methyltransferase (*BoCOQ5–2*) expressed in Arabidopsis further demonstrates that manipulating Se metabolism can be achieved by targeting pathways independent of sulfur metabolism (Zhou et al. 2009). BoCOQ5–2 is involved in the biosynthesis of ubiquinone, which has a role in respiration; additionally, it is an antioxidant in plants (Ohara et al. 2004) and likely protects mitochondria during stress (Bergamini et al. 2012). Transgenic COQ5–2 plants had improved Se tolerance, which was associated with decreased levels of hydrogen peroxide and increased dimethyl diselenide volatilization. Ubiquinone levels were not elevated in these plants. The authors conclude that increased volatilization was unlikely to be a direct consequence of manipulating the ubiquinone pathway. Rather, increased dimethyl diselenide volatilization likely stemmed from an improved antioxidant status in the COQ5–2 plants. If that is the case, then it is possible that increased levels of other antioxidants- such as vitamin C and vitamin E- may also alter Se metabolism in plants (Zhou and Li 2010). In line with the hypothesis that improved oxidative stress tolerance can alter Se metabolism in plants, overexpression of GSH synthetase also increases Se tolerance and accumulation in *B. juncea* (Bañuelos et al. 2005). In Arabidopsis, tolerance to selenite correlates tightly with internal GSH concentrations (Grant et al. 2011). More recently, overexpression of a peroxidase implicated in drought and salt stress also protected Arabidopsis plants against Se (Jiang et al. 2015). In summary, Se metabolism can be altered by genetic engineering approaches aimed at improving oxidative stress tolerance.

10.6 Transcriptomics Reveal Additional Genes That May Alter Se Metabolism and Tolerance

The advent of high-throughput sequencing has allowed researchers to identify genes and pathways responsive to stress conditions. For example, the transcriptome of selenate-treated Arabidopsis revealed an upregulation of many transcripts involved in ethylene and abscisic acid synthesis and signaling (Van Hoewyk et al. 2008). Indeed, further genetic analysis demonstrated that decreased levels of these two hormones increase both selenate and selenite sensitivity (Tamaoki et al. 2008), likely by mediating an oxidative stress response. A more recent transcriptome study used RNA-seq to determine the effects of selenate in *Astragalus chrysochlorus*, a secondary Se accumulator (Çakır et al. 2015). This study revealed an upregulation of genes involved in ABC transport, plant pathogen interactions, and biosynthesis of secondary metabolites. Additionally, many putative transcription factors were upregulated, including: TCP13-like, bZIP, bHLH041-like, heat stress A-3-like, trihelix GT-3b-like, and WRKY32. Additional experimentation is needed to elucidate if manipulation of these identified genes play a role in Se tolerance and accumulation.

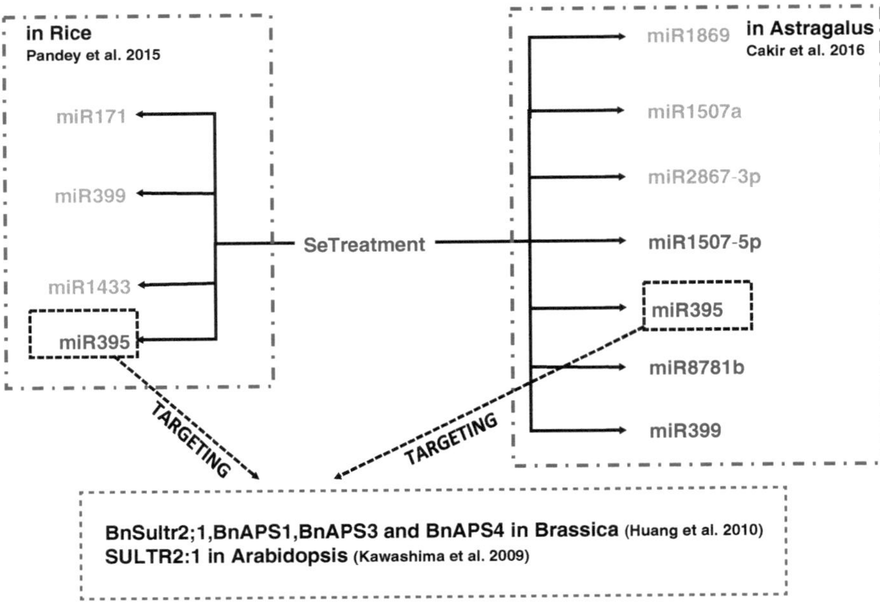

Fig. 10.2 Se treatment in plants alters the expression of miRNAs. *Green*: up-regulation. *Red*: down-regulation

Increased GSH concentration in plants is associated with improved tolerance to agents that induce oxidative stress, including Se (Noctor et al. 2012; Grant et al. 2011). Optimal glutamate and glutathione metabolism in Arabidopsis plants is maintained by glutamyl cyclotransferase (GGCT2; 1); this enzyme participates in the glutamyl cycle by recycling glutamate from GSH-conjugates, which can subsequently be used to make new GSH. Overexpression of GGCT2;1 in Arabidopsis improved arsenate tolerance; this phenotype was explained by the increased cytosolic breakdown of GSH conjugated to arsenic and decreased demand of *de novo* glutamate generated by the TCA cycle (Paulose et al. 2013). Selenate-treatment has been reported to decrease glutamate concentration in Arabidopsis (Van Hoewyk et al. 2008; Grant et al. 2011). It is possible that GGCT2;1 transgenics also confer Se tolerance, as suggested by a transcriptome study (Van Hoewyk et al. 2008). GGCT2;1 mRNA increased almost 100-fold in selenate-treated Arabidopsis (Van Hoewyk et al. 2008). Additionally, GGCT2;1 protein increases in *B. napus* treated with selenite (Dimkovikj and Van Hoewyk 2014), further implicating its involvement in a Se-stress response. Future studies may reveal that GGCT2;1 overexpression alleviates Se toxicity.

MicroRNAs (miRNAs) have also been recently implicated in mediating a Se response, as depicted in Fig. 10.2. Noncoding miRNAs post-transcriptionally regulate gene expression by participating in the degradation of target mRNAs (Bartel 2004), thereby inhibiting translation. miRNAs are known to function in many developmental and physiological processes (Zhang and Wang 2015). Two recent studies have explored how Se affects miRNA expression in plants. In one study, Se-induced miRNAs were identified in *A. chrysochlorus* using next generation sequencing analysis (Çakir et al. 2016). Computational studies revealed that Se induced miRNAs that target mRNAs controlling hormone signaling, plant-pathogen interactions, and sulfur metabolic pathways. The most significantly affected miRNAs were miR1507a, miR1869 and miR2867-3p, miR1507-5p and miR8781b; however, it is unknown what these miRNAs target or how they might mediate Se tolerance and accumulation. In another study performed in rice, Se increased expression of miR171, miR399 and miR1433, but decreased expression of miR395 (Pandey et al. 2015). miR395 targets ATP sulfurylases ATPS1 and ATPS4 and the sulfate transporter SULTR2;1 (Kawashima et al. 2009; Huang et al. 2010). These genes control sulfate accumulation and assimilation, and their expression was inversely correlated with decreased miR395 expression in rice. This result nicely coincides with transcriptome studies in Arabidopsis demonstrating that selenate induces genes involved in sulfur transport and assimilation (Van Hoewyk et al. 2008). In summary, manipulating miRNAs may also provide new approaches to alter Se metabolism.

References

Agalou A, Roussis A, Spaink HP (2005) The Arabidopsis selenium-binding protein confers tolerance to toxic levels of selenium. Funct Plant Biol 32:881–890

Bañuelos G, Terry N, LeDuc DL, Pilon-Smits EA, Mackey B (2005) Field trial of transgenic Indian mustard plants shows enhanced phytoremediation of selenium-contaminated sediment. Environ Sci Technol 39:1771–1777

Bañuelos G, LeDuc DL, Pilon-Smits EAH, Tagmount A, Terry N (2007) Transgenic Indian mustard overexpressing selenocysteine lyase or seleno- cysteine methyltransferase exhibit enhanced potential for selenium phytore- mediation under field conditions. Environ Sci Technol 41:599–605

Bartel DP (2004) MicroRNAs: genomics, biogenesis, mechanism, and function. Cell 116:281–297

Bergamini C, Moruzzi N, Sblendido A, Lenaz G, Fato R (2012) A water soluble CoQ10 formulation improves intracellular distribution and promotes mitochondrial respiration in cultured cells. PLoS One 7:e33712

Brown TA, Shrift A (1981) Exclusion of selenium from proteins of selenium-tolerant Astragalus species. Plant Physiol 67:1051–1053

Çakir O, Ari S (2013) Cloning and molecular characterization of selenocysteine methyltransferase (AchSMT) cDNA from Astragalus chrysochlorus. Plant Omics 6:100–106

Çakır O, Turgut-Kara N, Arı Ş (2012) Selenium metabolism in plants: molecular approaches. In: Montanaro G (ed) Advances in selected plant physiology aspects. InTech, Rijeka, pp 209–232

Çakır Ö, Turgut-Kara N, Arı Ş, Zhang B (2015) De novo transcriptome assembly and comparative analysis elucidate complicated mechanism regulating *Astragalus chrysochlorus* response to selenium stimuli. PloSOne 10:e0135677

Çakir O, Candar-Cakir B, Zhang BH (2016) Small RNA and degradome sequencing reveals important microRNA function in Astragalus chrysochlorus response to selenium stimuli. Plant Biotechnol J 14:543–556

Davis CD (2012) Selenium supplementation and cancer prevention. Curr Nutr Rep 1:16–23 de Souza MP, Pilon-Smits EAH, Lytle CM, Hwang S, Tai JC, Honma TSU, Yeh L, Terry N (1998) Rate-limiting steps in selenium volatilization by *Brassica juncea*. Plant Physiol 117:1487–1494

Dimkovikj A, Van Hoewyk D (2014) Selenite activates the alternative oxidase pathway and alters primary metabolism in *Brassica napus* roots: evidence of a mitochondrial stress response. BMC Plant Biol 14:259

Dimkovikj A, Fisher B, Hutchison K, Van Hoewyk D (2015) Stuck between a ROS and a hard place: analysis of the ubiquitin proteasome pathway in selenocysteine treated Brassica napus reveals different toxicities during selenium assimilation. J Plant Physiol 181:50–54

El Kassis E, Cathala N, Rouached H et al (2007) Characterization of a selenate-resistant Arabidopsis mutant: root growth as a potential target for selenate toxicity. Plant Physiol 143:1231–1241

El Mehdawi AF, Pilon-Smits EAH (2012) Ecological aspects of plant selenium hyperaccumulation. Plant Biol 14:1–10

Ellis DR, Sors TG, Brunk DG, Albrecht C, Orser C, Lahner B et al (2004) Production of Se-methylselenocysteine in transgenic plants expressing selenocysteine methyltransferase. BMC Plant Biol 4:1

Feng R, Chaoyang W, Tu S (2013) The roles of selenium in protecting plants against abiotic stresses. Environ Exp Bot 87:58–68

Fisher B, Yarmolinsky D, Abdel-Ghany S, Pilon M, Pilon-Smits EA, Sagi M, Van Hoewyk D (2016) Superoxide generated from the glutathione-mediated reduction of selenite damages the iron-sulfur cluster of chloroplastic ferredoxin. Plant Physiol Biochem 106:228–235

Grant K, Carey NM, Mendoza M, Schulze J, Pilon M, Pilon-Smits EA, Van Hoewyk D (2011) Adenosine 5′-phosphosulfate reductase (APR2) mutation in Arabidopsis implicates glutathione deficiency in selenate toxicity. Biochem J 438:325–335

Hallenbeck PC, George GN, Prince RC, Thorneley RN (2009) Characterization of a modified nitrogenase Fe protein from *Klebsiella pneumoniae* in which the 4Fe4S cluster has been replaced by a 4Fe4Se cluster. J Biol Inorg Chem 14:673–682

Hawkesford MJ, De Kok LJ (2006) Managing sulphur metabolism in plants. Plant Cell Environ 29:382–395

Huang SQ, Xiang AL, Che LL, Chen S, Li H, Song JB, Yang ZM (2010) A set of miRNAs from *Brassica napus* in response to sulphate deficiency and cadmium stress. Plant Biotechnol J 8(8):887–899

Hugouvieux V, Dutilleul C, Jourdain A, Reynaud F, Lopez V, Bourguignon J (2009) Arabidopsis putative selenium-binding protein1 expression is tightly linked to cellular sulfur demand and can reduce sensitivity to stresses requiring glutathione for tolerance. Plant Physiol 151:768–781

Institute of Medicine (2000) Dietary reference intakes for vitamin C, vitamin E, selenium, and carotenoids. National Academies Press, Washington, DC

Jiang L, Gao QC, Chen ZP, Zhang JJ, Bai XY, He XL, Xu QX (2015) Selenium tolerance of an Arabidopsis drought-resistant mutant csm1-1. Rus J Plant Physiol 62:625–631

Kawashima CG, Yoshimoto N, Maruyama-Nakashita A, Tsuchiya YN, Saito K, Takahashi H, Dalmay T (2009) Sulphur starvation induces the expression of microRNA-395 and one of its target genes but in different cell types. Plant J 57:313–321

Kessi J, Hanselmann KW (2004) Similarities between the abiotic reduction of selenite with glutathione and the dissimilatory reaction mediated by *Rhodospirillum rubrum* and *Escherichia coli*. J Biol Chem 279:50662–50669

LeDuc DL, Tarun AS, Montes-Bayon M, Meija J, Malit MF, Wu CP, Abdel Samie M, Chiang CY, Tagmount A, Neuhierl B, Böck A (2004) Overexpression of selenocysteine methyltransferase in Arabidopsis and Indian mustard increases selenium tolerance and accumulation. Plant Physiol 135:377–383

LeDuc DL, Abdel Samie M, Montes-Bayon M, Wu CP, Reisinger SJ, Terry N (2006) Overexpressing both ATP sulfurylase and selenocysteine methyltransferase enhances selenium phytoremediation traits in Indian mustard. Environ Pollut 144:70–76

Lyi SM, Heller LI, Rutzke M, Welch RM, Kochian LV, Li L (2005) Molecular and biochemical characterization of the selenocysteine Se-methyltransferase gene and Se-methylselenocysteine synthesis in broccoli. Plant Physiol 138:409–420

Neuhierl B, Bock A (1996) On the mechanism of selenium tolerance in selenium-accumulating plants. Purification and characterization of a specific selenocysteine methyltransferase from cultured cells of *Astragalus bisulcatus*. Eur J Biochem 239:235–238

Ng BH, Anderson JW (1979) Light-dependent incorporation of selenite and sulphite into selenocysteine and cysteine by isolated pea chloroplasts. Phytochemistry 18:573–580

Noctor G, Mhamdi A, Chaouch S et al (2012) Glutathione in plants: an integrated overview. Plant Cell Environ 35:454–484

Ohara K, Kokado Y, Yamamoto H, Sato F, Yazaki K (2004) Engineering of ubiquinone biosynthesis using the yeast coq2 gene confers oxidative stress tolerance in transgenic tobacco. Plant J 40:734–743

Pandey C, Raghuram B, Sinha AK, Gupta M (2015) miRNA plays a role in the antagonistic effect of selenium on arsenic stress in rice seedlings. Metallomics 7:857–866

Papp LV, Lu J, Holmgren A, Khanna KK (2007) From selenium to selenoproteins: synthesis, identity, and their role in human health. Antiox Red Signal 9:775–806

Paulose B, Chhikara S, Coomey J, H-i J, Vatamaniuk O, Dhankher OP (2013) A g-glutamyl cyclotransferase protects arabidopsis plants from heavy metal toxicity by recycling glutamate to maintain glutathione homeostasis. Plant Cell 25:4580–4595

Pilbeam DJ, Greathead HMR, Drihem K (2015) Selenium. In: Barker AV, Pilbeam DJ (eds) A handbook of plant nutrition, 2nd edn. CRC Press, Boca Raton, pp 165–198

Pilon M, Owen JD, Garifullina GF, Kurihara T, Mihara H, Esaki N, Pilon-Smits EAH (2003) Enhanced selenium tolerance and accumulation in transgenic Arabidopsis expressing a mouse selenocysteine lyase. Plant Physiol 131:1250–1257

Pilon-Smits EA (2015) Selenium in plants. Luttge U, Beyschlag W, ed. Progress in botany. Springer International Publishing. Vancouver, In, pp 93–107

Pilon-Smits EA, LeDuc DL (2009) Phytoremediation of selenium using transgenic plants. Curr Opin Biotechnol 20:207–212

Pilon-Smits EAH (2005) Phytoremediation. Annu Rev Plant Biol 56: 15–39

Pilon-Smits EAH, Hwang SB et al (1999) Overexpression of ATP sulfurylase in *Brassica juncea* leads to increased selenate uptake, reduction and tolerance. Plant Physiol 119:123–132

Sabbagh M, Van Hoewyk D (2012) Malformed selenoproteins are removed by the ubiquitin-proteasome pathway in Stanleya pinnata. Plant Cell Physiol 53:555–564

Schild F, Kieffer-Jaquinod S, Palencia A, Cobessi D, Sarret G, Zubieta C, Jourdain A, Dumas R, Forge V, Testemale D, Bourguignon J (2014) Biochemical and biophysical characterization of the selenium-binding and reducing site in *Arabidopsis thaliana* homologue to mammals selenium-binding protein 1. J Biol Chem 289:31765–31776

Seko Y, Saito T, Kitahara J, Imura N (1989) Active oxygen generation by the reaction of selenite with reduced glutathione in vitro. In: Wendel A (ed) Proceedings of the fourth international symposium on selenium in biology and medicine. Springer, Heidelburg, pp 70–73

Sors TG, Ellis DR, Na GN (2005) Analysis of sulfur and selenium assimilation in Astragalus plants with varying capacities to accumulate selenium. Plant J 42:785–797

Sors TG, Martin CP, Salt DE (2009) Characterization of selenocysteine methyltransferases from Astragalus species with contrasting selenium accumulation capacity. Plant J 59:110–122

Spallholz JE (1994) On the nature of selenium toxicity and carcinostatic activity. Free Radic Biol Med 17:45–64

Stadtman TC (1990) Selenium biochemistry. Annu Rev Biochem 59:111–127

Stadtman TC (1996) Selenocysteine. Annu Rev Biochem 65:83–100

Tamaoki M, Freeman JL, Pilon-Smits EA (2008) Cooperative ethylene and jasmonic acid signaling regulates selenite resistance in Arabidopsis. Plant Physiol 146:1219–1230

Terry N, Zayed A, de Souza P, Tarun A (2000) Selenium in higher plants. Annu Rev Plant Physiol Plant Mol Biol 51:401–432

Vallentine P, Hung CY, Xie J, Van Hoewyk D (2014) The ubiquitin-proteasome pathway protects Chlamydomonas reinhardtii against selenite toxicity, but is impaired as reactive oxygen species accumulate. AoB Plants 6:plu062

Van Hoewyk D (2013) A tale of two toxicities: malformed selenoproteins and oxidative stress both contribute to selenium stress in plants. Ann Bot 112:965–972

Van Hoewyk D (2016) Defects in endoplasmic reticulum-associated degradation (ERAD) increase selenate sensitivity in Arabidopsis. Plant Signal Behav. doi:10.1080/15592324.2016.1171451

Van Hoewyk D, Garifullina GF, Ackley AR, Abdel-Ghany SE, Marcus MA, Fakra S, Ishiyama K, Inoue E, Pilon M, Takahashi H, Pilon-Smits EA (2005) Overexpression of AtCpNifS enhances selenium tolerance and accumulation in Arabidopsis. Plant Physiol 139:1518–1528

Van Hoewyk D, Takahashi H, Inoue E, Hess A, Tamaoki M, Pilon-Smits EAH (2008) Transcriptome analyses give insight into selenium-stress responses and selenium tolerance mechanisms in Arabidopsis. Physiol Plant 132:236–253

van Huysen T, Abdel-Ghany S, Hale KL, LeDuc D, Terry N, Pilon-Smits EAH (2003) Overexpression of cystathionine-g-synthase in Indian mustard enhances selenium volatilization. Planta 218:71–78

Whanger PD (2002) Selenocompounds in plants and animals and their biological significance. J Am Coll Nutr 21:223–232

White PJ (2015) Selenium accumulation by plants. Ann Bot 117:217–235

Ying Q, Ansong E, Diamond AM, Yang W (2015) A critical role for cysteine 57 in the biological functions of selenium binding protein-1. Internat J Mol Sci 16:27599–27608

Zayed A, Lytle CM, Terry N (1998) Accumulation and volatilization of different chemical species of selenium by plants. Planta 206:284–292

Zhang BH (2015) MicroRNA: a new target for improving plant tolerance to abiotic stress. J Exp Bot 66:1749–1761

Zhang B, Wang Q (2015) MicroRNA-based biotechnology for plant improvement. J Cell Physiol 230:1–5

Zhang LH, Byrne PF, Pilon-Smits EAH (2006) Mapping quantitative trait loci associated with selenate tolerance in *Arabidopsis thaliana*. New Phytol 170:33–42

Zhang L, Hu B, Li W et al (2014) OsPT2, a phosphate transporter, is involved in the active uptake of selenite in rice. New Phytol 20:1183–1191

Zhao XQ, Mitani N, Yamaji N, Shen RF, Ma JF (2010) Involvement of silicon influx transporter OsNIP2; 1 in selenite uptake in rice. Plant Physiol 153:1871–1877

Zhou X, Li L (2010) Think outside the box: Selenium volatilization altered by a broccoli gene in the ubiquinone biosynthetic pathway. Plant Signal Behav 5:76–77

Zhou X, Yuan Y, Yang Y, Rutzke M, Thannhauser TW, Kochian LV, Li L (2009) Involvement of a broccoli COQ5 methyltransferase in the production of volatile selenium compounds. Plant Physiol 151:528–540

Zhu L, Jiang CJ, Deng WW, Gao X, Wang RJ, Wan XC (2008) Cloning and expression of selenocysteine methyltransferase cDNA from *Camellia sinensis*. Acta Physiol Plant 30:167–174

Zhu YG, Pilon-Smits EA, Zhao FJ, Williams PN, Meharg AA (2009) Selenium in higher plants: understanding mechanisms for biofortification and phytoremediation. Trends Plant Sci 14:436–442

Chapter 11
Ecology of Selenium in Plants

Colin F. Quinn, Ali F. El Mehdawi, and Elizabeth A.H. Pilon-Smits

Abstract Selenium (Se) is both essential at low levels and toxic at higher levels to most organisms. Plant Se accumulation therefore may affect interactions with ecological partners positively or negatively. The ecological implications of plant Se accumulation are especially intriguing for Se hyperaccumulator species, which have evolved the capacity to take up Se to extraordinarily high levels, around 1% of dry weight. In this chapter, we summarize ecological aspects of Se in plants, including how Se can act as a defense mechanism against herbivores, how some herbivores have disarmed this defense, how Se can be transferred to higher trophic levels, how Se hyperaccumulating plants alter soil Se distribution and speciation around them and how this affects other plant species. The effects of plant Se on plant-microbe interactions are not reviewed here, since they are covered elsewhere. Insight into ecological implications of plant Se accumulation sheds light on evolutionary pressures that led to Se hyperaccumulation, and the importance of plant Se (hyper)accumulation for Se cycling. In addition, better understanding of the ecological impacts of Se in plants can help manage seleniferous habitats and optimize crop Se biofortification and the use of plants in phytoremediation to clean up Se polluted areas.

Keywords Hyperaccumulation • Elemental defense • Allelopathy • Ecology

11.1 Introduction to the Ecology of Selenium in Plants

Selenium (Se), an essential element for animals and toxic to both plants and animals at higher concentrations, is found in different concentrations in soils throughout the world (Kabata-Pendias 1998). Many plant species, either inadvertently or advertently, take up Se when growing on soils with elevated concentrations of Se, called seleniferous soils. Plants can be separated into three categories, Se nonaccumulators, Se accumulators and Se hyperaccumulators (Baker et al. 2000). All of these categories of plants take up and assimilate Se using the sulfur (S) assimilation pathway, but to different extent: when growing in natural habitats, Se nonaccumulating plants have less than 100 mg Se per kg

C.F. Quinn • A.F. El Mehdawi • E.A.H. Pilon-Smits (✉)
Biology Department, Colorado State University, Fort Collins, CO 80523-1878, USA
e-mail: colin.quinn15@gmail.com; a_mezo_10@yahoo.com; epsmits@colostate.edu

E.A.H. Pilon-Smits et al. (eds.), *Selenium in plants*, Plant Ecophysiology 11,
DOI 10.1007/978-3-319-56249-0_11

dry weight (DW) in their tissues, while Se accumulating plants have 100–1000 mg Se per kg DW and hyperaccumulating plants have more than 1000 mg Se per kg DW (Baker et al. 2000). Hyperaccumulators often have Se concentrations 100 times higher than other species growing at the same site, which would be toxic to other plant species and may confer an ecological advantage for Se hyperaccumulating plants (Baker et al. 2000). Hyperaccumulation is not specific to Se. This phenomenon is found for a number of elements (including As, Co, Cu, Pb, Mn, Ni, Se and Zn) and occurs in over 500 species of plants (Cappa and Pilon-Smits 2014; Pollard et al. 2014). Selenium hyperaccumulating plants, which are found in approximately 45 plant taxa, are found on seleniferous soils such as those in the Western United States which was covered by an ocean during the Cretaceous Period (approximately 65 million years ago) that deposited large quantities of Se. For example, many species of *Astragalus* as well as some *Stanleya*, *Symphyotrichum*, *Xylorhiza* and *Oonopsis* species native to this region have evolved to become Se hyperaccumulators (Beath et al. 1939; Galeas et al. 2007).

Since there is no known essential function for Se in plants, and the beneficial effects of Se occur typically at very low tissue levels, it is intriguing why plants would accumulate so much of this element. There are several theories for why some species have evolved to take up elevated concentrations toxic elements like Se. These include: to serve as a mechanism to increase drought resistance, to use as an allelopathic chemical to prevent other plants from competing for the same resources, to protect plants from herbivores and pathogens (the elemental defense hypothesis), as a tolerance mechanism to the toxic element, or as an inadvertent side effect (Boyd and Martens 1992). The bulk of the research investigating the ecological aspects of Se in plants has focused on, and supported, the elemental defense hypothesis. There is also some evidence for elemental allelopathy (El Mehdawi et al. 2011a). In this chapter we examine the ecological significance of Se in plants by reviewing research that investigates how plants with elevated levels of Se interact with their environment. A summary of Se effects on ecological interactions is shown in Fig. 11.1.

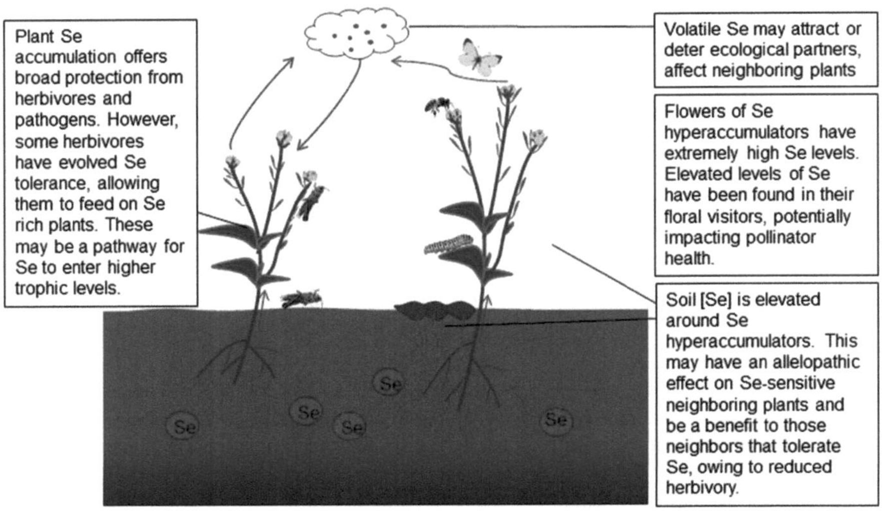

Fig. 11.1 Ecological interactions involving Se hyperaccumulator plants

11.2 Selenium in Plants Impacts Plant-Animal Interactions

As a result of Se being toxic to animals, it is expected that elevated levels of Se in plants will protect plants from herbivores and pathogens. However, it should be noted that plant species vary in the forms of Se accumulated, and these different forms of Se have different toxicity levels. As compared to other species, Se hyperaccumulators store and distribute Se differently, which may result in different effects of the accumulated Se on ecological interactions (Freeman et al. 2006a; Prins et al. 2011; Quinn et al. 2011b). Hyperaccumulators typically show preferential uptake of Se over S (higher tissue Se/S ratio), translocate more Se to the shoot and to sink organs, assimilate Se more readily into organic forms and store Se preferentially in specific tissues such as the epidermis. The elemental defense hypothesis proposed by Boyd and Martens (1992) states that toxic elements are hyperaccumulated to protect plants from herbivory and/or pathogen attacks. Research has supported the elemental defense hypothesis for a number of elements, including Se. Plants that (hyper)accumulate Se are protected from a wide variety of herbivores, as reviewed below, and by El Mehdawi and Pilon-Smits (2012).

As early as the 1930s, ranchers in the Western U.S. knew that Se was a toxic compound in certain "locoweeds", which, when ingested, can cause toxicity and death in livestock (Wilber 1980). More recently, controlled feeding studies have helped determine how Se can act as a defensive mechanism against animals. In choice and non-choice laboratory and field studies, plants with elevated levels of Se were shown to be protected from a variety of herbivores with different feeding mechanisms, both due to herbivore avoidance of high-Se plants, and to herbivore Se toxicity after feeding. Some varieties of saltbush (*Atriplex*), a Se accumulator, were protected from beet armyworm (*Spodoptera exigua*) herbivory due to toxicity (Vickerman et al. 2002). Selenium accumulator Indian mustard (*Brassica juncea*) was protected by Se from cabbage looper (*Trichoplusi ni*) and cabbage white butterfly (*Pieris rapae*) herbivory due to both toxicity and deterrence (Banuelos et al. 2002; Hanson et al. 2003). Both Indian mustard and the related Se hyperaccumulator Prince's plume (*Stanleya pinnata*) were protected from herbivory by orthopteran species due to both toxicity and deterrence (Freeman et al. 2007). Aphids, a phloem feeding insect, also preferred to feed on Indian mustard with lower levels of Se and suffered toxicity after feeding on plants with high levels of Se (Hanson et al. 2004). Furthermore, the Se hyperaccumulators two-grooved milkvetch (*Astragalus bisulcatus*) and Prince's plume containing elevated levels of Se both suffered less herbivory from spider mites and thrips, both cell-disruptor herbivores, than plants of the same species with low levels of Se (Quinn et al. 2010). A field study found that plants containing elevated tissue Se levels harbored fewer arthropods and fewer arthropod species than other plants at the same site with lower levels of Se (Galeas et al. 2008). Even vertebrate herbivores, such as prairie dogs, avoided plants with elevated levels of Se in the field (Quinn et al. 2008). When plants grown in the greenhouse with high- and low levels of Se were placed in the field, prairie dogs fed more on the plants with less Se. Furthermore, plants with elevated levels of Se that were placed in the field exposed to prairie dogs and other herbivores survived better over a two-year time period than plants of the same species containing less Se (Freeman et al. 2009).

The threshold where Se in plants starts to provide protection from herbivores is low. Levels as low as 50 mg Se per kg DW were shown to protect Prince's plume from prairie dog herbivory, and 230–447 mg Se per kg DW was shown to protect Indian mustard from Lepidoptera (El Mehdawi and Pilon-Smits 2012). In natural environments it is common for Se accumulating plants to have >100 mg Se per kg DW and hyperaccumulators commonly contain >1000 mg Se per kg DW. Thus, both Se accumulator and hyperaccumulator plants can receive ecological benefit in the form of decreased herbivory, with higher levels likely protecting more effectively and from more herbivore species. Both inorganic and organic forms of Se apparently offer herbivore protection, since Indian mustard accumulates mostly selenate while Prince's plume accumulates mainly methyl-SeCys. These findings gives insight into the ecological selection pressures that may have led to Se hyperaccumulation. Hyperaccumulators likely evolved to take up Se to serve as an ecological advantage, potentially as protection from herbivores and pathogens.

The particular location where plants sequester Se may also significantly affect the degree of elemental defense conferred. Hyperaccumulators, such as Prince's plume and two-grooved milkvetch store Se more in young leaves and reproductive organs as compared to older leaves, and they store the Se mainly in the periphery of leaves and in pollen and ovules (Freeman et al. 2006a; Quinn et al. 2011b). Interestingly, these specific Se sequestration patterns in hyperaccumulators are different from those of sulfur. These findings suggest that hyperaccumulators purposely store Se in specific areas, potentially to protect their most valuable structures as well as the likely entry points for herbivores and pathogens. In Se accumulators such as Indian mustard, Se is more evenly distributed and follows the same patterns as S, which suggest that these plants have not evolved ways to specifically use Se for the same ecological function as Se hyperaccumulators. The high Se levels in flowers may also have ecological consequences for plant-pollinator interactions, as discussed in more detail later in this chapter.

The form of Se found in nonaccumulators and accumulators is different than the form found in Se hyperaccumulators. The primary form of Se in many non-hyperaccumulators is selenate, the same form found in most seleniferous soils, while the primary form of Se in hyperaccumulators is methyl SeCys (Freeman et al. 2010). Both selenate and methyl SeCys are toxic to animals; while inorganic Se causes oxidative stress, methyl SeCys (when demethylated) disrupts protein function if it replaces Cys in proteins (Van Hoewyk 2013). Perhaps as an evolutionary response to the form of Se stored in hyperaccumulators, methyl SeCys, some herbivores of these species have evolved Se tolerance and some may even prefer to feed on Se hyperaccumulating plants. Some case studies are reviewed below.

In a laboratory study, a variety of diamondback moth (*Plutella xylostella*) from a non-seleniferous area, known to be sensitive to Se was compared with a variety of the same species collected at a seleniferous field site feeding on Se hyperaccumulators. The moth larvae collected from the seleniferous field site (referred to henceforth as the Se tolerant variety) showed no toxic effects when fed Prince's plume with elevated concentrations of Se, while the Se-sensitive variety suffered toxicity and death when forced to feed on Prince's plume leaves with elevated levels of Se (Freeman et al. 2006b). The Se tolerant variety of moth was found to store the ingested Se as methyl SeCys, the same form found in the plant, while the Se sensitive variety stored SeCys. If methyl-SeCys is not demethylated to SeCys (as in the Se tolerant moth) it likely is not toxic to animals because it is not incorporated into proteins. In contrast, the SeCys accumulating in the Se-sensitive moth likely caused toxicity due to incorporation into proteins (Stadtman 1990). Thus, the difference in Se tolerance between the two moth varieties appears to be that the Se tolerant variety has lost the capacity to demethylate methyl SeCys, leading to Se tolerance. Another apparent difference between the two moth varieties was that the Se tolerant moth sequestered Se in its hindgut while the Se sensitive moth did not; this may also contribute to Se tolerance (Freeman et al. 2006b). Interestingly, the Se-sensitive variety of diamondback moth preferred to oviposit (lay eggs) and feed on plants with lower Se levels, while the Se tolerant variety showed no preference in this respect between high- and low- Se plants (Freeman et al. 2006b). The findings that the Se tolerant variety of this diamondback moth has lost its aversion to feed and oviposit on high-Se plants and can feed on Se hyperaccumulators without ill effects suggest that this herbivore is able to occupy the exclusive niche offered by the high-Se plant host, and may even have become a Se specialist.

Another Se hyperaccumulator, two-grooved milkvetch (*A. bisulcatus*) was also found to be subject to herbivory by moth larvae from two species (*Apamea sordens* and a Gelechiidae) while growing in its natural seleniferous habitat (Valdez et al. 2012). Seeds of the same plant species were colonized by two herbivores, seed beetle *Acanthoscelides fraterculus* and seed chalcid *Bruchophagus mexicanus* (Freeman et al. 2012). Both seed herbivores apparently excluded Se, since they contained around 1000-fold lower Se levels than was in the seed.

While it has not been tested directly, it is likely that herbivores and other animals can identify plants with high levels of Se due to the odorous volatile Se that they release. Indeed, aphids were observed in choice studies to prefer a low-Se leaf disk over a high-Se leaf disk of the same species, solely based on smell (Hanson et al. 2004). Even for humans it is easy to identify a stand of Se hyperaccumulating plants in the field, based on the strong smell of volatile Se. Herbivores that can exclusively feed on high-Se plants may actually use the volatile Se as a cue to find their host – this has not been investigated for Se, but has been found for other herbivores that are specialist feeders on toxic plants. The fascinating arms race of plant defense versus disarmament may also impact ecological processes at higher trophic levels, which is discussed later in this chapter. In addition to plant herbivore interactions, elevated Se in plants may impact pollinators, as will be discussed next.

Like leaves, flowers can contain substantial Se levels, particularly in Se hyperaccumulators. Selenium levels up to 4000 mg Se per kg DW were found in the flower parts of the Se hyperaccumulator Prince's plume (Quinn et al. 2011b). This could negatively impact reproduction if floral Se deters or is toxic to pollinators, or if Se inhibits plant physiological processes involved with reproduction. In fact, when Indian mustard grown in a greenhouse and supplied with selenate accumulated over 500 mg Se per kg DW, pollen and seed germination decreased (Quinn et al. 2011b). This suggests a reproductive disadvantage when Se accumulator plants sequester Se to levels higher than 500 mg Se per kg DW. However, accumulators in the field rarely contain more than 100 mg Se per kg DW, making it unlikely Se impacts pollen and seed germination in the field. The Se hyperaccumulator Prince's plume, in contrast, did not have decreased pollen or seed germination rates, even at 4000 mg Se kg^{-1} DW (Quinn et al. 2011b). Within Prince's plume flowers, the pistils and anthers contained the highest levels of Se, with Se highly concentrated in the pollen and ovules, areas that are targeted by pollinators such as bees. Nectar of Prince's plume, which is collected by bees, also contained significant Se levels: 200 mg Se per kg DW. This amount is ecologically relevant because it has been shown to be an order of magnitude higher than those toxic to some insects (Hanson et al. 2003). The majority of Se in floral issues of Prince's plume was methyl-SeCys, the same form found in the plant's leaves (Quinn et al. 2011b). Indian mustard contained somewhat lower levels of Se in floral parts as compared to the rest of the plant. Interestingly, the form of Se in Indian mustard flowers was around two-thirds methyl-SeCys, while its leaves contained primarily selenate (Freeman et al. 2006a; Quinn et al. 2011b).

To help determine the ecological impacts of elevated levels of Se in plants on pollination and reproductive fitness, Indian mustard and Prince's plume containing high and low levels of Se were placed in the field and pollinator visits were monitored. Even when plants contained up to 4000 mg Se per kg DW, pollinator visitation was not influenced (Quinn et al. 2011b). Honey bees and bumble bees did not appear to be deterred by high-Se plants, and actively collected high-Se pollen and nectar (Quinn et al. 2011b). There remains much to study about how Se impacts pollination and pollinators. For example, if Se is accumulated by pollinators foraging on naturally occurring high-Se plants, this could have a positive or negative impact on pollinator health. If Se is ingested and is toxic to pollinators it could have a negative impact, but if Se is ingested and tolerated by pollinators, it may play a role in protecting these pollinators from predators or serve as a nutrient. A limited survey showed that honeybees collected on seleniferous sites in the Western United States, where honeybees are not native, contained around 20 mg Se per kg DW while bumble bees native to the seleniferous areas contained 250 mg Se per kg DW (Quinn et al. 2011b). More studies are needed to explain this difference and to investigate the effect of Se on pollinator health. Furthermore, if Se is accumulated in honey from Se-rich habitats then this honey could potentially be used as a commercial or dietary product as Se fortified food, to supplement Se-poor human diets. Honey collected from northern Colorado, a seleniferous area, contained, on average, 1 mg Se per kg DW, a level at which a few tablespoons per day correspond with the suggested daily dietary Se requirement of humans (Quinn et al. 2011b).

11.3 Effects of Plant Se Accumulation at Multiple Trophic Levels

As discussed above, some species of herbivores have evolved Se tolerance, feed on hyperaccumulators and accumulate high levels of Se, which may be toxic to predators and impact higher trophic levels. Indeed, a generalist predator, the spined soldier bug (*Podisus maculiventris*) grew slower and had higher mortality rates when fed Lepidopteran that had elevated Se concentrations (Vickerman and Trumble 2003). Thus, Se tolerant herbivores may enjoy similar protection from their accumulated Se from predators as Se hyperaccumulating plant species enjoy from herbivores. However, some predators may overcome such protection, as suggested by observations of the Se tolerant diamondback moth in the field hosting the parasitic wasp *Diadegma insulare* (Freeman et al. 2006b). Further investigation revealed that this wasp had similar Se concentrations to its host, the Se tolerant diamondback moth, and was able to tolerate such high concentrations of Se using the same mechanism as its host. *Diadegma insulare* accumulated Se in the form of methyl-SeCys, the same form found in the Se tolerant diamondback moth and Se hyperaccumulating plant species (Freeman et al. 2006b). Similarly, a Se tolerant moth that feeds on hyperaccumulator *A. bisulcatus* in its natural seleniferous habitat was found to be parasitized by a wasp (Valdez et al. 2012). These findings indicate that in seleniferous areas there is a gateway for Se to enter the local ecosystem through Se tolerant plant, herbivore and predator species. Little is known about the impact of Se on higher trophic levels and it would be interesting to conduct future studies.

11.4 Plant Se Hyperaccumulation Impacts Neighboring Plants

Another fascinating area of study that is receiving increasing attention in the literature is the ecological role of Se in plant-plant interactions. Since Se hyperaccumulators contain up to 100 times more Se than other plants growing on the same soils, it is possible that this Se impacts ecological interactions between plant species growing in close proximity.

The process by which plants concentrate high levels of certain elements in their surrounding soil, possibly as the result of deposition of litter, root exudation and root turn-over is called phytoenrichment (Morris et al. 2009). Since Se hyperaccumulators typically concentrate Se to around 1000-fold higher levels than those in the soil, and are perennials that shed their leaves annually, they may be expected to phytoenrich their surrounding soil. In support of this hypothesis, the Se concentration in soil surrounding hyperaccumulator plants such as *A. bisulcatus* and *S. pinnata* was 7- to 13-fold higher (up to 266 mg/kg) than Se in soil surrounding non-hyperaccumulator species *Medicago sativa* and *Helianthus pumilus* growing on the same site (El Mehdawi et al. 2011a, b). Also, Se hyperaccumulators were found to exude selenocompounds from their roots (El Mehdawi et al. 2012) and soil Se levels were enhanced when Se-rich hyperaccumulator leaf material was allowed to decompose for 12 month in a seleniferous habitat (Quinn et al. 2011a). If hyperaccumulators indeed enrich their surrounding soil with Se, and perhaps also convert the form of Se from inorganic to organic (which is taken much more readily by plants), this could impact Se concentration, growth and competition in neighboring plants. Indeed, tissue levels of Se were up to 20-fold higher in plant species *Artemisia ludoviciana* and *Symphyotrichum ericoides* growing within 1 m of Se hyperaccumulators than when growing more than 4 m away from these hyperaccumulators (El Mehdawi et al. 2011a, b) (Fig. 11.2). While it is plausible to assume that these enhanced Se levels in soil and plants surrounding hyperaccumulator plants are the result of phytoenrichment by the hyperaccumulator, it cannot be excluded that soil Se distribution is simply heterogeneous and that Se hyperaccumulators are more abundant in Se 'hot spots'.

The high Se levels in plants that neighbor Se hyperaccumulators may have a negative effect on their germination and growth, if they are Se-sensitive species. Indeed, when *Arabidopsis thaliana,* a Se sensitive species, was sown in soil collected adjacent to hyperaccumulator plants it showed significantly reduced germination and growth compared to when it was sown on soil collected from around non-hyperaccumulators (El Mehdawi et al. 2011a) (Fig. 11.3). This may be indicative of elemental allelopathy, a process by which plants concentrate toxic elements as a means to better compete with neighboring plants, and one of the hypothesized functions of hyperaccumulation (Boyd and Martens 1992). The allelopathy hypothesis was further supported by the finding that ground cover was about 10% less and species diversity slightly lower around Se hyperaccumulator species *A. bisulcatus* and *S. pinnata* than around non-accumulators (El Mehdawi et al. 2011a). Thus, it is possible that Se hyperaccumulator plant species benefit from their accumulated Se through decreased competition from surrounding vegetation. This may also influence the species composition in these plant communities, and with that, species composition at higher trophic levels. This will be interesting to further investigate.

As mentioned, Se sensitive plants germinate and grow slower on soils collected under the canopy of Se hyperaccumulators. This can create reduced competition, and may increase growth of Se tolerant plants in close proximity to hyperaccumulators. Indeed, the increased soil Se concentration around hyperaccumulators was shown to facilitate growth of Se-tolerant plant species through reduced herbivory and potentially enhanced physiological fitness (El Mehdawi et al. 2011b). *Artemisia ludoviciana* and *S. ericoides*, when growing in the sphere of influence of Se hyperaccumulators, thrived and grew bigger than individuals from the same species growing further away from hyperaccumulators despite, or perhaps even because of, their elevated Se levels (El Mehdawi et al. 2011b). *Artemisia ludoviciana* and *S. ericoides* growing less than a meter from hyperaccumulators were twofold larger, harbored fewer arthropods and showed less herbivory damage, as compared to plants of the same species growing further than four meters from hyperaccumulators. These Se-enriched neighbors of hyperaccumulator plant species, some of which contained over 1000 mg Se per kg DW, were used in controlled herbivory studies in comparison with their low-Se counterparts collected next to non-accumulators. In choice experiments, grasshoppers gathered from the same area preferred to feed on low-Se *A. ludoviciana* and *S. ericoides* plants collected farther away from hyperaccumulators, and when given no choice, the grasshoppers showed high Se accumulation in their tissues and increased mortality after feeding on the high-Se plants. Hence, Se phytoenrichment associated with growing next to Se hyperaccumulator plants appears to offer a competitive advantage Se-tolerant neighboring

Fig. 11.2 *Arabidopsis thaliana* growth is impaired on soil collected around hyperaccumulator species *S. pinnata* and *A. bisulcatus* (*left*) as compared to soil collected around non-hyperaccumulator species *Helianthus pumilus* and *Medicago sativa* (*right*)

Fig. 11.3 When growing next to Se hyperaccumulators plants (HA) in the field, *A. ludoviciana* and *S. ericoides* are bigger in size compared to when growing next to non-hyperaccumulators (non-HA) species (**a**, **c**). In addition, plants growing next to hyperaccumulators have higher concentrations of Se (data not shown) and are better protected against grasshopper herbivory, likely as a result of the elevated Se (**b**, **d**) (Adapted from El Mehdawi et al. 2015)

plants via reduced herbivory (El Mehdawi et al. 2011b). In addition to the ecological benefit, *A. ludoviciana* and *S. ericoides* plants may experience a physiological benefit from their hyperaccumulator mediated Se enrichment. Selenium has been shown to enhance growth for a variety of plant species (Pilon-Smits et al. 2009), and the growth of *S. ericoides* clearly responded favorably to selenate treatment in herbivore-free greenhouse experiments (El Mehdawi et al. 2015). Therefore, it appears that Se hyperaccumulator plant species facilitate their Se-tolerant neighbors *A. ludoviciana* and *S. ericoides* via enhanced Se concentration, which promotes their growth and protects them from herbivory.

11.5 Practical Implications and Research Gaps

The state of the knowledge of the ecology of Se in plants has important implications for how seleniferous habitats are managed and for the phytoremediation of Se polluted lands. For example, understanding the ecological roles of Se hyperaccumulators in seleniferous ecosystems could help land owners manage seleniferous habitats in the Western United States. Understanding the ecological implications of Se accumulation in plants may also help optimize phytoremediation, via cocropping or intercropping with Se hyperaccumulator species, and reducing herbivore losses.

The study of the ecological effects of Se in plants is still in its infancy. Recent studies investigating how elevated plant Se impacts higher trophic levels and plant community composition have laid the groundwork for future research. There is much more to learn about how Se impacts various trophic levels by following the concentration and speciation of Se throughout the food web. In addition, more research into how Se hyperaccumulating plants change the distribution of Se in soils and associated ecological consequences will help expand our knowledge. Ultimately, it will be useful to look ecosystem-wide, and understand better how seleniferous habitats differ ecologically from similar, low Se habitats, and how seleniferous habitats with and without Se hyperaccumulators compare. Furthermore, the ecological impacts of Se accumulation in plants likely are similar to the impacts of other accumulated toxic elements, such as arsenic, nickel, lead and zinc. Through continued interdisciplinary approaches, the field of plant elemental ecology is expected to continue to expand and provide increasingly better insight into these fascinating and likely impactful ecological effects.

References

Baker AJM, McGrath SP, Reeves RD, Smith JAC (2000) Metal hyperaccumulator plants: a review of the ecology and physiology of a biological resource for phytoremediation of metal polluted soils. In: Terry N, Banuelos GS (eds) Phytoremediation of contaminated soil and water. CRC Press, Boca Raton, pp 85–107

Banuelos GS, Vickerman DB, Trumble JT, Shannon MC, Davis CD, Finley JW, Mayland HF (2002) Biotransfer possibilities of selenium from plants used in phytoremediation. Int J Phytoremediat 4:315–329

Beath OA, Gilbert CS, Eppson HF (1939) The use of indicator plants in locating seleniferous soils in the Western United States. I. General. Am J Bot 26:257–269

Boyd RS, Martens SN (1992) The raison d'e^tre for metal hyperaccumulation by plants. In: AJM B, Proctor J, Reeves RD (eds) The vegetation of ultramafic (serpentine) soils. Intercept, Andover, pp 279–289

Cappa JJ, Pilon-Smits EAH (2014) Evolutionary aspects of elemental hyperaccumulation. Planta 239:267–275

El Mehdawi AF, Pilon-Smits EAH (2012) Ecological aspects of plant selenium hyperaccumulation. Plant Biol 14:1–10

El Mehdawi AF, Quinn CF, Pilon-Smits EAH (2011a) Effects of selenium hyperaccumulation on plant–plant interactions: evidence for elemental allelopathy. New Phytol 191:120–131

El Mehdawi AF, Quinn CF, Pilon-Smits EAH (2011b) Selenium hyperaccumulators facilitate selenium-tolerant neighbors via phytoenrichment and reduced herbivory. Curr Biol 21:1440–1449

El Mehdawi AF, Cappa JJ, Fakra SC, Self J, Pilon-Smits EAH (2012) Interactions of selenium and non-accumulators during co-cultivation on seleniferous or non-seleniferous soil – the importance of having good neighbors. New Phytol 194:264–277

El Mehdawi AF, Paschke M, Pilon-Smits EAH (2015) *Symphyotrichum ericoides* populations from seleniferous and non- seleniferous soil display striking variation in selenium accumulation. New Phytol 206:231–242

Freeman JL, Zhang LH, Marcus MA, Fakra S, Pilon-Smits EAH (2006a) Spatial imaging, speciation and quantification of selenium in the hyperaccumulator plants *Astragalus bisulcatus*and *Stanleya pinnata*. Plant Physiol 142:124–134

Freeman JL, Quinn CF, Marcus MA, Fakra S, Pilon-Smits EAH (2006b) Selenium-tolerant diamondback moth disarms hyperaccumulator plant defense. Curr Biol 16:2181–2192

Freeman JL, Lindblom SD, Quinn CF, Fakra S, Marcus MA, Pilon-Smits EAH (2007) Selenium accumulation protects plants from herbivory by Orthoptera via toxicity and deterrence. New Phytol 175:490–500

Freeman JL, Quinn CF, Lindblom SD, Klamper EM, Pilon-Smits EAH (2009) Selenium protects the hyperaccumulator Stanleya pinnata against black-tailed prairie dog herbivory in native seleniferous habitats. Am J Bot 96:1075–1085

Freeman JL, Tamaoki M, Stushnoff C, Quinn CF, Cappa JJ, Devonshire J, Fakra S, Marcus MA, McGrath S, Van Hoewyk D, Pilon-Smits EAH (2010) Molecular mechanisms of selenium tolerance and hyperaccumulation in *Stanleya pinnata*. Plant Physiol 153:1630–1652

Freeman JL, Marcus MA, Fakra SC, Devonshire J, McGrath SP, Pilon-Smits EAH (2012) Seeds of selenium hyperaccumulators *Stanleya pinnata and Astragalus bisulcatus* are colonized by Se-resistant, Se-excluding wasp and beetle herbivores. PLoS ONE 7(12) e50516

Galeas ML, Zhang LH, Freeman JL, Wegner M, Pilon-Smits EAH (2007) Seasonal fluctuations of selenium and sulfur accumulation in selenium hyperaccumulators and related non-accumulators. New Phytol 173:517–525

Galeas ML, Klamper EM, Bennett LE, Freeman JL, Kondratieff BC, Pilon-Smits EAH (2008) Selenium hyperaccumulation affects plant arthropod load in the field. New Phytol 177:715–724

Hanson B, Garifullina GF, Lindblom SD, Wangeline A, Ackley A, Kramer K, Norton AP, Lawrence CB, Pilon Smits EAH (2003) Selenium accumulation protects Brassica juncea from invertebrate herbivory and fungal infection. New Phytol 159:461–469

Hanson B, Lindblom SD, Loeffler ML, Pilon-Smits EAH (2004) Selenium protects plants from phloem feeding aphids due to both deterrence and toxicity. New Phytol 162:655–662

Kabata-Pendias A (1998) Geochemistry of selenium. J Environ Pathol Toxicol Oncol 17:173–177

Morris C, Grossl PR, Call CA (2009) Elemental allelopathy: processes, progress, and pitfalls. Plant Ecol 202:1–11

Pilon-Smits EAH, Quinn CF, Tapken W, Malagoli M, Schiavon M (2009) Physiological functions of beneficial elements. Curr Opin Plant Biol 12:267–274

Pollard AJ, Reeves RD, Baker AJ (2014) Facultative hyperaccumulation of heavy metals and metalloids. Plant Sci 217–218:8–17

Prins CN, Hantzis LJ, Quinn CF, Pilon-Smits EAH (2011) Effects of Selenium accumulation on reproductive functions in Brassica juncea and Stanleya pinnata. J Exp Bot 62:5233–5239

Quinn CF, Freeman JL, Galeas ML, Klamper EM, Pilon-Smits EAH (2008) The role of selenium in protecting plants against prairie dog herbivory: implications for the evolution of selenium hyperaccumulation. Oecologia 155:267–275

Quinn CF, Freeman JL, Reynolds RJB, Cappa JJ, Fakra SC, Marcus MA, Lindblom SD, Quinn EK, Bennett LE, Pilon-Smits EAH (2010) Selenium hyperaccumulation offers protection from cell disruptor herbivores. Plant Physiol 153:1630–1652

Quinn CF, Wyant K, Wangeline AL, Shulman J, Galeas ML, Valdez JR, Paschke MW, Pilon-Smits EAH (2011a) Enhanced decomposition of selenium hyperaccumulator litter in a seleniferous habitat – evidence for specialist decomposers. Plant and Soil 341:51–61

Quinn CF, Prins CN, Gross AM, Hantzis L, Reynolds RJB, Freeman JL, Yang SI, Covy PA, Bañuelos GS, Pickering IJ, Fakra SF, Marcus MA, Arathi HS, Pilon-Smits EAH (2011b) Selenium accumulation in flowers and its effects on pollination. New Phytol 192:727–737

Stadtman TC (1990) Selenium biochemistry. Annu Rev Biochem 59:111–127

Van Hoewyk D. 2013. A tale of two toxicities: malformed selenoproteins and oxidative stress both contribute to selenium stress in plants. Ann Bot 112:965–972

Vickerman DB, Trumble JT (2003) Biotransfer of Selenium: effects on an insect predator, *Podisus maculiventris*. Ecotoxicology 12:497–504

Vickerman DB, Shannon MC, Banuelos GS, Grieve CM, Trumble JT (2002) Evaluation of *Atriplex* lines for selenium accumulation, salt tolerance and suitability for a key agricultural insect pest. Environ Pollut 120:463–473

Valdez Barillas JR, Quinn CF, Freeman JL, Lindblom SD, Marcus MS, Fakra SC, Gilligan TM, Alford ER, Wangeline AL, Pilon-Smits EAH (2012) Selenium distribution and speciation in hyperaccumulator *Astragalus bisulcatus* and associated ecological partners. Plant Physiol 159:1834–1844

Wilber CG (1980) Toxicology of selenium: a review. Clin Toxicol 17:171–230

Chapter 12
Evolutionary Aspects of Plant Selenium Accumulation

R. Jason B. Reynolds, Jennifer J. Cappa, and Elizabeth A.H. Pilon-Smits

Abstract Essential selenium (Se) metabolism, as found in some photosynthetic cyanobacteria and algae, appears to have been lost in plants. Although not essential, Se is readily taken up by plants due to its similarity to sulfur (S), and typically plant accumulation of Se parallels that of S. In contrast, some plant species appear to preferentially take up Se over S, translocate and sequester Se and S independently, and accumulate Se to levels above 0.1% of dry matter. This so-called Se hyperaccumulation trait occurs in different plant lineages and likely has evolved independently multiple times. The variation in plant Se accumulation, particularly the phenomenon of hyperaccumulation, leads to some intriguing evolutionary questions: What may be the physiological and ecological benefits and constraints of Se hyperaccumulation? What sequence of events led to Se hyperaccumulation? Did tolerance and accumulation evolve simultaneously or sequentially, and what were the physiological, biochemical and genetic steps involved? These questions are explored in this chapter.

Keywords Selenium • Evolution • Hyperaccumulation • *Stanleya* • *Astragalus*

12.1 Variation in Plant Selenium Accumulation – Evolutionary Questions

Selenium (Se) is not uniformly distributed in the earth's crust: soil Se status can range from almost no Se to as much as 100 mg/kg, with 0.05 mg/kg as the estimated world average (Kabata-Pendias et al. 2001). Soils with elevated Se levels (> 0.1 mg/kg) are known as seleniferous, because they harbor vegetation with tissue Se concentrations that are potentially toxic for consumers (Dhillon and Dhillon 2001). Some plants native to seleniferous soils even concentrate Se upwards of 1000 mg/kg dry weight (DW), and are known as Se hyperaccumulators (Boyd and Martens

R.J.B. Reynolds • J.J. Cappa • E.A.H. Pilon-Smits (✉)
Biology Department, Colorado State University, Fort Collins, CO, USA
e-mail: Jason.Reynolds@colostate.edu; jjlarson@rams.colostate.edu; epsmits@colostate.edu

E.A.H. Pilon-Smits et al. (eds.), *Selenium in plants*, Plant Ecophysiology 11,
DOI 10.1007/978-3-319-56249-0_12

1992). Many of these seleniferous soils likely derive the high Se content from parent sedimentary rock originating in the cretaceous period when atmospheric Se from volcanic emissions was deposited in seas by rainfall, becoming a component of the sediment (Kabata-Pendias et al. 2001). Other soils are thought to derive their Se mainly from deposition and precipitation (Winkel et al. 2015). The predominant form of bioavailable Se in soils is SeO_4^{2-} (selenate), but Se can also be found as SeO_3^{2-} (selenite) in anoxic soils (Zhang et al. 2014). Selenium accumulation in plants is directly connected to the way in which plants take up, translocate and assimilate sulfur (S). This connection results from the chemical similarity these elements share as part of group 16 elements (chalcogens). Sulfur has long been known to be an essential macronutrient and is primarily taken up from soils as sulfate (SO_4^{2-}). Analogous with sulfate, selenate enters the plant via sulfate transporters, and is subsequently translocated and assimilated into seleno-amino acids via sulfate assimilation enzymes, and may also enter proteins and other S compounds (Terry et al. 2000).

Plant Se accumulation from soils high in bioavailable Se may be subject to positive or negative selection pressures. At low tissue concentration, Se can offer plants physiological benefits, and at higher tissue levels Se also offers ecological benefits (El Mehdawi and Pilon-Smits 2012). However, when Se is accumulated above a critical toxicity level it causes plant toxicity due to oxidative damage and to protein malfunction if Se displaces S in proteins (Van Hoewyk 2013).

Based on what is known about plant mechanisms of Se accumulation and toxicity, plant Se hyperaccumulation and hypertolerance may be hypothesized to involve increased expression of S transporters, S enzymes that convert Se to less toxic forms, and antioxidant enzymes. Apart from differences in expression level, there may also be differences in kinetic properties of S transporters or enzymes. Because of the chemical similarities between Se and S, most plant species cannot differentiate between the two elements, and their tissue Se concentration and Se/S ratio directly reflect that of the soil. Hyperaccumulator species, however, appear to disproportionally accumulate Se over S: their Se/S ratio is higher than that in their growth substrate (White 2016). Apparently, there is a selection pressure that favors the excessive accumulation of this toxic, nonessential element.

There have been type designations for different angiosperm species, based on the differences in their accumulation and tolerance of Se in natural environments. The early designations (Rosenfeld and Beath 1964; Brown and Shrift 1982), while useful, have been altered in later years to attempt to categorize angiosperms based on current understanding. Plant species that accumulate and tolerate Se at relatively low levels (below 100 mg/kg DW) when growing on seleniferous soil are called non-accumulators; these plants also typically do not tolerate higher tissue Se levels, and some may not be able to grow on seleniferous soils (Brown and Shrift 1982; Sors et al. 2005a, b). Other species accumulate and tolerate Se to levels between 100 and 1000 mg/kg DW and are known as Se accumulators or secondary accumulators (Terry et al. 2000). Finally, there are the Se hyperaccumulators (Boyd and Martens 1992; Salt et al. 1998), a phylogenetically diverse group of around 50 species that accumulate Se to tissue levels upwards of 1000 mg/kg DW, some as high as 15,000 mg/kg DW (Knight and Beath 1937). Selenium hyperaccumulators differ from other plant species in several respects. As mentioned, they have higher overall levels of Se, preferentially take up selenate over sulfate, and also convert inorganic Se to non-toxic non-protein amino acids, which are sequestered in specific tissues and organs, particularly in the pollen and ovules of reproductive organs and in the epidermis of young leaves (Freeman et al. 2006a).

Variation in Se tolerance and accumulation has been found not only between species, but also between populations and individuals within populations (Feist and Parker 2001; Zhang et al. 2006; El Mehdawi et al. 2015). This genetic variation is the basis upon which natural selection may act, to either drive or constrain the evolution of enhanced Se accumulation and Se tolerance. The observed variation in plant Se accumulation, especially the phenomenon of hyperaccumulation, leads to several intriguing evolutionary questions. Is Se hyperaccumulation a primitive or derived trait within the plant kingdom and within families and genera? Has the trait ever been lost after first evolving? Is hyperaccumulation a monophyletic or polyphyletic trait in the plant kingdom and within families and genera? What are the physiological and ecological benefits of Se hyperaccumulation, and are there any physiological or ecological constraints? How did hyperaccumulation evolve? Was it gradual, from non-accumulators via accumulators to hyperaccumulators? Did tolerance and accumulation evolve simultaneously or subsequently, and in what order? How do hyperaccumulators differ from other plants in terms of physiological and biochemical processes? How did these differences come about at the DNA level? These are questions addressed in this chapter.

12.2 Phylogenetic and Geographic Patterns of Se Hyperaccumulation

The phylogenetic distribution of Se hyperaccumulation may give insight into whether it is a primitive or derived trait, and whether it is monophyletic or polyphyletic. Selenium (Se) hyperaccumulation has been documented in 14 genera and 45 species (White 2016). These 14 genera are found in six eudicot orders; Fabales, Brassicales, Caryophyllales, Gentianales, Lamiales and Asterales (White 2016). The Asterales have the largest number of genera that hyperaccumulate Se, while the Fabales contain the most hyperaccumulator species, primarily because of the many Se hyperaccumulating *Astragalus* species (Cappa and Pilon-Smits 2014). The phylogenetic distribution of Se hyperaccumulation across six orders indicates that Se hyperaccumulation is a derived, polyphyletic trait that evolved convergently within at least six eudicot clades. Convergent evolution of Se hyperaccumulation is in some cases also evident at the family level: the Se hyperaccumulators in the Asterales are in the same tribe, Astereae, but there are multiple origins of Se hyperaccumulation within the Astereae (Fig. 12.1). Similar convergent patterns of

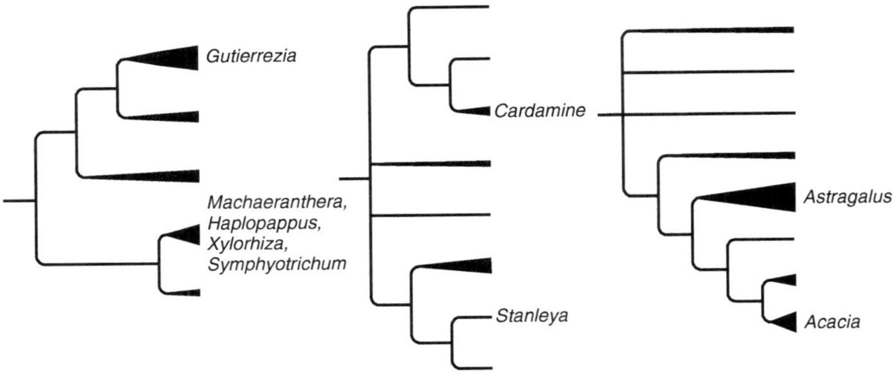

Fig. 12.1 Phylogenies of Astereae, Brassicaceae and Fabaceae showing selenium hyperaccumulator distribution. Branch widths represent relative number of taxa within each tree (not comparable across trees). (**a**) Tribe Astereae based off Urbatsch et al. 2003. Grindelia (not shown) is also a member of tribe Astereae. (**b**) Brassicaceae tribal level phylogeny based off BrassiBase (Koch et al. 2012; Kiefer et al. 2014). (**c**) Fabaceae subfamily phylogeny based off Angiosperm Phylogeny Group III (Stevens 2001)

evolution in angiosperms have been found for hyperaccumulation of other elements (Cappa and Pilon-Smits 2014).

If soil Se concentration is assumed to be the main driver for the evolution of hyperaccumulation, then most hyperaccumulators of Se may be expected to occur on seleniferous soil. Soil Se concentration in the United States has a very heterogeneous distribution with the highest levels occurring in the western half. According to a review of the literature by White (2016), the vast majority of reports of hyperaccumulators are indeed from the western United States.

In other parts of the world there are few reports of Se hyperaccumulators, even in areas where soil Se levels are high (e.g. parts of China). Of course, this may be due to a lack of study on hyperaccumulators, not an actual lack of hyperaccumulators in these areas. It may also be that some seleniferous areas are of more recent origin than others, so that less time has gone by during which Se hyperaccumulation could evolve. In the cases where hyperaccumulators have been found, they have been found where there is Se in the soil. Peterson and Butler (1962, 1971) reported two species that hyperaccumulate Se, they are *Neptunia amplexicaulis* found in north-western Queensland and *Morinda reticulata* found in Cape York, both in Australia. Furthermore, there has been reported a secondary Se accumulator species named *Cardamine hupingshanesis* found on seleniferous mine-tailings in western Hubei province in Enshi, China (Yuan et al. 2013).

12.3 Physiological and Ecological Benefits and Constraints of Se Hyperaccumulation

The possible abiotic and biotic selection pressures leading to hyperaccumulation of toxic elements were discussed by Boyd and Martens (1992), who proposed ways in which elemental hyperaccumulation may benefit a plant, and thus the selective pressures by which hyperaccumulation may evolve. Four of these hypotheses are relevant to the evolution of Se hyperaccumulation and will be discussed in the next paragraphs.

First, the *tolerance/disposal hypothesis* states that the sequestration of toxic elements like Se keeps it from interfering with essential processes, and the subsequent disposal by shedding of the tissues containing the element, confers tolerance. There is currently little evidence that hyperaccumulation has evolved as a means of tolerance/disposal. Assuming that the evolution toward hyperaccumulation started from a non-accumulator, there would need to be variation in tolerance in such a population, and this tolerance would have to be positively correlated with accumulation. Selenium tolerance and accumulation were not found to be correlated in the non-accumulator *Arabidopsis thaliana* (Zhang et al. 2006). Similarly, a study on the zinc hyperaccumulator *Arabidopsis halleri* indicated that tolerance and hyperaccumulation are genetically independent traits (Macnair et al. 1999).

Second, *drought tolerance* is another hypothesized benefit for elemental hyperaccumulation. This has not been well studied in Se hyperaccumulators, but there is some evidence that at low levels Se can protect plants (other than hyperaccumulators) from drought stress (Yao et al. 2009; Hasanuzzaman and Fujita 2011; Hasanuzzaman et al. 2011; Nawaz et al. 2015). This protection seems linked to increased overall antioxidant capacity, as well as increased production of chemicals that aid osmoregulation in plants. Specifically, Se treatment under drought stress was found to be associated with an increase in ascorbic acid and reduced glutathione content, and an associated reduction in malondialdehyde levels (Hasanuzzaman and Fujita 2011). Other drought stress studies reported Se-induced increases in peroxidase and catalase activities and higher levels of osmoprotectants including total soluble sugars, total free amino acids and total soluble proteins (Yao et al. 2009; Nawaz et al. 2015). Selenium treatment was also reported to result in increased biomass under drought conditions (Yao et al. 2009; Han et al. 2013). Any drought protection conferred by Se has a narrow range of effectiveness. At a soil Se concentration of 3 mg/kg as sodium selenite an opposite effect was found: wheat (*Triticum aestivum*) showed decreased levels of antioxidants (except glutathione and superoxide dismutase), decreased concentrations of osmoprotectants and reduced shoot biomass (Yao et al. 2009). In a study with tobacco a similar negative effect was found after treatment with 11.1 mg/kg sodium selenite (Han et al. 2013). For more information on Se and drought stress, see Ahmad et al. (2016) for a recent review. The effect of Se on drought stress resistance in Se hyperaccumulators is still unknown, and will be an interesting area of future research. Based on current information, any benefits conferred by Se in drought conditions seem unlikely to have been solely responsible for the evolution of Se hyperaccumulation. Beneficial plant tissue Se concentrations are very low (4–10 mg/kg DW), and the benefit of Se is lost at plant Se concentrations that are much lower compared to those in Se hyperaccumulators. However, it is possible that synergistic effects, combining the benefits of Se under drought conditions with those under other environmental stresses (as discussed in the next paragraphs) have collectively driven the evolution of Se hyperaccumulation.

Another hypothesized benefit of elemental hyperaccumulation is elemental *allelopathy* (Boyd and Martens 1998). In regard to Se, this would involve increased concentrations of Se in the soil near the hyperaccumulator, deposited there through turnover of hyperaccumulator shoot and root tissues or Se exudation by roots. Indeed, there is evidence that Se hyperaccumulators increase soil Se concentration around them, although it is generally difficult to distinguish whether a hyperaccumulator colonized a pre-existing high-Se spot, or rather created the Se hot spot itself by concentrating Se in their associated soil over time. After a 12-month litter bag study, the soil below hyperaccumulator litter contained approximately twice the Se concentration of soil below non-accumulator litter (Quinn et al. 2011b). While the Se levels were not very high, over time this process may increase soil Se concentrations near hyperaccumulators to toxic levels. Indeed, on a seleniferous field site, soils around Se hyperaccumulators *Astragalus bisulcatus* and *Stanleya pinnata* contained 7–11 fold higher concentrations of Se relative to soils around three non-hyperaccumulator species growing in the same area (El Mehdawi et al. 2011). These increased soil Se concentrations may result in inhibition of germination and growth of Se-sensitive neighboring plants, while conferring a competitive advantage to Se-resistant plants (El Mehdawi et al. 2011). The higher soil Se, as a negative allelopathic factor, may also explain the reduction in vegetative cover found around Se hyperaccumulator plants as compared to non-accumulator plants (El Mehdawi et al. 2011). In the same study, El Mehdawi et al. (2011) observed for two plant species up to 20-fold increased leaf Se concentration when growing near Se hyperaccumulators, as compared to further away from Se hyperaccumulators. Interestingly, the high-Se individuals growing next to hyperaccumulators were bigger in size and displayed less herbivory damage. Thus, hyperaccumulators not only are associated with higher surrounding soil Se levels, but also with higher Se levels in the surrounding vegetation. The soil Se around hyperaccumulator plants is likely in large part organic Se (El Mehdawi et al. 2015), a reflection of the high organic Se fraction in hyperaccumulators, but not non-accumulators (Freeman et al. 2006b). Organic Se is taken up more readily by vegetation than inorganic Se (Zayed et al. 1998). The observed increased soil Se concentrations associated with Se hyperaccumulators and the respective negative and positive growth responses of Se-sensitive and Se-resistant plant species when growing near Se hyperaccumulators give insight into the selective pressures exerted by Se in soils associated with hyperaccumulators. Likely, the increase in soil Se concentration associated with the presence of Se hyperaccumulator species can create special niches in which Se-resistant species may out-compete their Se-sensitive neighbors. This apparent environmental alteration brought about by Se hyperaccumulators, in combination with the variability in

Se tolerance in neighbors of Se hyperaccumulators may give rise to differences in plant species composition. This will be an interesting topic of future study.

The fourth hypothesized benefit for elemental hyperaccumulation is *elemental defense* (or more recently called the "defensive enhancement hypothesis" (Boyd 2012). For Se, this hypothesis predicts that a plant is protected from biotic stress as the concentration of Se in plant tissues increases to a point where it becomes toxic to herbivores or pathogens. Prior to Boyd and Martens' review (1992) it had already been suggested that by hyperaccumulating Ni, plants may be defended from insects and fungal pathogens (Reeves et al. 1981). Since then, several different hyperaccumulated elements have been shown to be able to protect plants from herbivory and/or pathogens (Boyd 2012). There is also abundant support for a protective effect of Se against biotic attack, both by means of deterrence and toxicity. A field survey and a variety of choice and non-choice feeding studies have shown protection by Se of both Se hyperaccumulators as well as non-hyperaccumulators from different types of herbivores and some fungal pathogens (Quinn et al. 2008, 2010, 2011a, b; Freeman et al. 2007, 2009; Galeas et al. 2008; El Mehdawi and Pilon-Smits 2012; Wu et al. 2015; Hanson et al. 2003, 2004; Vickerman and Trumble 1999; Trumble et al. 1998). The levels at which Se protects a plant have been found to be much lower than what a Se hyperaccumulator can tolerate and concentrate in its tissues (Hanson et al. 2004; Trumble et al. 1998). In fact, 2 mg/kg DW in leaf tissue was already toxic to aphids, and 10 mg/kg DW was an effective deterrent to aphids when given a choice of plants treated with or without Se (Hanson et al. 2004). The LC_{50} ("lethal concentration" required to kill 50% of subjects) for an army worm varied with the form of Se: for selenate and selenomethionine it was below 50 mg/kg DW, while it was below 20 mg/kg DW for selenite and selenocysteine (Trumble et al. 1998). When the effective concentrations of Se were supplied in pure chemical form to herbivores and pathogens, they typically result in comparable toxicity (Hanson et al. 2003). Nevertheless, it is possible that an additional protective effect of Se is derived from Se-induced upregulation of plant defense pathways, as observed in non-accumulator *A. thaliana* and Se hyperaccumulator *S. pinnata* (Van Hoewyk et al. 2008; Freeman et al. 2010).

Judged from the abundant evidence for a protective effect of Se against biotic stress, protection from herbivores and pathogens is likely to be an important driving force for the evolution of Se hyperaccumulation. The finding that the protective effect of Se starts at much lower tissue levels than the Se concentrations found in many Se hyperaccumulators raises the question: why do Se hyperaccumulators accumulate such high Se concentrations when low levels already offer protection? The Defensive Enhancement Hypothesis (Boyd 2012) suggests that, after an initial defensive benefit accrued from a relatively low initial concentration, increased concentration of an element provides enhanced plant fitness and drives the evolution of higher elemental concentrations until hyperaccumulation is achieved. Indeed, the results from plant-herbivore studies have shown that different herbivores show different Se sensitivity, indicating enhanced plant fitness with increasing Se accumulation capacity. Even at hyperaccumulator concentrations as high as 10,000 mg/kg DW, some Se-resistant herbivores and pathogens are able to occupy the plants

(Freeman et al. 2006b, 2012; Valdez Barillas et al. 2012). Some of these species may be specialists that exclusively feed on hyperaccumulators. Thus, it is feasible that hyperaccumulators are co-evolving with some of their herbivores or pathogens in an evolutionary arms race, where the plant is driven to evolve increasingly high Se hyperaccumulation, and the herbivore or pathogen is driven to increasingly high Se resistance. Likely, similar processes occur for other elemental hyperaccumulators. Thus, the evolution of Se hyperaccumulation may serve as a model for plant adaptation to metalliferous soils, in which the plant evolves mechanisms to accumulate and tolerate toxic elements and then optimizes the ecological advantages of the toxic elements they accumulate, meanwhile coevolving with ecological partners.

In order for Se hyperaccumulation to evolve, there has to be a selective advantage without too much of a cost to fitness. One potential constraint for plants that accumulate toxins is compromised symbiosis with mutualist partners, particularly pollinators. Given that *S. pinnata* preferentially sequesters Se in its reproductive organs, Quinn et al. 2011a tested *S. pinnata* flowers by x-ray microprobe analysis and found Se to be localized in the pollen and ovules, mainly as a C-Se-C forms, whereas Se was diffusely distributed throughout the flower of *B. juncea* and was found to be a mix of chemical Se species. When the seeds of all species of *Stanleya* were tested, it was found that regardless of *Stanleya* species the chemical form was C-Se-C, the Se was found in the embryo, absent from the endosperm, with the only difference between species being that *S. pinnata* had Se in the seed coat (Cappa et al. 2015). Prins et al. (2011) tested *B. juncea* and *S. pinnata* grown with and without Se for possible reproductive costs associated with Se accumulation. It was found that *B. juncea* plants upon exceeding tissue Se levels of 1000 mg/kg DW exhibited a decrease in overall reproductive functions including seed production, individual seed weight, and germination. In contrast, in *S. pinnata* there was no decrease in the same fitness measurements. Additionally, Quinn et al. 2011a, b found that at similar tissue Se concentrations *B. juncea* had a decrease in pollen germination, while *S. pinnata* showed no decrease. Another potential fitness cost of Se accumulation would be if pollinators would be deterred by high-Se flowers. However, this is not the case: pollinators showed no visitation preference between high- and low-Se *S. pinnata* or *B. juncea* individuals, and readily collected high-Se pollen and nectar (Quinn et al. 2011a, b). More research is needed to assess the effect of plant Se on the health of the pollinators, which may be positive or negative.

Another potential constraint of hyperaccumulation is reduced growth, if the hyperaccumulation processes are energetically costly. Hyperaccumulators do tend to grow more slowly than crop relatives, but this is not necessarily associated with the hyperaccumulation trait, and could also be a result of e.g. drought adaptation. More research is needed to address this issue.

12.4 Mechanisms of Se Hyperaccumulation

12.4.1 Competition Between Selenium and Sulfur

Selenium is atomically similar to sulfur (S) and therefore utilizes the S pathway for uptake into the plant and assimilation into organic molecules. In selenate:sulfate competition experiments, the related Brassicaceae *B. juncea* (non-hyperaccumulator) and *S. pinnata* (Se hyperaccumulator) showed remarkably different Se:S interactions. While 5 mM sulfate supply almost completely abolished selenate uptake (from 20 μM) in the non-hyperaccumulator, it had relatively little effect on selenate uptake in the hyperaccumulator (Harris et al. 2013; Schiavon et al. 2015). Furthermore, the Se/S ratio in the tissues in *B. juncea* largely reflected the Se/S ratio in the media, while *S. pinnata* was found to enrich itself with Se relative to S, as compared to the media (Schiavon et al. 2015). A similar result was found in Se hyperaccumulator *Astragalus racemosus*, which enriched itself with Se, as evidenced from a higher Se/S ratio compared to the supplied media (DeTar et al. 2015).

The key to the apparent ability of Se hyperaccumulators to discriminate between selenate and sulfate and preferentially take up selenate, may be a mutation in a sulfate/selenate transporter. The sulfate transporter family (SULTR) is a large ubiquitous family of transport proteins. In *Arabidopsis thaliana* the two high-affinity sulfate transporters *SULTR1;1* and *SULTR1;2* mediate sulfate uptake into the root, SULTR1;2 being the predominant transporter (Barberon et al. 2008). Schiavon et al. (2015) found *S. pinnata* to have much higher transcript levels of a *Sultr1;2*-like gene, relative to *B. juncea*. This is in agreement with an earlier macroarray study comparing transcript abundance between *S. pinnata* and *S. albescens* (a secondary accumulator) by Freeman et al. (2010), which reported *S. pinnata* to have constitutively elevated transcript abundance for *Sultr1;2*. Similarly, when Se hyperaccumulators *A. bisculatus* and *A. racemosus* were compared to non-hyperaccumulator *Astragalus* species, it was shown that under regular S status the hyperaccumulators had elevated transcript levels for sulfate transporters (Cabannes et al. 2011). The increased *SULTR1;2* transcript abundance in these hyperaccumulator species may be caused by differential regulation (e.g. higher expression of a transcription factor), by mutations in the promoter region of the *Sultr1;2* gene, by gene duplication events leading to increased gene copy number, or higher transcript stability. In either case, the increased transcript levels would be expected to be associated with higher transporter protein abundance and uptake capacity. In addition, mutations in the coding region may have given rise to altered substrate specificity, favoring selenate over sulfate. Given that Se hyperaccumulation is a convergent trait among eudicots, the underlying molecular mechanisms may differ and be lineage-specific. For other hyperaccumulated elements, an increase in copy number has been the most reported cause of increased transporter abundance (Hanikenne et al. 2008; Ueno et al. 2011; Craciun et al. 2012).

12.4.2 Assimilation of Selenate to Non-protein Organic Forms

Selenium is assimilated via the sulfate assimilation pathway. Because of this, many plant Se studies have examined gene expression of key enzymes involved in sulfate assimilation (Van Hoewyk et al. 2008; Sors et al. 2009; Freeman et al. 2010; Cakr et al. 2015). The first enzyme in the sulfate assimilation pathway is ATP-sulfurylase (ATPS). ATPS was found to be constitutively upregulated in the root of *S. pinnata* compared to *S. albescens* or *B. juncea*, which may explain the capacity of *S. pinnata* to more efficiently convert selenate to organic Se (Freeman et al. 2010; Schiavon et al. 2015). Indeed, ATPS may be a rate-limiting step for selenate assimilation, since overexpression of ATPS in *B. juncea* resulted in enhanced selenate reduction, as well as enhanced Se accumulation (Pilon-Smits et al. 1999). In contrast, ATPS was not found to be correlated with Se hyperaccumulator activity in *Astragalus* species when overexpressed in shoot tissue (Sors et al. 2005a). Intriguingly, an *ATPS2* homologue was found to be constitutively upregulated in the roots of *S. pinnata* (Schiavon et al. 2015); this isoform is the only cytosolic ATPS in *A. thaliana*, where it is targeted to both the cytosol and plastids (Bohrer et al. 2015). The finding that the only cytosolic form of ATPS is strongly upregulated in the hyperaccumulator is particularly interesting given that the current paradigm of plant sulfate assimilation is that it occurs in the shoot plastids, but for Se hyperaccumulator *S. pinnata* a significant portion was found to be assimilated in the roots and transported to the shoots as methyl-selenocysteine (Freeman 2006). This form was also found to be the exclusive Se species in the roots of *S. pinnata* and *A. bisulcatus*, adding further evidence for the reduction of Se in the roots of hyperaccumulators (Lindblom et al. 2013).

The enzyme APS reductase (APR) is responsible for the reduction of activated sulfate (adenosine phosphosulfate) to sulfite. Freeman et al. (2010) found that *APR1, APR2* and *APR3* all had constitutively higher transcript abundance in *S. pinnata* as compared to *S. albescens; APR2* transcript abundance in the shoots was induced by Se treatment and *APR2* and *APR3* were induced in the roots. Together, these results indicate that the evolution of hyperaccumulation involves constitutive upregulation of sulfate assimilation pathway genes. The underlying molecular mechanisms involved may be the same as described above for *Sultr1;2* expression, and all genes may even be regulated by a common transcription factor. In this context it is interesting to note that the transcription factor WRKY, involved in several biotic and abiotic stresses including the abiotic stresses heat, drought, cold, osmotic pressure and salt (Wang et al. 2016) has been found to be induced with Se in several species. In *Astragalus chrysochlorus* (a secondary Se accumulator) it was found that the WRKY transcription factor was upregulated with Se treatment in a transcriptome study (Cakir et al. 2015). WRKY was also found to be upregulated when *A. thaliana* was treated with selenate (Van Hoewyk et al. 2008). A specific transcription factor induced by Se uptake could be one of the "master switches" that leads to Se tolerance and eventually Se hyperaccumulation. Several studies also point to a role for defense- and stress hormones jasmonic acid (JA) and ethylene in Se responses and induction of genes involved in sulfate/selenate uptake and assimilation (Van Hoewyk et al. 2008; Freeman et al. 2010). More elaborate genomic studies are needed to fully unravel the connections between these processes.

12.4.3 Selenium Localization and Speciation, and Plant Fitness

A major question researchers have addressed is: where in plant tissues does Se reside and in which chemical forms? In a field survey, Galeas et al. (2007) found that Se hyperaccumulators *S. pinnata* and *A. bisculatus* translocated Se and S differently, and had different seasonal Se fluctuation patterns than related non-hyperaccumulator species. *Stanleya pinnata* and *A. bisculatus* remobilized Se from the leaves to the reproductive tissues in the summer, resulting in lower leaf tissue Se concentrations while non-hyperaccumulator species had the greatest Se concentrations in the leaf (Galeas et al. 2007). Similar differences were found between *S. pinnata* and non-accumulator *S. elata* when tested in the field. *Stanleya pinnata* had greater Se concentrations in the fruit than the leaves, while *S. elata* had greater Se concentrations in the leaves relative to the fruit (Cappa et al. 2014).

Using X-ray microprobe analysis, it has been shown that different localization patterns exist between hyperaccumulators and nonhyperaccumulators. *Stanleya pinnata* was found to sequester Se in the periphery of the leaf, in epidermal vacuoles (Freeman et al. 2010; Cappa et al. 2015) while *S. elata* (Cappa et al. 2015) and *B. juncea* (Freeman et al. 2010) localized Se in the vascular tissue. *Astragalus bisculatus* has been shown to localize Se in trichomes of the leaves (Freeman et al. 2006a). Additionally, the chemical form in which the Se is found is different between Se hyperaccumulators and nonhyperaccumulators. The majority of Se is found in an organic C-Se-C configuration in *S. pinnata*, which was identified as 80% methyl-selenocysteine and 20% selenocystathionine (Freeman et al. 2006b; Cappa et al. 2015). In comparison, a substantial fraction of the Se was found in inorganic form in non-hyperaccumulators *S. elata* ~ 30%, (Cappa et al. 2015) and *B. juncea* (close to 100%, Freeman et al. 2010), and *S. albescens* accumulated exlusively selenocystathionine (Freeman et al. 2006b). Freeman et al. (2006a) also reported that the Se in *A. bisculatus* was predominantly found in the C-Se-C form in the trichomes (identified as methyl-selenocysteine) with a fraction of the Se in an inorganic form in the regions of the leaf other than trichome. Therefore, the evolution of Se hyperaccumulation likely involves the synthesis of methyl-selenocysteine and the specific peripheral sequestration of this aminoacid in leaves. The enzyme responsible for methyl-selenocysteine synthesis in Se hyperaccumulators is selenocysteine methyltransferase (SMT) (Neuhierl and Böck 1996). Overexpression of this enzyme from *A. bisulcatus* in *A. thaliana* or *B. juncea* indeed enhanced accumulation of methyl-selenocysteine and total Se, as well as Se tolerance (Ellis et al. 2004; LeDuc et al. 2004). This enzyme also exists in secondary Se accumulator broccoli (*B. oleracea*,

Lyi et al. 2005), which indeed accumulates methyl-selenocysteine. Its homologue in *A. thaliana* is a homoserine methyltransferase (HMT), which may suggest that SMT evolved from HMT. The aminoacid transporter for methyl-selenocysteine is still unknown.

12.5 Evolution of Se Hyperaccumulation in *Stanleya*

Stanleya is a small genus consisting of eight species. Feist and Parker (2001) documented substantial ecotypic variation in Se accumulation among *S. pinnata* populations throughout the western United States, both in the field and in a common garden experiment. Cappa et al. (2014) examined Se hyperaccumulation in the field and greenhouse and found substantial variation in Se accumulation between *Stanleya* species. In field-collected samples, *S. pinnata* was the only hyperaccumulator collected with Se levels >0.1% DW (Cappa et al. 2014). Beath et al. (1940) also reported *S. bipinnata* to be a Se hyperaccumulator; in a molecular phylogenetic analysis this was shown by Cappa et al. (2015) to be a variety of *S. pinnata*. Among the four varieties of *S. pinnata*, var. *pinnata* showed the highest Se accumulation properties. Interestingly, a geographic pattern was found in *S. pinnata*, var. *pinnata* that was related to Se hyperaccumulation: the highest Se accumulating populations were found on the eastern side of the Continental Divide (Cappa et al. 2014). There was also a polyploidy event within *S. pinnata var. pinnata* where the majority of individuals on the east side of the continental divide were diploid while all individuals on the west side were tetraploid (Cappa et al. 2014). While this ploidy difference did coincide with field Se levels, a common garden experiment did not find a genetic difference between diploids and tetraploids. Cappa et al. (2015) used leaves collected from *Stanleya* field populations (at least three populations per species or variety) for a phylogenetic analysis and conducted an ancestral-reconstruction analysis to predict the ancestral states for Se accumulation and Se tolerance, mapped onto the tree. Using the phylogeny, it was hypothesized that hyperaccumulation evolved once in *Stanleya*, within the *S. pinnata/bipinnata* clade (Cappa et al. 2015). Furthermore, it was found that tolerance was more widespread, found in all but three taxa within *Stanleya* and in three of the four outgroups tested (Cappa et al. 2015). Because hyperaccumulation was found in only three closely related taxa within *Stanleya pinnata*, the authors hypothesized that tolerance most likely preceded hyperaccumulation (Cappa et al. 2015). The same result was found in *Noccaea* (Brassicaceae) metal hyperaccumulators, when compared to closely related taxa (Broadley et al. 2007).

Taking the above evidence into account, we suggest that the evolution of hyperaccumulation is likely preceded by genetic variance within populations leading to tolerance to a given element, followed by genomic changes in transporter abundance, assimilation and localization capabilities, eventually leading to hyperaccumulation in a subset of tolerant genera/species/individuals. The convergent evolution of Se hyperaccumulation in different clades may involve different molecular evolutionary pathways, but likely involves many shared mechanisms. The scenario leading to this adaptation in its current form likely includes many intermediate steps. These steps can be surmised by looking at the current variability within non-accumulator taxa (e.g. *Arabidopsis*), secondary accumulator taxa (e.g. *Brassica*) and hyperaccumulator taxa (e.g. *Stanleya*). Ideally, closely related non-accumulator, accumulator and hyperaccumulator taxa can be further analyzed using powerful genomic approaches, to fully elucidate the evolutionary patterns and mechanisms associated with Se hyperaccumulation.

References

Ahmad R et al (2016) Selenium (Se) improves drought tolerance in crop plants – a myth or fact? J Sci Food Agric 96(2):372–380. Available at: http://doi.wiley.com/10.1002/jsfa.7231

Barberon M, Berthomieu P, Clairotte M, Shibagaki N, Davidian JC, Gosti F (2008) Unequal functional redundancy between the two Arabidopsis thaliana high-affinity sulphate transporters SULTR1; 1 and SULTR1; 2. New Phytol 180(3):608–619

Barillas JRV, Quinn CF, Freeman JL, Lindblom SD, Fakra SC, Marcus MA, Gilligan TM, Alford ÉR, Wangeline AL, Pilon-Smits EA (2012) Selenium distribution and speciation in the hyperaccumulator *Astragalus bisulcatus* and associated ecological partners. Plant Physiol 159(4):1834–1844

Beath OA, Gilbert CS, Eppson HF (1940) The use of indicator plants in locating seleniferous areas in western United States. III. Further studies. Am J Bot 27:564–573

Bohrer AS, Yoshimoto N, Sekiguchi A, Rykulski N, Saito K, Takahashi H (2015) Alternative translational initiation of ATP sulfurylase underlying dual localization of sulfate assimilation pathways in plastids and cytosol in *Arabidopsis thaliana*. Front Plant Sci 5:750

Boyd RS (2012) Plant defense using toxic inorganic ions: conceptual models of the defensive enhancement and joint effects hypotheses. Plant Sci 195:88–95

Boyd RS, Martens SN (1992) The *raison d'être* for metal hyperaccumulation by plants. In: AJM B, Proctor J, Reeves RD (eds) The vegetation of ultramafic (Serpentine) soils. Intercept, Andover, pp 279–289

Boyd RS, Martens SN (1998) The significance of metal hyperaccumulation for biotic interactions. Chemoecology 8(1):1–7

Broadley MR, White PJ, Hammond JP, Zelko I, Lux A (2007) Zinc in plants. New Phytol 173:677–702

Brown TA, Shrift A (1982) Selenium: toxicity and tolerance in higher plants. Biol Rev 57(1):59–84

Cabannes E, Buchner P, Broadley MR, Hawkesford MJ (2011) A comparison of sulfate and selenium accumulation in relation to the expression of sulfate transporter genes in *Astragalus* species. Plant Physiol 157:2227–2239

Cakir O, Turgut-Kara N, Ari S, Zhang B (2015) De novo transcriptome assembly and comparative analysis elucidate complicated mechanism regulating *Astragalus chrysochlorus* response to selenium stimuli. PLoS One 10(10):1–16

Cappa JJ, Pilon-Smits EAH (2014) Evolutionary aspects of elemental hyperaccumulation. Planta 239:267–275

Cappa JJ, Cappa PJ, El Mehdawi AF, McAleer JM, Simmons MP, Pilon-Smits EAH (2014) Characterization of selenium and sulfur accumulation across the genus *Stanleya* (Brassicaceae): a field survey and common-garden experiment. Am J Bot 101:830–839

Cappa JJ, Yetter C, Fakra S, Cappa PJ, DeTar R, Landes C, Pilon-Smits EAH, Simmons MP (2015) Evolution of selenium hyperaccumulation in *Stanleya* (Brassicaceae) as inferred from phylogeny, physiology and X-ray microprobe analysis. New Phytol 205:583–595

Craciun A, Meyer C-L, Chen J, Roosens N, De Groodt R, Hilson P, Verbruggen N (2012) Variation in HMA4 copy number and expression among Noccaea caerulescens populations presenting different levels of Cd tolerance and accumulation. J Exp Bot 63:4179–4189

DeTar R, Alford E, Pilon-Smits EAH (2015) Molybdenum accumulation, tolerance and molybdenum–selenium–sulfur interactions in *Astragalus* selenium hyperaccumulator and nonaccumulator species. J Plant Physiol 183:32–40

Dhillon KS, Dhillon SK (2001) Distribution and management of seleniferous soils. Adv Agron 79:119–184

El Mehdawi AF, Pilon-Smits EAH (2012) Ecological aspects of plant selenium hyperaccumulation. Plant Biol 14(1):1–10

El Mehdawi AF, Quinn CF, Pilon-Smits EAH (2011) Effects of selenium hyperaccumulation on plant-plant interactions: evidence for elemental allelopathy? New Phytol 191(1):120–131

El Mehdawi AF, Paschke MW, Pilon-Smits EAH (2015) *Symphyotrichum ericoides* populations from seleniferous and nonseleniferous soil display striking variation in selenium accumulation. New Phytol 206(1):231–242

Ellis DR, Sors TG, Brunk DG, Albrecht C, Orser C, Lahner B, Wood KV, Harris HH, Pickering IJ, Salt DE (2004) Production of Se-methylselenocysteine in transgenic plants expressing selenocysteine methyltransferase. BMC Plant Biol 4(1):1

Feist LJ, Parker DR (2001) Ecotypic variation in selenium accumulation among populations of *Stanleya pinnata*. New Phytol 149:61–69

Freeman JL (2006) Spatial imaging, speciation, and quantification of selenium in the hyperaccumulator plants *Astragalus bisulcatus* and *Stanleya pinnata*. Plant Physiol 142(1):124–134

Freeman JL, Zhang LH, Marcus M, Fakra S, McGrath SP, Pilon-Smits EAH (2006a) Spatial imaging, speciation, and quantification of selenium in the hyperaccumulator plants *Astragalus bisulcatus* and *Stanleya pinnata*. Plant Physiol 142:124–134

Freeman JL, Quinn CF, Marcus MA, Fakra S, Pilon-Smits EA (2006b) Selenium-tolerant diamondback moth disarms hyperaccumulator plant defense. Curr Biol 16(22):2181–2192

Freeman JL, Lindblom SD, Quinn CF, Fakra S, Marcus MA, Pilon-Smits EA (2007) Selenium accumulation protects plants from herbivory by Orthoptera via toxicity and deterrence. New Phytol 175(3):490–500

Freeman JL, Tamaoki M, Stushnoff C, Quinn CF, Cappa JJ, Devonshire J, Fakra SF, Marcus MA, McGrath SP, Van Hoewyk D, Pilon-Smits EAH (2010) Molecular mechanisms of selenium tolerance and hyperaccumulation in *Stanleya pinnata*. Plant Physiol 153:1630–1652

Freeman JL, Quinn CF, Lindblom SD, Klamper EM, Pilon-Smits EA (2009) Selenium protects the hyperaccumulator *Stanleya pinnata* against black-tailed prairie dog herbivory in native seleniferous habitats. Am J Bot 96(6):1075–1085

Freeman JL, Marcus MA, Fakra SC, Devonshire J, McGrath SP, Quinn CF, Pilon-Smits EA (2012) Selenium hyperaccumulator plants *Stanleya pinnata* and *Astragalus bisulcatus* are colonized by Se-resistant, Se-excluding wasp and beetle seed herbivores. PLoS One 7(12):e50516

Galeas ML, Zhang LH, Freeman JL, Wegner M, Pilon-Smits EA (2007) Seasonal fluctuations of selenium and sulfur accumulation in selenium hyperaccumulators and related nonaccumulators. New Phytol 173(3):517–525

Galeas ML, Klamper EM, Bennett LE, Freeman JL, Kondratieff BC, Quinn CF, Pilon-Smits EA (2008) Selenium hyperaccumulation reduces plant arthropod loads in the field. New Phytol 177(3):715–724

Han D, Li X, Xiong S, Tu S, Chen Z, Li J, Xie Z (2013) Selenium uptake, speciation and stressed response of *Nicotiana tabacum* L. Environ Exp Bot 95:6–14

Hanikenne M, Talke IN, Haydon MJ, Lanz C, Nolte A, Motte P, Kroymann J, Weigel D, Krämer U (2008) Evolution of metal hyperaccumulation required cis-regulatory changes and triplication of HMA4. Nature 453:391–395

Hanson B, Garifullina GF, Lindblom SD, Wangeline A, Ackley A, Kramer K, Norton AP, Lawrence CB, Pilon-Smits EA (2003) Selenium accumulation protects *Brassica juncea* from invertebrate herbivory and fungal infection. New Phytol 159(2):461–469

Hanson B, Lindblom SD, Loeffler ML, Pilon-Smits EA (2004) Selenium protects plants from phloem-feeding aphids due to both deterrence and toxicity. New Phytol 162(3):655–662

Harris J, Schneberg KA, Pilon-Smits EAH (2013) Sulfur - selenium - molybdenum interactions distinguish selenium hyperaccumulator Stanleya pinnata from non-hyperaccumulator Brassica juncea (Brassicaceae). Planta 239:479–491

Hasanuzzaman M, Fujita M (2011) Selenium pretreatment upregulates the antioxidant defense and methylglyoxal detoxification system and confers enhanced tolerance to drought stress in rapeseed seedlings. Biol Trace Elem Res 143(3):1758–1776

Hasanuzzaman M, Hossain MA, Fujita M (2011) Selenium-induced up-regulation of the antioxidant defense and methylglyoxal detoxification system reduces salinity-induced damage in rapeseed seedlings. Biol Trace Elem Res 143(3):1704–1721

Kabata-Pendias AH, Pendias H, CRC Press (2001) Trace elements in soils and plants, 4th edn. CRC Press, Boca Raton

Kiefer M, Schmickl R, German DA, Mandáková T, Lysak MA, Al-Shehbaz IA, Franzke A, Mummenhoff K, Stamatakis A, Koch MA (2014) BrassiBase: introduction to a novel knowledge database on Brassicaceae evolution. Plant Cell Physiol 55(1):e3–e3

Koch, M.A., Kiefer, M., German, D.A., Al-Shehbaz, I.A., Franzke, A., Mummenhoff, K. and Schmickl, R., 2012. BrassiBase: tools and biological resources to study characters and traits in the Brassicaceae—version 1.1. Taxon, 61(5), pp.1001–1009.

LeDuc DL, Tarun AS, Montes-Bayon M, Meija J, Malit MF, Wu CP, AbdelSamie M, Chiang CY, Tagmount A, Neuhierl B, Böck A (2004) Overexpression of selenocysteine methyltransferase in Arabidopsis and Indian mustard increases selenium tolerance and accumulation. Plant Physiol 135(1):377–383

Lindblom SD, Valdez-Barillas JR, Fakra SC, Marcus MA, Wangeline AL, Pilon-Smits EAH (2013) Influence of microbial associations on selenium localization and speciation in roots of *Astragalus* and *Stanleya* hyperaccumulators. Environ Exp Bot 88:33–42

Lyi SM, Heller LI, Rutzke M, Welch RM, Kochian LV, Li L (2005) Molecular and biochemical characterization of the selenocysteine Se-methyltransferase gene and Se-methylselenocysteine synthesis in broccoli. Plant Physiol 138:409–420

Macnair MR, Bert V, Huitson SB, Saumitou-Laprade P, Petit D (1999) Zinc tolerance and hyperaccumulation are genetically independent characters. Proc R Soc Lond B Biol Sci 266(1434):2175–2179

Nawaz F, Ashraf MY, Ahmad R, Waraich EA, Shabbir RN, Bukhari MA (2015) Supplemental selenium improves wheat grain yield and quality through alterations in biochemical processes under normal and water deficit conditions. Food Chem 175:350–357

Neuhierl B, Böck A (1996) On the mechanism of selenium tolerance in selenium-accumulating plants: purification and characterization of a specific selenocysteine methyltransferase from cultured cells of *Astragalus bisulcatus*. Eur J Biochem 239:235–238

Peterson PJ, Butler GW (1962) The uptake and assimilation of selenite by higher plants. Aust J Biol Sci 24(2):175–178

Peterson PJ, Butler GW (1971) The occurrence of selenocystathionine in *Morinda reticulata* benth., a toxic seleniferous plant. Aust J Biol Sci 24(1):175–178

Pilon-Smits EAH, Hwang S, Lytle CM, Zhu Y, Tai JC, Bravo RC, Chen Y, Leustek T, Terry N (1999) Overexpression of ATP sulfurylase in Indian mustard leads to increased selenate uptake, reduction, and tolerance. Plant Physiol 119(1):123–132

Prins CN, Hantzis LJ, Quinn CF, Pilon-Smits EAH (2011) Effects of selenium accumulation on reproductive functions in *Brassica juncea* and *Stanleya pinnata*. J Exp Bot 62:5633–5640

Quinn CF, Freeman JL, Galeas ML, Klamper EM, Pilon-Smits EA (2008) The role of selenium in protecting plants against prairie dog herbivory: implications for the evolution of selenium hyperaccumulation. Oecologia 155(2):267–275

Quinn CF, Freeman JL, Reynolds RJ, Cappa JJ, Fakra SC, Marcus MA, Lindblom SD, Quinn EK, Bennett LE, Pilon-Smits EA (2010) Selenium hyperaccumulation offers protection from cell disruptor herbivores. BMC Ecol 10(1):1

Quinn CF, Prins CN, Freeman JL, Gross AM, Hantzis LJ, Reynolds RJB, Yang SI, Covey PA, Bañuelos GS, Pickering IJ, Fakra SC, Marcus MA, Arathi HS, Pilon-Smits EAH (2011a) Selenium accumulation in flowers and its effects on pollination. New Phytol 192:727–737

Quinn CF, Wyant KA, Wangeline AL, Shulman J, Galeas ML, Valdez JR, Self JR, Paschke MW, Pilon-Smits EA (2011b) Enhanced decomposition of selenium hyperaccumulator litter in a seleniferous habitat—evidence for specialist decomposers? Plant Soil 341(1–2):51–61

Reeves RD, Brooks RR, Macfarlane RM (1981) Nickel uptake by Californian Streptanthus and Caulanthus with particula reference to the hyperaccumulator *S. polygaloides* gray (Brassicaceae). Am J Bot 68(5):708

Rosenfeld I, Beath OA (1964) Selenium. Geobotany, biochemistry, toxicity and nutrition. Annu Rev Plant Physiol 20:475

Salt DE, Smith RD, Raskin I (1998) Phytoremediation. Annu Rev Plant Biol 49(1):643–668

Schiavon M, Pilon M, Malagoli M, Pilon-Smits EA (2015) Exploring the importance of sulfate transporters and ATP sulphurylases for selenium hyperaccumulation—a comparison of *Stanleya pinnata* and *Brassica juncea* (Brassicaceae). Front Plant Sci 6:2

Sors TG, Ellis DR, Na GN, Lahner B, Lee S, Leustek T, Pickering IJ, Salt DE (2005a) Analysis of sulfur and selenium assimilation in Astragalus plants with varying capacities to accumulate selenium. Plant J 42(6):785–797

Sors T, Ellis D, Salt D (2005b) Selenium uptake, translocation, assimilation and metabolic fate in plants. Photosynth Res:373–389

Sors TG, Martin CP, Salt DE (2009) Characterization of selenocysteine methyltransferases from Astragalus species with contrasting selenium accumulation capacity. Plant J Cell Mol Biol 59(1):110–122

Stevens PF (2001 onwards) Angiosperm Phylogeny website. version 12, July 2012 (and more or less continuously updated since). http://www.mobot.org/MOBOT/research/APweb/

Terry N, Zayed AM, De Souza MP, Tarun AS (2000) Selenium in higher plants. Annu Rev Plant Biol 51(1):401–432

Trumble JT, Kund GS, White KK (1998) Influence of form and quantity of selenium on the development and survival of an insect herbivore. Environ Pollut 101(2):175–182

Ueno D, Milner MJ, Yamaji N, Yokosho K, Koyama E, Zambrano MC, Kaskie M, Ebbs S, Kochian LV, Ma JF (2011) Elevated expression of TcHMA3 plays a key role in the extreme Cd tolerance in a Cd-hyperaccumulating ecotype of Thlaspi caerulescens. Plant J 66:852–862

Urbatsch LE, Roberts RP, Karaman V (2003) Phylogenetic evaluation of *Xylothamia*, *Gundlachia*, and related genera (Asteraceae, Astereae) based on ETS and ITS nrDNA sequence data. Am J Bot 90:634–649

Van Hoewyk D (2013) A tale of two toxicities: malformed selenoproteins and oxidative stress both contribute to selenium stress in plants. Ann Bot 112(6):965–972

Van Hoewyk D, Takahashi H, Inoue E, Hess A, Tamaoki M, Pilon-Smits EA (2008) Transcriptome analyses give insights into selenium-stress responses and selenium tolerance mechanisms in Arabidopsis. Physiol Plant 132(2):236–253

Vickerman D, Trumble J (1999) Feeding preferences of *Spodoptera exigua* in response to form and concentration of selenium. Arch Insect Biochem Physiol 42(1):64–73

Wang H, Wang H, Shao H, Tang X (2016) Recent advances in utilizing transcription factors to improve plant abiotic stress tolerance by transgenic technology. Front Plant Sci 7:67

White PJ (2016) Selenium accumulation by plants. Ann Bot 117:mcv180

Winkel LH, Vriens B, Jones GD, Schneider LS, Pilon-Smits E, Bañuelos GS (2015) Selenium cycling across soil-plant-atmosphere interfaces: a critical review. Nutrients 7(6):4199–4239

Wu Z, Yin X, Bañuelos GS, Lin ZQ, Zhu Z, Liu Y, Yuan L, Li M (2015) Effect of selenium on control of postharvest gray mold of tomato fruit and the possible mechanisms involved. Front Microbiol 6:1441

Wyoming Agricultural Experiment Station, Beath OA, Knight SH (1937) Occurrence of Selenium and Seleniferous vegetation in Wyoming

Yao X, Chu J, Wang G (2009) Effects of selenium on wheat seedlings under drought stress. Biol Trace Elem Res 130(3):283–290

Yuan L, Zhu Y, Lin ZQ, Banuelos G, Li W, Yin X (2013) A novel selenocystine-accumulating plant in selenium-mine drainage area in Enshi, China. PLoS One 8(6):e65615

Zayed A, Lytle CM, Terry N (1998) Accumulation and volatilization of different chemical species of selenium by plants. Planta 206(2):284–292

Zhang LH, Abdel-Ghany SE, Freeman JL, Ackley AR, Schiavon M, Pilon-Smits EA (2006) Investigation of selenium tolerance mechanisms in *Arabidopsis thaliana*. Physiol Plant 128(2):212–223

Zhang L, Hu B, Li W, Che R, Deng K, Li H, Yu F, Ling H, Li Y, Chu C (2014) OsPT2, a phosphate transporter, is involved in the active uptake of selenite in rice. New Phytol 201(4):1183–1191

Part IV
The Societal Relevance of Se for Human and Environmental Health: Biofortification and Phytoremediation

Chapter 13
Overview of Selenium Deficiency and Toxicity Worldwide: Affected Areas, Selenium-Related Health Issues, and Case Studies

André Rodrigues dos Reis, Hassan El-Ramady, Elcio Ferreira Santos, Priscila Lupino Gratão, and Lutz Schomburg

Abstract Selenium (Se) is an essential micronutrient for human and animal healthy due to its capabilities to support antioxidant defence systems. However, problems related to the deficiency of Se are emerging issue for human health worldwide and plant species differ considerably in their susceptibility to high concentrations of Se, and certain plant species can be able to accumulate Se to astonishingly high concentrations. Many factors can affect the content of Se in different foods, including different uptake rate by plants, which can be related to plant type, soil, pH, microbial activity, rainfall and a number of other biogeochemical parameters. Humans Se intake and Se status in the population depends firstly on Se concentrations in soils, and hence the Se concentrations in the harvested edible plants in these soils. Thus, this chapter aims to compile some information about research work on essentiality of Se for humans and other mammals, and the need for a sufficient daily Se intake.

Keywords Agronomic bio-fortification • Human health • Oxidative stress • Selenium toxicity • Keshan disease

---

A.R. dos Reis (✉)
UNESP – São Paulo State University, Tupã, SP 17602-496, Brazil
e-mail: andrereis@tupa.unesp.br

H. El-Ramady
Faculty of Agriculture, Kafrelsheikh University, Kafr el-Sheikh 33516, Egypt
e-mail: ramady2000@gmail.com

E.F. Santos
Center of Energy in Agriculture, University of São Paulo, Piracicaba, SP 13416-000, Brazil
e-mail: elciokw@yahoo.com.br

P.L. Gratão
UNESP – São Paulo State University, Jaboticabal, SP 14884-900, Brazil
e-mail: plgratao@fcav.unesp.br

L. Schomburg
Institute for Experimental Endocrinology, Charité – University Medical School, Suedring 10, CVK, D-13353 Berlin, Germany
e-mail: lutz.schomburg@charite.de

E.A.H. Pilon-Smits et al. (eds.), *Selenium in plants*, Plant Ecophysiology 11,
DOI 10.1007/978-3-319-56249-0_13

13.1 Introduction

Selenium (Se) is an essential micronutrient for human and animal health due to its capabilities to support antioxidant defence systems, but is harmful in excess (Fordyce 2013; Mora et al. 2015; Wrobel et al. 2016). Compared to other micronutrients, Se has one of the narrowest ranges between its toxic dose (> 400 μg/day) and dietary deficiency (< 40 μg/day), as reviewed by Kabata-Pendias and Mukherjee (2007) and Fordyce (2013). Therefore, both deficiency and toxicity of Se are global emerging problems. Several studies have investigated Se deficiency and toxicity in humans (Sah et al. 2013; Sun et al. 2014; Yang and Jia 2014; Zhu et al. 2015; Nagy et al. 2015; Oropeza-Moe et al. 2015; Krohn et al. 2016; Wrobel et al. 2016; Manzanares and Hardy 2016), algae (Gojkovic et al. 2015), yeast (Kieliszek et al. 2016), bacteria (Nancharaiah and Lens 2015; Ye et al. 2016; Lampis et al. 2016) and higher plants (Saidi et al. 2014a, b; Yusuf et al. 2016). Recently, two comprehensive books were published that reviewed the diversity of Se functions in health and disease (Brigelius-Flohé and Sies 2016) and global advances in Se research, from theory to application (Bañuelos et al. 2016).

While it is clear that Se is required for several essential biological functions for human health, many questions still need to be answered. The most important Se research findings are summarized below, along with some attempts to answer remaining urging questions: (1) Selenium is required by humans and other mammals. It is well established that Se is an essential nutrient for human health (Rayman 2000; Combs 2001; Surai 2006), and an insufficient supply causes or predisposes to disease; (2) No clear definition is currently given as to how much Se is required by humans. The amount minimally needed likely depends on a number of anthropometric characteristics: body weight, age, sex, health status. The recommendations given cover a wide range of reference intakes spanning a daily uptake of 25–125 μg/day (Hurst et al. 2013; Combs 2016); (3) The dietary intake level associated with Se deficiency for humans is reported to be <50 μg/day (Fairweather-Tait et al. 2011; Combs 2016). However, as mentioned above, this recommendation depends on a number of personal characteristics; (4) Research indicates that Se can reduce cancer risk when the concentration of plasma Se ranges from 70 to 106 ng/ml (Combs 2016). Lower Se concentrations seem to increase the risk of several types of cancer, while no positive chemopreventive effects have been observed above this level, which likely corresponds to the amount of Se needed for full expression of the human selenoproteins.

Concerning the importance of Se for human and animal health, many studies have investigated this relationship, such as El-Ramady et al. (2015b), Niedzielski et al. (2016), Han et al. (2016), Hauser-Davis et al. (2016), Krohn et al. (2016), Hoffmann (2016), Bissardon et al. (2016), Schomburg (2011), Wang et al. (2016a, b), Menezes et al. (2016) and Zanetti et al. (2016). On the other hand, the toxicity of Se to higher plants also has been documented in many studies including Molnárová and Fargasová (2009), Akbulut and Çakır (2010), Aggarwal et al. (2011), Madaan and Mudgal (2011), Srivastava et al. (2012), Hladu et al. (2013), Talukdar (2013),

Zhao et al. (2013), Chen et al. (2014), Sharma et al. (2014b), Mechora et al. (2015), El-Ramady et al. (2015a, b), Lehotai et al. (2015), Nawaz et al. (2015), Handa et al. (2016), Pilon et al. (2016), Pilon-Smits et al. (2016), and White (2016).

In general, the effects of Se deficiency on humans include muscle weakness and inflammation, fragile red blood cells, abnormal skin coloration, heart muscle dysfunction, susceptibility to cancer, Keshan and Kashin-Beck diseases, whereas Se toxicity includes liver and kidney damage, blood clotting, necrosis of heart and liver, hair and nail loss and nausea and vomiting (Kabata-Pendias and Mukherjee 2007). Selenium in not essential for plants, but has beneficial effects on plant growth and stress tolerance. Many studies have indicated that already small Se doses are sufficient for improving plant health (e.g. Kong et al. 2005; Eiche et al. 2015). High Se levels tend to induce different toxic effects in plants, including reduced photosynthetic efficiency and growth, chlorosis and finally plant death (Van Hoewyk 2013; Eiche et al. 2015). However, plant species differ considerably in their susceptibility to high dosages of Se, and certain plant species even show stimulation of growth on soils with high Se and are able to accumulate Se to astonishingly high concentrations (Pilon-Smits et al. 2016).

Problems related to the deficiency of Se are an emerging issue for human health worldwide. A solution for this problem can be achieved through Se biofortification of different crops, as reviewed by several authors using rice (Boldrin et al. 2013; Wang et al. 2014; Reis et al. 2014; Sharma et al. 2014b, c; Pandey and Gupta 2015; Li et al. 2016), maize (Chilimba et al. 2012; Longchamp and Castrec-Rouelle 2014; Longchamp et al. 2013, 2015), wheat (Acuña et al. 2013; Galinha et al. 2013; Fenech et al. 2013; Gong et al. 2014; Zhu et al. 2014; Li et al. 2014; Poblaciones et al. 2014; Yasin et al. 2014; Galinha et al. 2015; Lazo-Vélez et al. 2015) and cruciferous vegetables (Harris et al. 2014; Avila et al. 2014; Yasin et al. 2015b; Bañuelos et al. 2015; Bachiega et al. 2016). Different forms of Se-biofortification have been tested, including supplementation of fertilizers, foliar spraying directly on the plants or using Se-accumulating plant leftovers for soil fortification, as recently reported by Bañuelos et al. (2015), El-Ramady et al. (2015c), Malagoli et al. (2015), Galinha et al. (2015), Yasin et al. (2015a, b), Bañuelos et al. (2016), Faria et al. (2016), Mao et al. (2016), Ortiz-Monasterio et al. (2016), Reis et al. (2016), dos Reis (2016), El-Ramady et al. (2016a), Li et al. (2016), Domingues et al. (2016), and Sharma et al. (2016).

13.2 Global Areas Related to Se Deficiency and Toxicity

Selenium can be found in all agroecosystem components including soil, plants, rocks or water. In animal feed, the critical Se concentrations for Se adequacy and toxicity are 0.05–0.10 mg/kg and 4–5 mg/kg, respectively (Zanetti et al. 2016). Several global locations have been monitored where livestock may experience Se toxicity, including areas in the western U.S.A. such as the San Joaquin Valley in California (Frankenberger and Benson 1994), Colorado and Wyoming (El Mehdawi

Table 13.1 Soil, crop and water Se concentrations in different seleniferous areas worldwide

Country (region)	Total Se in soil (mg/kg)	Total Se in cultivated plants (mg/kg)	Total Se in water (µg/L)	References
China (Enshi, Yutangba)	0.10–42.3 (4.75)	Maize (seeds): 0.17–4.82 (1.48)	Stream water: 58.4	Zhu et al. (2008)
India (Punjab)	2.7–6.55 (3.63)	Wheat, rice, maize and mustard 13–670	Ground water: 479 (170)	Sharma et al. (2009)
China (Enshi, Yutangba)	3.76–79.08 (27.81)	*Adenocaulon himalaicum* (leaf) 299–2278 (760)	Stream water: 15.13–192.7 (52.66)	Yuan et al. (2012)
China (Enshi, Jianshi)	2.89–87.3 (9.36)	Maize: 0.39–37.2 (3.76)	Surface water: 2.0–519 (46)	Qin et al. (2013)
USA (Pine Ridge Fort Collins, CO)	8.2	*Brassica juncea* (leaf): 711		Yasin et al. (2015c)
India (Punjab)	0.024–3.06 (0.449)	Cultivated and naturally growing weed plants: 0.01–6.60 (0.2795)	Ground water: 0.01–35.6 (0.972)	Dhillon and Dhillon (2016a)

and Pilon-Smits 2012), Enshi, Hubei Province, China (Wang and Gao 2001), and Australia (Thomson 2004) and Punjab Nawanshahr–Hoshiarpur region India (Dhillon and Dhillon 2014). Due to the importance of Se (Table 13.1), several studies have been conducted to quantify the Se levels in soil, water and crops from different areas, including those published by Bañuelos and Lin (2007), Dhillon et al. (2008), Dhillon and Dhillon (2009a, b), (2014), (2016b), Sharma et al. (2009), (2014a), Yuan et al. (2012), Wang et al. (2012), Eiche et al. (2015), Schilling et al. (2015), Yasin et al. (2015c), (2016), Chawla et al. (2016), and Prakash (2016).

On the other hand, Se deficiency predisposes to certain endemic diseases, as has been well described for Se-poor areas in China, where Se deficiency predisposes people to Keshan disease, which is associated with childhood cardiomyopathy (Xia et al. 2005). There are more than 40 countries described as having areas with very low soil Se content, associated with human Se intakes of 10 µg/day or even less, such as in areas of China (Moreno-Reyes et al. 1998; Tan et al. 2002; Li et al. 2007; Han et al. 2016). In fact, in China Se deficient areas are reported to represent 72% of the country's total area; these areas are often not intensively populated. Besides these Se-deficient areas, there are also Chinese areas with very high Se levels in soil and in the agricultural products, thereby causing a high daily Se intake (Gao et al. 2011; Han et al. 2016).

13.3 Selenium Deficiency and Toxicity in Soils and Plants in Middle East and Europe

It has been documented that Se occurs in different mammalian tissues ranging from 0.7 in heart tissue to 2.5 mg/kg in muscles, with an estimated average Se content in human soft tissues of 0.11 mg/kg (Kabata-Pendias and Mukherjee 2007). Concerning the Se intake and its status in Middle East and Europe, a suboptimal Se status was found throughout these regions with only few exceptions (Table 13.2). In general, it can be noticed that the intake status of Se across Europe is low. This means that the Se level in European soils is inadequate, particularly in Eastern Europe. There is no complete systematic review of the soil Se quality and Se status in subjects living in the Middle East (Stoffaneller and Morse 2015; Sharma et al. 2009).

13.4 Selenium Status in Brazilian Soils and Crops

Selenium is one of various compounds and chemical elements important to ensure the quality of food, together with proteins, carbohydrates, fats, vitamins, iron (Fe), iodine (I), and zinc (Zn) (Rayman et al. 2012). Genetic breeding programs can contribute positively to the development of improved crop varieties. So far, crop breeding has focused almost exclusively on higher productivity, i.e. crop quantity rather than quality. However, problems of nutritional deficiencies are experienced by almost half the world's population, especially Fe, I, Se, vitamin A and Zn in developing countries (Rayman et al. 2012). These deficiencies happen basically due to two reasons: 1) low concentration of micronutrients in the soils, which are affected by texture class, mineral composition and soil pH; 2) a dilution effect of the essential micronutrients and vitamins for human health in the most productive varieties or cultivars. While Se concentrations in agricultural products (food) depend primarily on their concentrations in the soil, genotypic variation can also influence the absorption capacity of Se by plants.

The two major inorganic forms of Se in soils are selenate and selenite. In comparison to selenate, selenite forms usually are more strongly retained in the soil colloids, a process which depends on environment characteristics such as pH, ionic strength, ion concentration and other effects. Considering that the range between essential and toxic levels of Se in plants and animals is very narrow (Lyons et al. 2003), the study of Se levels in agricultural soils and their sorption behavior under different conditions of pH, ionic strength and concentration of competing ions becomes highly relevant for a better understanding of the dynamics of Se in soils.

In many areas of Brazil, agricultural products have low levels of Se, and thus it is important to understand the behavior of Se in soils and to assess the mechanisms of its transfer to edible parts of plants, which are primary sources of food for the population and animals. Table 13.3 summarizes what is known so far regarding the Se and sulfur concentrations in soils collected from different regions of Brazil.

Table 13.2 Survey for investigating human Se status in some Arabian and European countries (as a reference: a desirable plasma Se level is 70–106 μg/L)

Country	Subject details and number	Mean Se status in human (μg L^{-1})	References
Arabian countries			
Egypt	67 patient children and 60 healthy children	Serum: 40.1 in patient children and 83.3 in control	Saad et al. (2014)
Egypt	80 obese children and 80 healthy children	Serum: 63.6 in the obese compared to 78.3 in controls	Azab et al. (2014)
Egypt	108 patients children and 60 healthy children	Serum: 31.5 in patients and 65.9 in control	Sherief et al. (2014)
Jordan	Subjects: 73 total; 56 smokers; 17 non-smokers	Blood: 332 in smokers and 187 in case of non-smokers	Massadeh et al. (2010)
KSA	42 Saudi in 45–60 years and 34 Saudi in 20–30 years	Serum: 91.24 and 86.63, resp.	Stoffaneller and Morse (2015)
KSA	170 diabetics with an equal number of control	Urinary: 31.1 for diabetics and 39.1 for control	El-Yazigi and Legayada (1996)
KSA	513 children	Serum: < 56 from 53.4% of total	Al-Saleh et al. (2006)
Kuwait	66 obese female patients and 44 female control	Serum: 86.08 in obese group and 101.14 in the control group	Alasfar et al. (2011)
Lebanon	159 healthy men and 284 women; age 18–65 years	Plasma: 151.2 for men and 135.0 for women	Obeid et al. (2008)
Yemen	75 patient children and 74 healthy control	Serum: 78.96 in cases and 94.75 in controls	Elemraid et al. (2011)
European countries			
Austria	Patients with autoimmune thyroiditis and control	Serum: 98.0 in the patients and 103.2 in control	Wimmer et al. (2014)
Denmark	97 patients and 830 control	Serum: 89.9 for patients and 98.8 in controls	Bülow Pedersen et al. (2013)
Denmark	3333 males (53–74 years)	Low serum: 31.58–78.96 and high serum 102.65–236.88	Suadicani et al. (2012)
Estonia	404 subjects (19.5–52 years)	Serum: 26–116 (mean: 75)	Rauhamaa et al. (2008)
Germany	60 patients (aged 65 year)	From 89.05 to 70.84	Stoppe et al. (2011)
Germany	104 cardiac surgical patients	Blood: 89.05 and 70.84 pre- and post-surgery, respectively	Stoppe et al. (2013)
Germany	44 trauma patients	Plasma: 62.38	Blass et al. (2013)

(continued)

Table 13.2 (continued)

Country	Subject details and number	Mean Se status in human (μg L^{-1})	References
Greece	47 singleton pregnant women in age 30 + 5 years	Urine: 91, 82 and 69 for the 1st trimester, 2nd and 3rd trimester, res.	Koukkou et al. (2014)
Finland	60 adults	Plasma: 70.27 in the 1970s to 110.54 after Se-fertilizers in 1984	Alfthan et al. (2015)
France	1389 subjects aged 59–71 years followed for 9 years	Plasma: 16.58 in men and 15.79 in women	Akbaraly et al. (2010)
Hungary	197 consecutive patients	Blood: in non-survivors 102.2 compared with survivors 111.1	Koszta et al. (2012)
Italy	54 melanoma patients and 56 control	Plasma: 99 in the cases and 89 in the control	Vinceti et al. (2012)
Poland	95 lung cancer cases, 113 laryngeal cancer cases	Serum: 63.2 compared to 74.6 control	Jaworska et al. (2013)
Poland	80 children (age 6–17; 40 boys, 40 girls)	Serum: 102.3 and 111.1 in control girls and boys, respectively	Błażewicz et al. (2015)
Portugal	136 women (20–44 years)	Serum: 81	Lopes et al. (2004)
Slovenia	15 recruits	Plasma: 71.75–82 (mean 76.87)	Pograjc et al. (2012)
Spain	84 healthy adults (31 males and 53 females	Plasma: 87.3 in males and 67.3 in females	Millán Adame et al. (2012)
Spain	340 subjects	86.5% had plasma Se below 125	Sánchez et al. (2010)
The Netherlands	1197 pregnant women from 12 weeks gestation	Serum: at 12 weeks and after 75.80 and 80.54, respectively	Rayman et al. (2011)
UK	501 elderly volunteers	Plasma: 90.71 at baseline	Rayman et al. (2012)
UK	1042 subjects (19–64 years)	Plasma: 86.86	Stranges et al. (2010)

There still is a lack of comprehensive information about the distribution of Se in Brazilian soils; this is an aim of current research. Better knowledge of the of the levels and dynamics of Se in soils throughout Brazil are expected to greatly contribute to future research aiming to provide optimal levels of Se to crops, applied through fertilizer (agronomic bio-fortification) to increase the natural intake of Se by the Brazilian population.

Ferreira et al. (2002) observed that food consumed in Brazil has significantly low concentrations of Se. This observation likely is due to low Se concentrations in Brazilian soils. Similar results were reported by Faria (2009), showing very low Se

Table 13.3 Concentrations of Selenium and Sulfur in Brazilian soils

City	State	Se (μg/kg)	S (g/kg)	Geographic coordinates	References
Sena Madureira - Acre	Acre	184	5	9° 25′ 54″ S 68° 35′ 42″ W	Silva Junior (2016)
Itacoatiara - Amazonas	Amazonas	530	17	3° 6′ 31″ S 58° 26′ 33″ W	Silva Junior (2016)
Silvânia	Goiás	49	–	16° 39′ 32″ S 48° 36′ 29″ W	Carvalho (2011)
Itaúba	Mato Grosso	174	7	11° 06′ 00″ S 55° 02′ 06″ W	Silva Junior (2016)
Pirapora	Minas Gerais	44	–	17° 20′ 42″ S 44° 56′ 06″ W	Carvalho (2011)
Capinópolis	Minas Gerais	50	–	18° 40′ 55″ S 49° 34′ 11″ W	Carvalho (2011)
Caracaraí - Roraima	Roraima	182	10	1° 28′ 10″ S 60° 44′ 16″ W	Silva Junior (2016)
Alvinlândia	São Paulo	10	–	22° 26′ 00″ S 49° 45′ 00″ W	Nogueira et al. (2013)
Analândia	São Paulo	70	–	22° 07′ 00″ S 47° 39′ 00″ W	Nogueira et al. (2013)
Araras	São Paulo	60	–	22° 19′ 00″ S 47° 10′ 00″ W	Nogueira et al. (2013)
Bonfim Paulista	São Paulo	200	–	21° 05′ 00″ S 47° 08′ 00″ W	Nogueira et al. (2013)
Capivari	São Paulo	50	–	22° 59′ 01″ S 47° 30′ 00″ W	Nogueira et al. (2013)
Capivari	São Paulo	110	–	22° 59′ 10″ S 47° 30′ 10″ W	Nogueira et al. (2013)
Conchal	São Paulo	50	–	22° 19′ 00″ S 47° 00′ 10″ W	Nogueira et al. (2013)
Cosmópolis	São Paulo	110	–	22° 38′ 00″ S 47° 11′ 00″ W	Nogueira et al. (2013)
Gália	São Paulo	10	–	22° 17′ 00″ S 49° 33′ 00″ W	Nogueira et al. (2013)
Garça	São Paulo	10	–	22° 12′ 00″ S 49° 56′ 00″ W	Nogueira et al. (2013)
Garça	São Paulo	10	–	22° 12′ 00″ S 49° 39′ 10″ W	Nogueira et al. (2013)
Ibaté	São Paulo	70	–	21° 57′ 00″ S 47° 59′ 00″ W	Nogueira et al. (2013)
Ibituruna	São Paulo	300	–	21° 8′ 36″ S 44° 44′ 24″ W	Nogueira et al. (2013)
Itirapina	São Paulo	70	–	22° 15′ 10″ S 47° 00′ 49″ W	Nogueira et al. (2013)
Itirapina	São Paulo	70	–	22° 15′ 00″ S 47° 49′ 00″ W	Nogueira et al. (2013)
Itirapina	São Paulo	80	6	22° 15′ 54″ S 47° 52′ 44″ W	Faria (2009)

(continued)

Table 13.3 (continued)

City	State	Se (µg/kg)	S (g/kg)	Geographic coordinates	References
Marília	São Paulo	10	–	22° 13′ 15″ S 49° 56′ 55″ W	Nogueira et al. (2013)
Matão	São Paulo	98	5	21° 35′ 27″ S 48° 26′ 54″ W	Faria (2009)
Miguelópolis	São Paulo	30	–	20° 10′ 00″ S 48° 02′ 00″ W	Nogueira et al. (2013)
Mogi Mirim	São Paulo	70	–	22° 22′ 00″ S 46° 56′ 00″ W	Nogueira et al. (2013)
Mogi-Guaçu	São Paulo	100	–	22° 22′ 00″ S 46° 56′ 00″ W	Nogueira et al. (2013)
Pariquera Açu	São Paulo	670	–	24° 43′ 00″ S 47° 52′ 00″W	Nogueira et al. (2013)
Pariquera Açu	São Paulo	650	–	24° 43′ 10″ S 47° 52′ 10″ W	Nogueira et al. (2013)
Piracicaba	São Paulo	60	–	22° 43′ 10″ S 47° 38′ 10″ W	Nogueira et al. (2013)
Piracicaba	São Paulo	560	–	22° 43′ 15″ S 47° 38′ 16″ W	Nogueira et al. (2013)
Piracicaba	São Paulo	30	–	22° 43′ 10″ S 47° 38′ 20″ W	Nogueira et al. (2013)
Piracicaba	São Paulo	320	–	22° 43′ 18″ S 47° 38′ 23″ W	Nogueira et al. (2013)
Piracicaba	São Paulo	68	6	22° 38′ 36″ S 47° 49′ 52″ W	Faria (2009)
Piracicaba	São Paulo	220	12	22° 42′ 40″ S 47° 37′ 43″ W	Faria (2009)
Piracicaba	São Paulo	108	17	22° 38′ 40″ S 47° 49′ 24″ W	Faria (2009)
Pirassununga	São Paulo	160	8	22° 04′ 60″ S 47° 34′ 36″ W	Faria (2009)
Pirassununga	São Paulo	78	8	21° 56′ 30″ S 47° 28′ 50″ W	Faria (2009)
Pirassununga	São Paulo	197	9	21° 57′ 60″ S 47° 26′ 60″ W	Faria (2009)
Ribeirão Preto	São Paulo	40	–	21° 10′ 00″ S 47° 48′ 00″ W	Nogueira et al. (2013)
Ribeirão Preto	São Paulo	110	–	21° 10′ 00″ S 47° 48′ 00″ W	Nogueira et al. (2013)
São Carlos	São Paulo	60	–	22° 01′ 00″ S 47° 53′ 00″ W	Nogueira et al. (2013)
São Pedro	São Paulo	10	–	22° 32′ 15″ S 47° 54′ 00″ W	Nogueira et al. (2013)
São Pedro	São Paulo	10	–	22° 32′ 23″ S 47° 54′ 16″ W	Nogueira et al. (2013)
Deficiency Se range		**100–600**	**–**		**Lyons et al.** (2003)

concentrations in pasture grass (Brachiaria sp. and Stylosanthes sp.) ranging from 40 to 66 μg/kg. On the other hand, Brazilian nuts growing in the North-West of Brazil are considered the richest food source for Se, but its concentrations range dramatically from 0.03 to 512 mg/kg, likely reflecting soil Se. There is evidence of Se deficiency in the Brazilian human population; however, no extensive research data on the subject are available.

13.5 Selenium Status in Soils in Relation to Plant and Human Health

The relationship between Se content in soils and plants as well as human health can be followed through many recent studies (Hatfield et al. 2012; Yuan et al. 2012; Fordyce 2013; Hurst et al. 2013; El-Ramady et al. 2015b,c; Alfthan et al. 2015; Mora et al. 2015; Winkel et al. 2015; El-Ramady et al. 2016b; Wang et al. 2016a; White 2016). It should be noted that human Se intake and Se status in the population depends firstly on Se concentration in soils, and hence the Se concentrations in the harvested edible plants in these soils. In other words, human Se intake and Se status start from the Se concentration in soils. Many factors can affect the content of Se in different foods, including different uptake rate by plants, which is related to plant type, soil pH, microbial activity, rainfall and a number of other biogeochemical parameters (Stoffaneller and Morse 2015). Therefore, human Se intake and Se status can be largely controlled by manipulating Se concentrations in plants, which are a function of soil Se concentration, speciation and bioavailability, as well as the activity of soil microorganisms (Winkel et al. 2015).

The interrelation of soil, crop and human Se status has been impressively shown in a recent Chinese study analyzing the importance of Se for the risk of thyroid diseases. Wu and colleagues studied the neighboring Se-rich and Se-poor regions of Ziyang and Ningshan counties, where average soil Se concentrations were 4–33 mg/kg and 0.17 mg/kg, respectively. This difference directly translated into the average blood Se concentrations of the farmers living in these areas, with the subjects from Ziyang displaying 103.6 μg/L (IQR 79.7, 135.9) versus the farmers from the Se-poor area of Ningshan, who had an average of only 57.4 μg/L (IQR: 39.4, 82.1). A lower Se status is known to increase thyroid disease risks (Schomburg 2011), and consequently, the farmers in Ningshan showed an almost twice as high prevalence of hypothyroidism and autoimmune thyroid diseases than those from Ziyang (Wu et al. 2015). Extrapolating this concept, it would be highly fascinating and interesting trying to calculate the number of Se-dependent diseases worldwide that could be prevented by a better Se supply. Of course, due to the different life styles, differences in genotype, environment, nutrition and activity patterns and the complex and multifactorial reasons for human diseases, this is very complicated to do.

One way to test the importance of Se for human health is via a retrospective observational study. This is a longitudinal human study where serum or plasma samples are collected, analyzed and stored over long periods of time. Then, e.g. 20 years later, when some of the participants have developed a certain disease, their blood samples are analyzed together with samples from comparable control subjects from within the same study. One instructive example has just been published analyzing the importance of the Se status for preventing colorectal cancer (Hughes et al. 2015), which clearly provided evidence that within the same population, the subjects with relatively low Se status had a significantly increased risk for this devastating disease.

A second very powerful way of testing the importance of Se for human health is by conducting randomized controlled supplementation studies. Here, subjects are recruited and asked to take a daily supplement containing Se, while a control group takes a placebo. The participants do not know into which group they have been recruited, an assay design called "blind". In the high quality trials also the medical doctors are unaware of the *verum* or placebo status of a given patient, in which case the study is denoted as a "double blind" study. These studies run over several years and are relatively expensive.

With respect to cancer, two most important double-blind randomized controlled trials (RCT) have been conducted in the US. The nutritional prevention of cancer trials (NPC trial) yielded an impressive reduction of cancer cases by Se supplementation over a study period of around 5 years (Clark et al. 1996). The more recent SELECT (Selenium and Vitamin E Cancer Prevention Trial) failed to replicate these impressive chemopreventive effects of Se (Klein et al. 2011). The most likely reason for this discrepancy lies in the baseline Se status of the participants, which were already very high at the start of SELECT. The comparison of these two RCT is supporting the notion that health benefits of Se supplements are restricted to those human subjects who have an insufficient intake and a submaximal expression of the biologically active selenoproteins. Similar results have been obtained in a number of respective animal experiments. Together, these studies highlight the essentiality of Se for humans and other mammals, and the need for a sufficient daily Se intake.

In conclusion, a dramatic number of humans worldwide likely fall into the Se deficient category. Lyons et al. (2003) estimated that around a billion people are Se deficient, and it might even be more. The fraction of sub-optimally Se supplied humans currently may include a large part of the European population (with the exception of Finland, where a nation-wide Se supplementation effort is in place), large parts of Africa and Asia (including China), and also Australia, New Zealand and large parts of South America. This "hidden hunger" may translate to higher incidence of infections (e.g. in Africa), osteopathy problems (in China), and cancer and thyroid problems (in Europe). Selenium toxicity is a problem of smaller magnitude, but has its own set of devastating effects in different areas across the world. A solution to both of these problems is to focus on development of Se-enriched dietary plant material.

13.6 Roles of Plants in Alleviating Se Deficiency and Toxicity

It is well documented that Se has a vital role in alleviating toxic effects in plants of heavy metals and other oxidative stresses, and can promote plant growth when supplied at low concentration. However, a phytotoxic effect and inhibition of plant growth has also been reported for many plant species when grown under high Se concentration (Feng et al. 2013). All plants readily take up Se, a property that may be used for cleanup of excess environmental Se (phytoremediation) or for biofortification of crops with Se. Phytoremediation of Se-enriched soil or water can be considered an emerging field. Some studies have been published concerning the role of plants in alleviating environmental Se toxicity, such as Gupta and Gupta (2015) and Hawrylak-Nowak et al. (2015). Concerning the use of plants to solve the problem of Se deficiency, several plant species have been successfully biofortified with Se to alleviate this deficiency, including rice (Pandey and Gupta 2015), maize (Longchamp et al. 2013, 2015), wheat (Galinha et al. 2015; Lazo-Vélez et al. 2015), cucumber (Hawrylak-Nowak et al. 2015), lentil (Ekanayake et al. 2015), lettuce (Hawrylak-Nowak 2013) and cruciferous vegetables (Bañuelos et al. 2015; Bachiega et al. 2016). Selenium may be applied to soil as fertilizer or as foliar spray, in the forms of selenate or selenite.

Collectively, it is becoming more and more obvious that Se plays an important role in human health, and dietary Se intake worldwide largely depends on crop Se content. Plant Se accumulation depends on a given soil, as well as on plant species, as some plants are able to accumulate and tolerate high Se concentrations while other plant species do not take up much Se and are sensitive to it. Furthermore, Se bioavailability in soil has a direct impact on the Se concentrations of the plants that are locally produced and consumed, and thereby on the daily Se intake of humans. The results from an increasing number of clinical studies highlight the health risks associated with too low a daily Se intake, a problem that may affect a billion or more people. In order to improve this situation in the future, research needs to be intensified on the improvement of soil Se bioavailability, ways for controlling and optimizing Se uptake and accumulation in plants, and the many health effects that are related to Se status in humans. It is hoped that an increased awareness for this topic will in the long run improve human health in general and especially in the low-income countries where infectious and childhood diseases are a constant and deadly health threat to large parts of the population, and also to the general human community. Biofortification of crops with Se can be a relatively cost-effective and safe way to bring about important significant health benefits to the world population. This trace element can be analytically monitored fairly easily, and its levels controlled on its path from atmosphere to soil, to plant and finally to animal and human organisms.

References

Acuña JJ, Jorquera MA, Barr PJ, Crowley DE, de la Mora M (2013) Selenobacteria selected from the rhizosphere as a potential tool for Se biofortification of wheat crops. Biol Fertil Soils 49:175–185

Aggarwal M, Sharma S, Kaur N, Pathania D, Bhandhari K, Kaushal N, Kaur R, Singh K, Srivastava A, Nayyar H (2011) Exogenous proline application reduces phytotoxic effects of selenium by minimising oxidative stress and improves growth in bean (*Phaseolus vulgaris* L.) seedlings. Biol Trace Elem Res 140(3):354–367

Akbaraly TN, Arnaud J, Rayman MP, Hininger-Favier I, Roussel AM, Berr C, Fontbonne A (2010) Plasma selenium and risk of dysglycemia in an elderly French population: results from the prospective epidemiology of vascular ageing study. Nutr Metab 7:1–7

Akbulut M, Çakır S (2010) The effects of Se phytotoxicity on the antioxidant systems of leaf tissues in barley (*Hordeum vulgare* L.) seedlings. Plant Physiol Biochem 48(2–3):160–166

Alasfar F, Ben-Nakhi M, Khoursheed M, Kehinde EO, Alsaleh M (2011) Selenium is significantly depleted among morbidly obese female patients seeking bariatric surgery. Obes Surg 21:1710–1713

Alfthan G, Eurola M, Ekholm P, Venäläinen ER, Root T, Korkalainen K, Hartikainen H, Salminen P, Hietaniemi V, Aspila P, Aro A (2015) Effects of nationwide addition of selenium to fertilizers on foods, and animal and human health in Finland: from deficiency to optimal selenium status of the population. J Trace Elem Med Biol 31:142–147

Al-Saleh I, Billedo G, El-Doush I, El-Din M, Yosef G (2006) Selenium and vitamins status in Saudi children. Clin Chim Acta 368:99–109

Avila FW, Yang Y, Faquin V, Ramos SJ, Guilherme LRG, Thannhauser TW, Li L (2014) Impact of selenium supply on Se-methylselenocysteine and glucosinolate accumulation in selenium-biofortified brassica sprouts. Food Chem 165:578–586

Azab SF, Saleh SH, Elsaeed WF, Elshafie MA, Sherief M, Esh AM (2014) Serum trace elements in obese Egyptian children: a case-control study. Ital J Pediatr 10:1–7

Bachiega P, Salgado JM, de Carvalho JE, Ruiz ALTG, Schwarz K, Tezotto T, Morzelle MC (2016) Antioxidant and antiproliferative activities in different maturation stages of broccoli (*Brassica oleracea* Italica) biofortified with selenium. Food Chem 190:771–776

Bañuelos GS, Lin ZQ (2007) Acceleration of selenium volatilization in seleniferous agricultural drainage sediments amended with methionine and casein. Environ Pollut 150:306–312

Bañuelos GS, Arroyo I, Pickering IJ, Yang SI, Freeman JL (2015) Selenium biofortification of broccoli and carrots grown in soil amended with Se-enriched hyperaccumulator *Stanleya pinnata*. Food Chem 166:603–608

Bañuelos GS, Lin Z-Q, de Moraes MF, Guilherme LRG, Reis AR (2016) Global advances in selenium research from theory to application. CRC/Taylor & Francis Group, London

Bissardon C, Charlet L, Bohic S, Khan I (2016) Role of the selenium in articular cartilage metabolism, growth, and maturation. In: Bañuelos GS, Lin Z-Q, de Moraes MF, Guilherme LRG, Reis AR (eds) Global advances in selenium research from theory to application. CRC/Taylor & Francis Group, London, pp 77–78

Blass SC, Goost H, Burger C, Tolba RH, Stoffel-Wagner B, Stehle P, Ellinger S (2013) Extracellular micronutrient levels and pro−/antioxidant status in trauma patients with wound healing disorders: results of a cross-sectional study. Nutr J 12:1–7

Błażewicz A, Klatka M, Astel A, Korona-Glowniak I, Dolliver W, Szwerc W, Kocjan R (2015) Serum and urinary selenium levels in obese children: a cross-sectional study. J Trace Elem Med Biol 29:116–122

Boldrin PF, Faquin V, Ramos SJ, Boldrin KVF, Avila FW, Guilherme LRG (2013) Soil and foliar application of selenium in rice biofortification. J Food Compos Anal 31:238–244

Brigelius-Flohé R, Sies H (2016) Diversity of selenium functions in health and disease. Oxidative stress and disease series, vol 38. CRC/Press Taylor & Francis, Boca Raton

Bülow Pedersen I, Knudsen N, Carlé A, Schomburg L, Köhrle J, Jørgensen T, Rasmussen L (2013) Serum selenium is low in newly diagnosed graves' disease: a population-based study. Clin Endocrinol 79:584–590
Carvalho GS (2011) Selenium and mercury in cerrados soils of Brazil. PhD thesis in Soil Science. Federal University of Lavras, Brazil
Chawla R, Loomba R, Chaudhary RJ, Singh S, Dhillon KS (2016) Impact of high selenium exposure on organ function & biochemical profile of the rural population living in seleniferous soils in Punjab, India. In: Bañuelos GS, Lin Z-Q, de Moraes MF, Guilherme LRG, Reis AR (eds) Global advances in selenium research from theory to application. CRC/Taylor & Francis Group, London, pp 93–94
Chen Y, Mo HZ, Hu LB, Li YQ, Chen J, Yang LF (2014) The endogenous nitric oxide mediates selenium-induced phytotoxicity by promoting ROS generation in *Brassica rapa*. PLoS One 9(10):e110901
Chilimba ADC, Young SD, Black CR, Meacham MC, Lammel J, Broadley MR (2012) Agronomic biofortification of maize with selenium (Se) in Malawi. Field Crop Res 125:118–128
Clark RF, Strukle E, Williams SR, Manoguerra AS (1996) Selenium poisoning from a nutritional supplement. JAMA 275(14):1087–1088
Combs GF (2001) Selenium in global food systems. Brit J Nutr 85:517–547
Combs GF Jr (2016) Who can benefit from selenium? In: Brigelius-Flohé R, Sies H (eds) Diversity of selenium functions in health and disease, Oxidative stress and disease series, vol 38. CRC Press/Taylor & Francis, Boca Raton, pp 3–15
Dhillon KS, Dhillon SK (2009a) Accumulation and distribution of selenium in some vegetable crops grown in selenate-Se treated clay loam soil. Front Agric China 3(4):366–373
Dhillon KS, Dhillon SK (2009b) Selenium concentrations of common weeds and agricultural crops grown in the seleniferous soils of northwestern India. Sci Total Environ 407:6150–6156
Dhillon KS, Dhillon SK (2014) Development and mapping of seleniferous soils in northwestern India. Chemosphere 99:56–63
Dhillon KS, Dhillon SK (2016a) Selenium in groundwater and its contribution towards daily dietary Se intake under different hydrogeological zones of Punjab. J Hydrol 533:615–626
Dhillon KS, Dhillon SK (2016b) Phytoremediation of selenium contaminated soils: strategies and limitations. In: Bañuelos GS, Lin Z-Q, de Moraes MF, Guilherme LRG, Reis AR (eds) Global advances in selenium research from theory to application. CRC/Taylor & Francis Group, London, pp 201–202
Dhillon SK, Dhillon KS, Kohli A, Khera KL (2008) Evaluation of leaching and runoff losses of selenium from seleniferous soils through simulated rainfall. J Plant Nutr Soil Sci 171:187–192
Domingues CRS, Pascoalino JAL, Moraes MF, Santos CLR, Reis AR, Franco FA, Evangelista A, Scheeren PL (2016) Wheat biofortification: genotypic variation and selenium fertilization in Brazil. In: Bañuelos GS, Lin Z-Q, de Moraes MF, Guilherme LRG, Reis AR (eds) Global advances in selenium research from theory to application. CRC/Taylor & Francis Group, London, pp 183–184
dos Reis AR (2016) Selenium status in Brazilian soils and crops: agronomic biofortification as a strategy to improve food quality. In: Bañuelos GS, Lin Z-Q, de Moraes MF, Guilherme LRG, Reis AR (eds) Global advances in selenium research from theory to application. CRC/Taylor & Francis Group, London, pp 179–180
Eiche E, Bardelli F, Nothstein AK, Charlet L, Göttlicher J, Steininger R, Dhillon KS, Sadana US (2015) Selenium distribution and speciation in plant parts of wheat (*Triticum aestivum*) and Indian mustard (*Brassica juncea*) from a seleniferous area of Punjab. Sci Total Environ 505:952–961
Ekanayake LJ, Thavarajah D, Vial E, Schatz B, McGee R, Thavarajah P (2015) Selenium fertilization on lentil (*Lens culinaris Medikus*) grain yield, seed selenium concentration, and antioxidant activity. Field Crop Res 177:9–14

El Mehdawi AF, Pilon-Smits EA (2012) Ecological aspects of plant selenium hyperaccumulation. Plant Biol (Stuttg) 14(1):1–10

Elemraid MA, Mackenzie IJ, Fraser WD, Harper G, Faragher B, Atef Z, Al-Aghbari N, Brabin BJ (2011) A case-control study of nutritional factors associated with chronic suppurative otitis media in Yemeni children. Eur J Clin Nutr 65:895–902

El-Ramady H, Abdalla N, Alshaal T, El-Henawy A, Faizy S. E-D A, Shams M S, Shalaby T, Bayoumi Y, Elhawat N, Shehata S, Sztrik A, Prokisch J, Fári M, Pilon-Smits E A, Domokos-Szabolcsy É (2015a). Selenium and its role in higher plants. *In*: Lichtfouse E, Schwarzbauer J, Robertet D (eds.), Environ Chem Sustainable World 7:235–296.

El-Ramady H, Domokos-Szabolcsy É, Shalaby T A, Prokisch J, Fári M (2015b). Selenium in agriculture: water, air, soil, plants, food, animals and nanoselenium. *In*: Lichtfouse E (ed.), Environ Chem Sustainable World 5:153–232.

El-Ramady H, Abdalla N, Alshaal T, Domokos-Szabolcsy É, Elhawat N, Prokisch J, Sztrik A, Fári M, El-Marsafawy S, Shams MS (2015c) Selenium in soils under climate change, implication for human health. Environ Chem Lett 13(1):1–19

El-Ramady H, Alshaal T, Abdalla N, Prokisch J, Sztrik A, Fári M, Domokos-Szabolcsy É (2016a) Selenium and nano-selenium biofortified sprouts using micro-farm systems. In: Bañuelos GS, Lin Z-Q, de Moraes MF, Guilherme LRG, Reis AR (eds) Global advances in selenium research from theory to application. CRC/Taylor & Francis Group, London, pp 189–190

El-Ramady H, Abdalla N, Taha HS, Alshaal T, El-Henawy A, Faizy SE-DA, Shams MS, Youssef MS, Shalaby T, Bayoumi Y, Elhawat N, Shehata S, Sztrik A, Prokisch J, Fári M, Domokos-Szabolcsy É, Pilon-Smits EA, Selmar D, Haneklaus S, Schnug E (2016b) Selenium and nano-selenium in plant nutrition. Environ Chem Lett 14(1):123–147

El-Yazigi A, Legayada E (1996) Urinary selenium in healthy and diabetic Saudi Arabians. Biol Trace Elem Res 52:55–63

Fairweather-Tait SJ, Bao YP, Broadley MR, Collings R, Ford D, Hesketh JE, Hurst R (2011) Selenium in human health and disease. Antioxid Redox Signal 14:1337–1383

Faria LAF (2009) Overview Selenium Quantity in Soils and Plants of State São Paulo, and Selenium Application on Different Forages. PhD thesis in Animal Science, The University São Paulo, Brazil. 57p.

Faria L A, Machado M C, Abdalla A L, Righeto P P, Campos L L, Karp F H S, Kamogawa M Y (2016). Agronomic biofortification of Brachiaria with selenium along with urea. In: Bañuelos G S, Lin Z-Q, de Moraes MF, Guilherme L R G, and Reis A R (Eds.), Global advances in selenium research from theory to application, CRC/Taylor & Francis Group, London, pp: 159–160.

Fenech M, Wu J, Graham R, Lyons G (2013). Selenium biofortified wheat. In: Preedy V R, Srirajaskanthan R, Patel V B. Handbook of food fortification and health: from concepts to public health applications volume 1. Nutrition and health. Springer Science + Business Media New York, pp. 349–356.

Feng R, Wei C, Tu D (2013). The roles of selenium in protecting plants against abiotic stress. Environ Exp Bot 87:58–68.

Ferreira KS, Gomes JC, Bellato CR, Jordão CP (2002) Selenium content in Brazilian foods. Pan Am J Public Health 11(3):171–177

Fordyce FM (2013) Selenium deficiency and toxicity in the environment. In: Selinus O (ed.), Essentials of Medical Geology: Rev Edn, British Geological Survey, pp 375–419

Frankenberger WT Jr, Benson S (1994) Selenium in the environment. Marcel Dekkar, New York, 456p

Galinha C, Freitas MC, Pacheco AMG, Coutinho J, Macas B, Almeida AS (2013) Selenium supplementation of Portuguese wheat cultivars through foliar treatment in actual field conditions. J Radioanal Nucl Chem 297:227–231

Galinha C, Sánchez-Martínez M, Pacheco AMG, do Carmo Freitas M, Coutinho J, Maçãs B, Sofia Almeida A, Teresa Pérez-Corona M, Madrid Y, Wolterbeek HT (2015) Characterization of selenium-enriched wheat by agronomic biofortification. J Food Sci Technol 52(7):4236–4245

Gao J, Liu Y, Huang Y, Lin Z, Bañuelos GS, Lam MH-W, Yin X (2011) Daily selenium intake in a moderate selenium deficiency area of Suzhou. Food Chem 126(3):1088–1093
Gojkovic Ž, Garbayo I, Luis GJ, Ariza I, Márová C, Vílchez (2015) Selenium bioaccumulation and toxicity in cultures of green microalgae. Algal Res 7:106–116
Gong P, Li T, Wang A, Sun F, Gu S, Yin X, Guan W (2014) Screening wheat genotypes for selenium biofortification in Brazil. In: Bañuelos GS, Lin Z–Q, Yin XB (eds) Selenium in the environment and human health. Taylor & Francis Group, London, pp 142–143
Gupta S, Gupta M (2015) Alleviation of selenium toxicity in *Brassica juncea* L.: salicylic acid-mediated modulation in toxicity indicators, stress modulators, and sulfur-related gene transcripts. Protoplasma 253(6):1–14
Han J, Liang H, Yi J, Tan W, He S, Wu X, Shi X, Ma J, Guo X (2016) Selenium deficiency induced damages and altered expressions of metalloproteinases and their inhibitors (MMP1/3, TIMP1/3) in the kidneys of growing rats. J Trace Elem Med Biol 34:1–9
Handa N, Bhardwaj R, Poonam HK, Kapoor D, Rattan A, Kaur S, Thukral AK, Kaur S, Arora S, Kapoor N (2016) Selenium: an antioxidant protectant in plants under stress. In: Ahmed P (ed) Plant metal interaction. Elsevier, Amsterdam, pp 179–208
Harris J, Schneberg KA, Pilon-Smits EAH (2014) Sulfur–selenium–molybdenum interactions distinguish selenium hyperaccumulator *Stanleya pinnata* from non-hyperaccumulator *Brassica juncea* (Brassicaceae). Planta 239:479–491
Hatfield DL, Berry MJ, Gladyshev VN (2012) Selenium: its molecular biology and role in human health. Springer, New York
Hauser-Davis RA, Silva JAN, Rocha RCC, Saint'Pierre T, Ziolli RL, Arruda MAZ (2016) Acute selenium selenite exposure effects on oxidative stress biomarkers and essential metals and trace-elements in the model organism zebrafish (*Danio rerio*). J Trace Elem Med Biol 33:68–72
Hawrylak-Nowak B (2013) Comparative effects of selenite and selenate on growth and selenium accumulation in lettuce plants under hydroponic conditions. Plant Growth Regul 70(2):149–157
Hawrylak-Nowak B, Matraszek R, Pogorzelec M (2015) The dual effects of two inorganic selenium forms on the growth, selected physiological parameters and macronutrients accumulation in cucumber plants. Acta Physiol Plant 37:41
Hladu KR, Parker DR, Tran KD, Trumble JT (2013) Effects of selenium accumulation on phytotoxicity, herbivory, and pollination ecology in radish (*Raphanus sativus* L.) Environ Pollut 172:70–75
Hoffmann PR (2016) Selenium as a regulator of immune and inflammatory responses. In: Bañuelos GS, Lin Z-Q, de Moraes MF, Guilherme LRG, Reis AR (eds) Global advances in selenium research from theory to application. CRC/Taylor & Francis Group, London
Hughes DJ, Fedirko V, Jenab M, Schomburg L, Méplan C, Freisling H, Bueno-de-Mesquita HB, Hybsier S, Becker NP, Czuban M, Tjønneland A, Outzen M, Boutron-Ruault MC, Racine A, Bastide N, Kühn T, Kaaks R, Trichopoulos D, Trichopoulou A, Lagiou P, Panico S, Peeters PH, Weiderpass E, Skeie G, Dagrun E, Chirlaque MD, Sánchez MJ, Ardanaz E, Ljuslinder I, Wennberg M, Bradbury KE, Vineis P, Naccarati A, Palli D, Boeing H, Overvad K, Dorronsoro M, Jakszyn P, Cross AJ, Quirós JR, Stepien M, Kong SY, Duarte-Salles T, Riboli E, Hesketh JE (2015) Selenium status is associated with colorectal cancer risk in the European prospective investigation of cancer and nutrition cohort. Int J Cancer 136(5):1149–1161
Hurst R, Collings R, Harvey L, King M, Hooper L, Bouwman J, Gurinovic M, Fairweather-Tait SJ (2013) EURRECA – estimating selenium requirements four deriving dietary reference values. Crit Rev Food Sci 53:1077–1096
Jaworska K, Gupta S, Durda K, Muszyńska M, Sukiennicki G, Jaworowska E, Grodzki T, Sulikowski M, Waloszczyk P, Wójcik J (2013) A low selenium level is associated with lung and laryngeal cancers. PLoS One 8:e59051
Kabata-Pendias A, Mukherjee AB (2007) Trace elements from soil to human. Springer, Berlin/Heidelberg/New York

Kieliszek M, Błażejak S, Płaczek M (2016) Spectrophotometric evaluation of selenium binding by *Saccharomyces cerevisiae* ATCC MYA-2200 and *Candida utilis* ATCC 9950 yeast. J Trace Elem Med Biol 35:90–96

Klein EA, Thompson IM, Tangen CM, Crowley JJ, Lucia MS, Goodman FJ, Minasian LM, Ford LG, Parnes HL, Gaziano JM, Karp DD, Lieber MM, Walther PJ, Klotz L, Parsons JK, Chin JL, Darke AK, Lippman SM, Goodman GE, Meyskens FL, Baker LH (2011) Vitamin E and the risk of prostate cancer the selenium and vitamin E cancer prevention trial (SELECT). JAMA 306(14):1549–1556

Kong L, Wang M, Bi D (2005) Selenium modulates the activities of antioxidant enzymes, osmotic homeostasis, and promotes the growth of sorrel seedlings under salt stress. Plant Growth Regul 45:155–163

Koszta G, Kacska Z, Szatmári K, Szerafin T, Fülesdi B (2012) Lower whole blood selenium level is associated with higher operative risk and mortality following cardiac surgery. J Anesth 26:812–821

Koukkou E, Ilias I, Alexiou M, Mamali I, Nicopoulou S, Alevizaki M, Markou K (2014) Urine selenium changes during pregnancy do not correlate with thyroid autoantibodies in a mildly iodine deficient population. Biol Trace Elem Res 157:9–13

Krohn RM, Lemaire M, Silva LFN, Lemarié C, Bolt A, Mann KK, Smits JE (2016) High-selenium lentil diet protects against arsenic-induced atherosclerosis in a mouse model. J Nutr Biochem 27:9–15

Lampis S, Zonaro E, Bertolini C, Cecconi D, Monti F, Micaroni M, Turner RJ, Butler CS, Vallini G (2016) Selenite biotransformation and detoxification by Stenotrophomonas maltophilia SeITE02: novel clues on the route to bacterial biogenesis of selenium nanoparticles. J Hazard Mater S0304-3894(16):30162–30165

Lazo-Vélez MA, Chávez-Santoscoy A, Serna-Saldivar SO (2015) Selenium-enriched breads and their benefits in human nutrition and health as affected by agronomic, milling, and baking factors. Cereal Chem 92(2):134–144

Lehotai N, Lyubenova L, Schröder P, Feigl G, Ördög A, Szilágyi K, Erdei L, Kolbert Z (2015) Nitro-oxidative stress contributes to selenite toxicity in pea (*Pisum sativum* L). Plant Soil 400:107–122

Li N, Gao ZD, Luo DG, Tang X, Chen D, Hu Y (2007) Selenium level in the environment and the population of Zhoukoudian area. Sci Total Environ 381(1):105–111

Li T, Wang A, Gong P, Gu S, Yuan L, Li F, Yin X, Guan W (2014) Microbial-enhanced selenium biofortification of wheat (*Triticum aestivum* L.) In: Bañuelos GS, Lin Z-Q, Yin X (eds) Selenium in the environment and human health. Taylor & Francis Group, London, pp 138–139

Li M Q, Hasan M K, Li C X, Ahammed G J, Xia X J, Shi K, Zhou Y H, Reiter R J, Yu J Q, Xu M X, Zhou J. (2016). Melatonin mediates selenium-induced tolerance to cadmium stress to tomato plants. J Pineal Res June 6.

Longchamp M, Castrec-Rouelle M (2014) Uptake of selenate versus selenite in *Zea mays*: biofortification of crops and forage. In: Banuelos GS, Lin Z-Q, Yin X (eds) Selenium in the environment and human health. Taylor & Francis Group, London, pp 118–119

Longchamp M, Angeli N, Castrec-Rouelle M (2013) Selenium uptake in *Zea mays* supplied with selenate or selenite under hydroponic conditions. Plant Soil 362:107–117

Longchamp M, Castrec-Rouelle M, Biron P, Bariac T (2015) Variations in the accumulation, localization and rate of metabolization of selenium in mature *Zea mays* plants supplied with selenite or selenate. Food Chem 182:128–135

Lopes PA, Santos MC, Vicente L, Rodrigues MO, Pavão ML, Nève J, Viegas-Crespo AM (2004) Trace element status (Se, Cu, Zn) in healthy Portuguese subjects of Lisbon population: a reference study. Biol Trace Elem Res 101:1–17

Lyons G, Stangoulis L, Graham R (2003) High-selenium wheat: biofortification for better health. Nutr Res Rev 16:45–60

Madaan N, Mudgal V (2011) Phytotoxic effect of selenium on the accessions of wheat and safflower. Res J Environ Sci 5:82–87

Malagoli M, Schiavon M, dall' Acqua S, Pilon-Smits EAH (2015) Effects of selenium biofortification on crop nutritional quality. Front Plant Sci 6:280

Manzanares W, Hardy G (2016) Can dietary selenium intake increase the risk of toxicity in healthy children? Nutrition 32(1):149–150

Mao H, Lyons GH, Wang ZH (2016) Using agronomic biofortification to reduce micronutrient deficiency in food crops on loess soil in China. In: Bañuelos GS, Lin Z-Q, de Moraes MF, Guilherme LRG, Reis AR (eds) Global advances in selenium research from theory to application. CRC/Taylor & Francis Group, London, pp 163–164

Massadeh A, Gharibeh A, Omari K, Al-Momani I, Alomary A, Tumah H, Hayajneh W (2010) Simultaneous determination of Cd, Pb, Cu, Zn, and Se in human blood of Jordanian smokers by ICP-OES. Biol Trace Elem Res 133:1–11

Mechora S, Stibilj V, Germ M (2015) Response of duckweed to various concentrations of selenite. Environ Sci Pollut Res 22:2416–2422

Menezes C, Marins A, Murussi C, Pretto A, Leitemperger J, Loro VL (2016) Effects of diphenyl diselenide on growth, oxidative damage, and antioxidant response in silver catfish. Sci Total Environ 542:231–237

Millán EA, Florea D, Sáez Pérez L, Molina JL, López-González B, Cruz PA, Planells del Pozo E (2012) Deficient selenium status of a healthy adult Spanish population. Nutr Hosp 27:524–528

Molnárová M, Fargasová A (2009) Se (IV) phytotoxicity for monocotyledonae cereals (*Hordeum vulgare* L., *Triticum aestivum* L.) and dicotyledonae crops (*Sinapis alba* L., *Brassica napus* L.) J Hazard Mater 172(2–3):854–861

Mora ML, Durán P, Acuña AJ, Cartes P, Demanet R, Gianfreda L (2015) Improving selenium status in plant nutrition and quality. J Soil Sci Plant Nutr 15(2):486–503

Moreno-Reyes R, Suetens C, Mathieu F, Begaux F, Zhu D, Boelaert MM, Nève J, Perlmutter N, Vanderpas J (1998) Kashin-Beck osteoarthropathy in rural Tibet in relation to selenium and iodine status. N Engl J Med 339:1112–1120

Nagy G, Benko I, Kiraly G, Voros O, Tanczos B, Sztrik A, Takács T, Pocsi I, Prokisch J, Banfalvi G (2015) Cellular and nephrotoxicity of selenium species. J Trace Elem Med Biol 30:160–170

Nancharaiah YV, Lens PNL (2015) Selenium biomineralization for biotechnological applications. Trends Biotechnol 33(6):323–330

Nawaz F, Ahmad R, Ashraf MY, Waraich EA, Khan SZ (2015) Effect of selenium foliar spray on physiological and biochemical processes and chemical constituents of wheat under drought stress. Ecotoxicol Environ Saf 113:191–200

Niedzielski P, Rudnicka M, Wachelka M, Kozak L, Rzany M, Wozniak M, Kaskow Z (2016) Selenium species in selenium fortified dietary supplements. Food Chem 190:454–459

Nogueira TAR, Alleoni LRF, He Z, Villanueva FCA, Poggere GC, Abreu Junior CH (2013) Teores naturais e valor de referência de qualidade para selênio em solos do estado de São Paulo. In: XXXIV CONGRESSO BRASILEIRO DE CIÊNCIA DO SOLO, 2013. Anais. Florianópolis. CD-ROM.

Obeid O, Elfakhani M, Hlais S, Iskandar M, Batal M, Mouneimne Y, Adra N, Hwalla N (2008) Plasma copper, zinc, and selenium levels and correlates with metabolic syndrome components of Lebanese adults. Biol Trace Elem Res 123:58–65

Oropeza-Moe M, Wisløff H, Bernhoft A (2015) Selenium deficiency associated porcine and human cardiomyopathies. J Trace Elem Med Biol 31:148–156

Ortiz-Monasterio I, Cárdenas ME, Lyons GH (2016) Biofortification of irrigated wheat with Se fertilizer: timing, rate, method and type of wheat. In: Bañuelos GS, Lin Z-Q, de Moraes MF, Guilherme LRG, Reis AR (eds) Global advances in selenium research from theory to application. CRC/Taylor & Francis Group, London, pp 167–168

Pandey C, Gupta M (2015) Selenium and auxin mitigates arsenic stress in rice (*Oryza sativa* L.) by combining the role of stress indicators, modulators and genotoxicity assay. J Hazard Mater 287:384–391

Pilon M, El Mehdawi AF, Cappa JJ, Wang J, Pilon-Smits EAH (2016) Molecular mechanisms of selenium hyperaccumulation in *Stanleya pinnata*: potential key genes *SpSultr1;2* and *SpAPS2*. In: Bañuelos GS, Lin Z-Q, de Moraes MF, Guilherme LRG, Reis AR (eds) Global advances in selenium research from theory to application. CRC/Taylor & Francis Group, London, pp 115–116

Pilon-Smits EAH, El-Mehdawi AF, Cappa JJ, Wang J, Cochran AT, Reynolds RJ, Sura de Jong M (2016) New insights into the multifaceted ecological and evolutionary aspects of plant selenium hyperaccumulation. In: Bañuelos GS, Lin Z-Q, de Moraes MF, Guilherme LRG, Reis AR (eds) Global advances in selenium research from theory to application. CRC/Taylor & Francis Group, London, pp 125–126

Poblaciones MJ, Rodrigo S, Santamaria O, Chen Y, McGrath SP (2014) Selenium accumulation and speciation in biofortified chickpea (*Cicer arietinum* L.) under Mediterranean conditions. J Sci Food Agric 94(6):1101–1106

Pograjc L, Stibilj V, Falnoga I (2012) Impact of intensive physical activity on selenium status. Biol Trace Elem Res 145:291–299

Prakash NT (2016) Quantification, speciation and bioaccessibility of selenium from Se-rich cereals cultivated in seleniferous soils of India. In: Bañuelos GS, Lin Z-Q, de Moraes MF, Guilherme LRG, Reis AR (eds) Global advances in selenium research from theory to application. CRC/Taylor & Francis Group, London, pp 119–120

Qin H-B, Zhu J-M, Liang L, Wang M-S, Su H (2013) The bioavailability of selenium and risk assessment for human selenium poisoning in high-Se areas. Environ Int 52:66–74

Rauhamaa P, Kantola M, Viitak A, Kaasik T, Mussalo-Rauhamaa H (2008) Selenium levels of Estonians. Eur J Clin Nutr 62:1075–1078

Rayman MP (2000) The importance of selenium to human health. Lancet 356:233–241

Rayman MP, Wijnen H, Vader H, Kooistra L, Pop V (2011) Maternal selenium status during early gestation and risk for preterm birth. CMAJ 183:49–555

Rayman MP, Blundell-Pound G, Pastor-Barriuso R, Guallar E, Steinbrenner H, Stranges S (2012) A randomized trial of selenium supplementation and risk of type-2 diabetes, as assessed by plasma adiponectin. PLoS One 7:e45269

Reis AR, Guilherme LRG, Moraes MF, Ramos SJ (2014) High-selenium upland rice: agronomic biofortification strategies to improve human nutrition. In: Bañuelos GS, Lin Z-Q, Yin X (eds) Selenium in the environment and human health. Taylor & Francis Group, London, pp 113–114

Reis HPG, Barcelos JPQ, Reis AR, Moraes MF (2016) Genotypic variation and agronomic biofortication of upland rice with selenium. In: Bañuelos GS, Lin Z-Q, de Moraes MF, Guilherme LRG, Reis AR (eds) Global advances in selenium research from theory to application. CRC/Taylor & Francis Group, London, pp 175–176

Saad K, Farghaly HS, Badry R, Othman HA (2014) Selenium and antioxidant levels decreased in blood of children with breath-holding spells. J Child Neurol 29:1339–1343

Sah S, Vandenberg A, Smits J (2013) Treating chronic arsenic toxicity with high selenium lentil diets. Toxicol Appl Pharmacol 272(1):256–262

Saidi I, Nawel N, Djebali W (2014a) Role of selenium in preventing manganese toxicity in sunflower (*Helianthus annuus*) seedling. S Afr J Bot 94:88–94

Saidi I, Chtourou Y, Djebali W (2014b) Selenium alleviates cadmium toxicity by preventing oxidative stress in sunflower (*Helianthus annuus*) seedlings. J Plant Physiol 171(5):85–91

Sánchez C, López-Jurado M, Aranda P, Llopis J (2010) Plasma levels of copper, manganese and selenium in an adult population in southern Spain: influence of age, obesity and lifestyle factors. Sci Total Environ 408:1014–1020

Schilling K, Johnson TM, Dhillon KS, Mason PRD (2015) Fate of selenium in soils at a seleniferous site recorded by high precision Se isotope measurements. Environ Sci Technol 49:9690–9698
Schomburg L (2011) Selenium, selenoproteins and the thyroid gland: interactions in health and disease. Nature 8:160–171
Sharma VK (2009) Aggregation and toxicity of titanium dioxide nanoparticles in aquatic environment—a review. Journal of Environmental Science and Health, Part A: Toxic/Hazardous Substances and Environmental Engineering 44(14):1485–1495
Sharma N, Prakash R, Srivastava A, Sadana US, Acharya R, Prakash NT, Reddy AVR (2009) Profile of selenium in soil and crops in seleniferous area of Punjab, India by neutron activation analysis. J Radioanal Nucl Chem 281:59–62
Sharma G, Sharma A R, Bhavesh R, Park J, Ganbold B, Nam J-S, Lee S-S. (2014a). Biomolecule-mediated synthesis of selenium nanoparticles using dried *Vitis vinifera* (Raisin) extract. Molecules, 19: 2761–2770.
Sharma S, Kaur N, Kaur S, Nayyar H (2014b) Ascorbic acid reduces the phytotoxic effects of selenium on rice (*Oryza sativa* L.) by up-regulation of antioxidative and metal-tolerance mechanisms. J Plant Physiol Pathol 2(3)
Sharma S, Goyal R, Sandana US (2014c) Selenium accumulation and antioxidant status of rice plants grown on seleniferous soil from northwestern India. Rice Sci 21(6):327–334
Sharma P, Aggarwal P, Kaur A (2016) Biofortification: a new approach to eradicate hidden hunger. Food Rev Int 33(1):1–21
Sherief LM, Abd El-Salam SM, Kamal NM, El Safy O, Almalky MA, Azab SF, Morsy HM, Gharieb AF (2014) Nutritional biomarkers in children and adolescents with Betathalassemia-major: an Egyptian center experience. Biomed Res Int 2014:1–7
Silva Junior ED (2016) Selenium in Brazilian nuts (*Bertholletia excelsa*) and soils from Amazon region. PhD thesis in Soil Science. Federal University of Lavras, Brazil. 75p.
Srivastava A, Pathania D, Swain KK, Ajith N, Acharya R, Reddy AVR, Nayyar H (2012) Application of INAA for phyto-accumulation study of selenium by chickpea plant. J Radioanal Nucl Chem 294(2):315–318
Stoffaneller R, Morse NL (2015) A review of dietary selenium intake and selenium status in Europe and the Middle East. Forum Nutr 7(3):1494–1537
Stoppe C, Schälte G, Rossaint R, Coburn M, Graf B, Spillner J, Marx G, Rex S (2011) The intraoperative decrease of selenium is associated with the postoperative development of multiorgan dysfunction in cardiac surgical patients. Crit Care Med 39:1879–1885
Stoppe C, Spillner J, Rossaint R, Coburn M, Schälte G, Wildenhues A, Marx G, Rex S (2013) Selenium blood concentrations in patients undergoing elective cardiac surgery and receiving perioperative sodium selenite. Nutrition 29:158–165
Stranges S, Laclaustra M, Ji C, Cappuccio FP, Navas-Acien A, Ordovas JM, Rayman M, Guallar E (2010) Higher selenium status is associated with adverse blood lipid profile in British adults. J Nutr 140:81–87
Suadicani P, Hein HO, Gyntelberg F (2012) Serum selenium level and risk of lung cancer mortality: a 16-year follow-up of the Copenhagen male study. Eur Respir J 39:1443–1448
Sun H, Rathinasabapathi JB, Wu B, Luo J, Pu L-P, Ma LQ (2014) Arsenic and selenium toxicity and their interactive effects in humans. Environ Int 69:148–158
Surai P (2006) Selenium in food and feed: selenomethionine and beyond. In: Selenium in nutrition and health. Nottingham University Press, Nottingham, pp 151–212
Talukdar D (2013) Selenium priming selectively ameliorates weed – induced phytotoxicity by modulating antioxidant defense components in lentil (*Lens culinaris* Medik.) and grass pea (*Lathyrus sativus* L.) Ann Rev Res in Biol 3(3):195–212
Tan JA, Zhu WY, Wang WY, Li RB, Wang DC, Yang LS (2002) Selenium in soil and endemic diseases in China. Sci Total Environ 284:227–235
Thomson CD (2004) Selenium and iodine intakes and status in New Zealand and Australia. Br J Nutr 91(5):661–672

Van Hoewyk D (2013) A tale of two toxicities: malformed selenoproteins and oxidative stress both contribute to selenium stress in plants. Ann Bot 112(6):965–972

Vinceti M, Crespi CM, Malagoli C, Bottecchi I, Ferrari A, Sieri S, Krogh V, Alber D, Bergomi M, Seidenari S, Pellacani G (2012) A case-control study of the risk of cutaneous melanoma associated with three selenium exposure indicators. Tumori 98:287–295

Wang Z, Gao Y (2001) Biogeochemical cycling of selenium in Chinese environments. Appl Geochem 16:1345–1351

Wang J, Xiao Y P, Liang X Q, Shao X H, Zhang K (2012). Determination of arsenic, mercury and selenium in *Gynostemma pentaphyllum* and rhizospheric soil samples collected from different regions by hydride generation atomic fluorescence spectrometry. Guang Pu Xue Yu Guang Pu Fen Xi 32(3):813–816 (in Chinese)

Wang X, Tam NF, Fu S, Ametkhan A, Ouyang Y, Ye Z (2014) Selenium addition alters mercury uptake, bioavailability in the rhizosphere and root anatomy of rice (*Oryza sativa*). Ann Bot 114(2):271–278

Wang Q, Zhang J, Zhao B, Xin X, Deng X, Zhang H (2016a) Influence of long-term fertilization on selenium accumulation in soil and uptake by crops. Pedosphere 26(1):120–129

Wang D, Liu Y, Liu D (2016b) Keshan disease and Kaschin-Beck disease in China: is there still selenium deficiency? In: Bañuelos GS, Lin Z-Q, de Moraes MF, Guilherme LRG, Reis AR (eds) Global advances in selenium research from theory to application. CRC/Taylor & Francis Group, London, pp 95–96

White PJ (2016) The genetics of selenium accumulation by plants. In: Bañuelos GS, Lin Z-Q, de Moraes MF, Guilherme LRG, Reis AR (eds) Global advances in selenium research from theory to application. CRC/Taylor & Francis Group, London, pp 111–112

Wimmer I, Hartmann T, Brustbauer R, Minear G, Dam K (2014) Selenium levels in patients with autoimmune thyroiditis and controls in lower Austria. Horm Metab Res 46:707–709

Winkel LHE, Vriens B, Jones GD, Schneider LS, Pilon-Smits EAH, Bañuelos G (2015) Selenium cycling across soil-plant-atmosphere interfaces: a critical review. Forum Nutr 7(6):4199–4239

Wrobel JK, Power R, Toborek M (2016) Biological activity of selenium: revisited. IUBMB Life 68(2):97–105

Wu Q, Rayman M, Lu H, Schomburg L, Cui B, Gao C, Chen P, Zhuang G, Zhang Z, Peng X, Li H, Zhao Y, He X, Zeng G, Qin F, Hou P, Shi B (2015) Low population selenium status is associated with increased prevalence of thyroid disease. J Clin Endocrinol Metab 100(11):4037–4047

Xia Y, Hill KE, Byrne DW, Xu J, Burk RF (2005) Effectiveness of selenium supplements in a low-selenium area of China. Am J Clin Nutr 81:829–834

Yang H, Jia X (2014) Safety evaluation of Se-methylselenocysteine as nutritional selenium supplement: acute toxicity, genotoxicity and subchronictoxicity. Regul Toxicol Pharmacol 70(3):720–727

Yasin M, Faisal M, Pilon-Smits EAH (2014) Selenium biofortification and its effects on flour paste viscosity properties in wheat. In: Bañuelos GS, Lin Z-Q, Yin X (eds) Selenium in the environment and human health. Taylor & Francis Group, London, pp 140–141

Yasin M, El-Mehdawi AF, Pilon-Smits EA, Faisal M (2015a) Selenium-fortified wheat: potential of microbes for biofortification of selenium and other essential nutrients. Int J Theor Phys 17(8):777–786

Yasin M, El-Mehdawi AF, Anwar A, Pilon-Smits EA, Faisal M (2015b) Microbial-enhanced selenium and iron biofortification of wheat (*Triticum aestivum* L.): applications in phytoremediation and biofortification. Int J Theor Phys 17(1–6):341–347

Yasin M, El Mehdawi AF, Jahn CE, Anwar A, Turner MFS, Faisal M, Pilon-Smits EAH (2015c) Seleniferous soils as a source for production of selenium-enriched foods and potential of bacteria to enhance plant selenium uptake. Plant Soil 386:385–394

Yasin M, Faisal M, El Mehdawi AF, Pilon-Smits EAH (2016) Microbe-assisted selenium phytoremediation and phytomanagement of natural seleniferous areas. In: Bañuelos GS, Lin Z-Q, de

Moraes MF, Guilherme LRG, Reis AR (eds) Global advances in selenium research from theory to application. CRC, Taylor & Francis Group, London, pp 199–200

Ye S, Zhang J, Liu Z, Zhang Y, Li J, Li YO (2016) Biosynthesis of selenium rich exopolysaccharide (Se-EPS) by *Pseudomonas* PT-8 and characterization of its antioxidant activities. Carbohydr Polym 142:230–239

Yuan L, Yin X, Zhu Y, Li F, Huang Y, Liu Y, Lin Z (2012) Selenium in plants and soils, and selenosis in Enshi, China: implications for selenium biofortification. In: Yin X, Yuan L (eds) Phytoremediation and biofortification. Springer, Dordrecht, pp 7–31

Yusuf M, Khan TA, Fariduddin Q (2016) Interaction of epibrassinolide and selenium ameliorates the excess copper in *Brassica juncea* through altered proline metabolism and antioxidants. Ecotoxicol Environ Saf 129:25–34

Zanetti MA, Correa LB, Saran Netto A, Cunha JA, Santana RSS, Cozzolino SMF (2016) Influence of canola oil, vitamin E and selenium on cattle meat quality and its effects on nutrition and health of humans. In: Bañuelos GS, Lin Z-Q, de Moraes MF, Guilherme LRG, Reis AR (eds) Global advances in selenium research from theory to application. CRC/Taylor & Francis Group, London, pp 97–98

Zhao J, Gao Y, Li Y-F, Hu Y, Peng X, Dong Y, Li B, Chen C, Chai Z (2013) Selenium inhibits the phytotoxicity of mercury in garlic (*Allium sativum*). Environ Res 125:75–81

Zhu J, Wang N, Li S, Li L, Su H, Liu C (2008) Distribution and transport of selenium in Yutangba, China: impact of human activities. Sci Total Environ 392(2–3):252–261

Zhu Y, Yin X, Liu S, Yuan S (2014) The selenium speciation in the seeds of the common wheat genotypes tending to accumulate high concentrations of selenium. In: Bañuelos GS, Lin Z-Q, Yin X (eds) Selenium in the environment and human health. Taylor & Francis Group, London, pp 136–137

Zhu X, Jiang M, Song E, Jiang X, Song Y (2015) Selenium deficiency sensitizes the skin for UVB-induced oxidative damage and inflammation which involved the activation of p38 MAPK signaling. Food Chem Toxicol 75:139–145

Chapter 14
Selenium Biofortification

Gary S. Bañuelos, Zhi-Qing Lin, and Martin Broadley

Abstract Selenium (Se) is an essential trace element for animals and humans, and thus, low dietary intake of Se can cause health disorders in humans and animals. The Se content of food is highly dependent on soil Se bioavailability and the ability of plants to take up and accumulate Se in edible tissues. Compared to the recommended daily Se allowance value of 55 μg per person, the estimated Se intake rate from food consumption is often lower than this recommended value in many parts of the world. To overcome the Se deficiency and its related public health issues, biofortification strategies have been applied to produce Se-enriched agricultural products through alternative new agronomic practices and the development of new biotechnologies in recent decades. For example, Se-amended soil fertilizers or foliar Se applications have been used to increase Se accumulation in crops, and genetically engineered plants have also been developed to increase the uptake of Se from soil. In addition, the use of Se-laden plant materials as organic Se fertilizers represents a unique environmentally-friendly strategy to implement the goal of Se biofortification. The importance of plant and soil microbial interaction and identification of selenoamino acids in plant tissues have also been documented for the enhancement of soil and biological Se bioavailability, respectively. This chapter has explored some major mechanisms underlying the Se biofortification process and potential benefits in promoting functional agricultural production. The authors have also addressed the economic and public acceptance aspects of Se biofortification.

Keywords Biofortication • Selenoamino acids • Se amendments • Bioavailability

G.S. Bañuelos (✉)
USDA, Agricultural Research Service, San Joaquin Valley Agricultural Sciences Center, 9611 South Riverbend Avenue, Parlier, CA 93648-9757, USA
e-mail: gary.banuelos@ars.usda.gov

Z.-Q. Lin
Environmental Sciences Program and Department of Biological Sciences, Southern Illinois University Edwardsville, Edwardsville, IL 62026-1651, USA
e-mail: zhlin@siue.edu

M. Broadley
University of Nottingham, School of Biosciences, Sutton Borrington Campus, Sutton Borrington, Leicestershire LE12 5RD, UK
e-mail: martin.broadley@nottingham.ac.uk

E.A.H. Pilon-Smits et al. (eds.), *Selenium in plants*, Plant Ecophysiology 11,
DOI 10.1007/978-3-319-56249-0_14

14.1 Introduction

14.1.1 Se in the Natural Environment

The metalloid selenium (Se) is ubiquitous in the environment with total concentrations in most soils ranging from 0.01 to 2.0 mg/kg; however, higher concentrations can occur in seleniferous areas (Fordyce 2013). Soil Se concentrations and Se bioavailability vary with environmental conditions, and the distribution of Se in soils is usually heterogeneous and site specific (Wang and Gao 2001), and varies geographically (Steinnes 2009; Winkel et al. 2011). Areas with high concentrations of Se are characteristic for igneous rock, sandstone, granite and limestone (Dhillon and Dhillon 2003; Van Metre and Callan 2001), while soils developed under temperate and humid/sub-humid conditions are quite poor in Se (Lyons et al. 2007). In general, Se in soils can originate from both local and regional sources. Local sources include rocks (geogenic sources) from which Se can be mobilized via weathering processes. In addition, atmospheric deposition (wet and dry) of Se is derived from both anthropogenic (fossil fuel burning, metal smelting, ship emissions) and natural sources (biomethylation, volcanic activity) (Blazina et al. 2014; Fernández-Martínez and Charlet 2009; Floor and Román-Ross 2012; Plant et al. 2004; Wen and Carignan 2007; Winkel et al. 2015). Regionally, shales are the primary sources of high Se soils in China (Shao and Zheng 2008), parts of California (Presser et al. 1994), Colorado (Seiler et al. 1999), South Dakota and Wyoming (Kulp and Pratt 2004). Conversely, Northern European countries are among the low-selenium regions, particularly the Scandinavian countries (Kápolna et al. 2009). Total soil Se is not a useful index of plant-available Se and it cannot be used as a reliable parameter in risk assessment or the determination of Se supplementation need (Chilimba et al. 2011). The mobility and plant availability of Se in soils is controlled by a number of chemical and biochemical processes: sorption, desorption, formation of inorganic and organic complexes, precipitation, dissolution and methylation to volatile compounds (Alfthan et al. 2011). Using Extended X-ray Absorption Fine Structure (EXAFS) data, Peak and Sparks (2002) showed that specific bonding (inner-sphere co-ordination) of selenate (Se^{VI}; $pKa_2 = 1.92$) on iron (Fe) hydrous oxides declined between pH 3.5 and pH 6. By contrast, selenite (Se^{IV}; $pKa_2 = 7.3$) was specifically adsorbed beyond the 'point of zero charge' of Fe oxides (pH 7–8), whereas the adsorption envelope of $HSeO_3^{1-}/SeO_3^{2-}$ on hematite showed a marked fall in sorption strength in the pH range 6–8 (Duc et al. 2006); expected from the second pKa value (7.3) of selenious acid (Vuori et al. 1989). The pH value at which Se uptake increases corresponds closely with the value at which selenate adsorption on Fe oxides ceases and the selenite sorption envelope declines. An additional factor may be the dependence of inorganic speciation on Eh-pH relations. Thus, it is clear from Eh-pH predominance diagrams (Seby et al. 2001) and recognized from studies of solubility (Masscheleyn et al. 1990) that selenate is the dominant form of available inorganic Se under oxic and alkaline soil conditions.

The speciation of Se in the soil is key for determining Se content in food and feed plants. Campbell et al. (1997) reported that amending soils with sulfur (S) increased Se uptake by wheat and canola in low Se soils, however, as soil S increased, plants accumulated less Se. Stroud et al. (2010b) found that when the plant S status is sufficient, grain Se increases but when the plant lacks S, grain Se decreases even when the plot is fertilized with S because the plants accumulate S over Se. Toler et al. (2007) reported that as Se increased, there was an upregulation of S accumulation as well, highlighting a protective mechanism of S against Se toxicity. Consequently, predictive Se uptake models need to include consideration of soil S content (Kikkert et al. 2013).

14.1.2 Se Essentiality in Human Health

Selenium is an essential trace element for animals and humans (Combs 2001; Fairweather-Tait et al. 2011). For this reason, the Se content in the human diet has become a topic of great interest in the public health systems around the world. Low dietary intake of Se can cause health disorders, including oxidative related- stress, epilepsy, fertility reduction and immune deficiency (Nazıroğlu 2009; Rayman 2012; Zeng and Combs 2008). For the human, the thyroid gland has the highest Se concentration of all the tissues (Schomburg and Köhrle 2008), and clearly needs an adequate supply of Se. Diet is a principal route of the daily intake of Se for animals and humans; however, plant-derived food products usually contain various amounts of Se (Diaz-Alarcon et al. 1994). Despite evidence from *in-vitro* and animal studies that Se is important to immunity (Hoffmann et al. 2010; Wood et al. 2000), there are few Se and human immunity studies. However, Se deficiency (serum or plasma Se < 85 μg/L) has been associated with decreased survival in HIV-infected patients (Rayman 2000).

14.1.3 Se-Deficiencies

Food is the major source of Se for the general population and Se deficiencies can arise if dietary Se supply and intake is not adequate. The Se content of food is highly dependent on the amount of bioavailable Se in the soil and on the ability of plants to take up and accumulate the element. In addition, the intake of Se is highly dependent on the area of residence and whether only local food is consumed or whether the diet is replenished with imported food products. Typical recommended daily allowance (RDA) values range from 40 to 75 μg/person/d (Fairweather-Tait et al. 2011). However, the estimated Se intake rate from food consumption is often lower than this range, due to insufficient Se content in the soil, resulting in low Se concentrations in food products (Jardine and Kidd 2011; Navarro-Alarcon and

Cabrera-Vique 2008). Data for Se intake for many parts of Africa, southern Asia, and South America are scarce, although studies using food balance sheets, food composition tables and dietary surveys indicate that dietary Se deficiencies are likely to be widespread in sub Saharan Africa (Joy et al. 2014, 2015a, b). Individuals with low protein intake will also take in lower amounts of Se, since a major source of Se in food products is from protein (Combs 2001). As a preventative strategy to minimize Se deficiencies in a country's population, Finland adopted Se fertilization as a national public health measure, leading to increases in dietary intake of Se, and has continued since the 1980s (Alfthan et al. 2011).

14.2 Biofortification Strategies

14.2.1 Relationship Between Se and S

Plant foods are the major dietary sources of Se in most countries around the world, followed by meats and seafood (ODS 2016). For this reason, it is vital to increase Se uptake by plants and to produce crops with higher Se concentrations and bioavailability in their edible tissues. One of the most promising approaches to mitigate a low transfer of Se and other nutrients from soil into the food chain involves a concept called biofortification (White and Broadley 2009). Excellent review articles have appeared on the concept of biofortification for reducing other micronutrient malnutrition (Bouis and Welch 2010; Carvalho and Vasconcelos 2013; Johns and Eyzaguirre 2007; Saltzman et al. 2013), as well as the creation of The Harvest Plus Challenge Program 2004, which was a large biofortification platform for reducing malnutrition in Asia and Africa (Carvalho and Vasconcelos 2013).

Biofortification, as an agronomic-based strategy, can be utilized to produce Se-enriched food products that may help reduce dietary deficiencies of Se occurring throughout susceptible regions of the world (Broadley et al. 2006, 2010). Before we can effectively develop a Se-biofortification strategy, it is important to understand the relationship between Se and S, especially since there are no pieces of conclusive evidence demonstrating Se as an essential nutrient required by plants (Pilon-Smits 2015). Because the chemical and physical properties of Se and S are very similar (Combs and Combs 1984), non-Se accumulating plants cannot effectively distinguish between absorbing Se as selenate and S as sulfate. Sulfate has been observed to compete with selenate by roots (Ulrich and Shrift 1968). In this regard, selenate (similar to sulfate) can transport across the plasma membrane of root epidermal cells against their electrochemical gradients (Hawkesford et al. 1993). The active transport of Se appears to occur via shared transporter-proteins: selenate via sulfate transporters (Terry et al. 2000) and selenite via phosphate transporters (Yonghua et al. 2008). There are considerable differences between the mechanisms involved in uptake and transport of selenate, selenite and organic compounds like selenomethionine (SeMet). Both selenate and organic Se compound absorption in plants from the soil solution are active processes, whereas selenite was also reported to be accumulated through passive diffusion (Abrams et al. 1990). Some studies show that absorbed selenite is readily reduced to organic compounds in plants and some plants are able to oxidize selenite back to selenate in small amounts (Shrift and Ulrich 1969). Earlier work indicate that selenate translocation in plants parallels that of sulfate, even though the proportion of oxidized and reduced forms of Se and S are different, they have the similar distribution pattern (Pilon-Smits et al. 1999). The better understanding of Se and S metabolism requires more detailed biochemical studies and Se/S flux analyses. Molecular studies and the overexpression of genes encoding proteins involved in the uptake, transport and assimilation of both S and Se will expand our understanding of the close relationship between these two elements and may provide useful information for developing effective Se biofortification strategies.

14.2.2 Application of Soil Se Fertilizers

One of the key issues in biofortification is to select the most appropriate method to biofortify plants with Se that can be effectively delivered to the plant. Using a meta-analysis approach, Ros et al. (2016) showed that selenate-based fertilizers have a high potential to increase Se uptake by crops, and subsequently the Se intake in animals and humans. Agronomic Se biofortification of food crops has been practiced commercially in Se-deficient regions by adding Se-amended inorganic fertilizers to soils, e.g. in Finland (Alfthan et al. 2011). Its application has also been studied in the field in many other soil/crop systems in different countries, e.g. United Kingdom (UK) (Hartikainen 2005; Lyons 2010; Stroud et al. 2010a, b), Europe (Poblaciones et al. 2013, 2014), New Zealand (Curtin et al. 2006), Africa (Chilimba et al. 2012a, b) and China (Wu et al. 2015). The most commonly added form of Se used in these studies, selenate, and to a lesser extent selenite as sodium or barium salts, can be applied in granular or blended forms directly to the soil, or as high volume liquid drenches, and much of which will enter the soil (Broadley et al. 2010; Iwashita and Nishi 2004; Rayman et al. 2008; Shrestha et al. 2006). Applying high quantities of Se fertilizers may not always be the most sustainable strategy to employ. In addition to potential leaching of excessive Se, a further drawback is the need for regular applications, which can make this approach costly (Hirschi 2009; White and Broadley 2009; Winkler 2011).

14.2.3 Foliar Application of Se

Effective enrichment of agricultural crops with Se using soil Se-enriched fertilizers can be challenging due to varying baseline soil Se concentrations, soil types, redox potential, pH, and microbiological activity (Hartfiel and Bahners 1988). As an alternative, foliar application of Se has been used for enriching Se in agricultural products (Smrkolj et al. 2006). With this method, a Se-containing solution is sprayed onto leaf surface of the crop. In this regard, soil chemistry and microbiological processes have less impact on Se ensuring a higher uptake efficacy with low volumes of applied Se solution. Factors such as the amount of Se applied, leaf area and surface structure, and differences in plant-specific metabolism of Se differ among crops and must be considered. Foliar application of Se (IV) or Se (VI) has successfully increased the Se concentration of many crops, including potato (Poggi et al. 2000), rice (Hu et al. 2002), soybean (Yang et al. 2003), cabbage, onion, garlic and radish (Slekovec and Goessler 2005), buckwheat (Smrkolj et al. 2006), and carrots (Kápolna et al. 2009). Ros et al. (2016) estimated that selenate foliar fertilization seems to be the most effective fertilizer strategy to enhance crop Se uptake in most arable crops. Kápolna et al. (2007) also reported a greater movement of Se to the carrot root when Se was applied in a foliar form as Se (VI). Importantly, when plants are exposed to high Se concentrations from foliar sprays, they may show symptoms of phytotoxicity, especially at Se concentrations of over 100 μg/ml (Kápolna et al. 2009).

Foliar application strategies require careful consideration of some of the following factors related to spraying the Se solution onto plants: (1) Se solutions must be carefully prepared and delivered using well-calibrated spraying equipment; (2) windy and/or rainy days should be avoided and plants must have adequate leaf surface area to ensure absorption of Se; and (3) growth stages need to be determined for timing of applications to ensure greatest Se absorption.

14.2.4 Application of Se-Enriched Organic Fertilizers

The use of Se-enriched organic or green fertilizers may be another effective alternative soil amendment to produce Se biofortified crops. Early work (Ajwa et al. 1998; Bañuelos et al. 1992) and more recently Bañuelos et al. (2015a) showed that Se applied via organic matter (green manure) can be taken up by various plant species. Freeman and Bañuelos (2011) first suggested the possibility of using Se-hyperaccumulating plant materials as an organic-Se enriched fertilizer for biofortifying food crops. In this regards, Bañuelos et al. (2015a) observed that more than 90% of organic Se added from Se-enriched *Stanleya pinnata* to grow Se-enriched broccoli and carrots under ideal soil moisture conditions was converted to inorganic selenate and selenite. Total Se concentrations in both broccoli florets and carrots was correlated with the amount of organic Se added to the soil. In addition, plant uptake of Se from Se-laden organic matter containing large proportions of organic compounds (e.g. SeMet, selenocysteine (SeCys)) can reportedly occur at rates greater than those from using inorganic sources of Se (Abrams et al. 1990; Kikkert and Berkelaar 2013). The uptake mechanisms for organic Se compounds are, however, still poorly known, but likely amino acid transporters might be involved (Kikkert and Berkelaar 2013).

Businelli et al. (2015) proposed using Se-enriched peat (10–20 mg Se/kg dry weight) during the pre-transplanting stage with cucumbers, lettuce and tomatoes as an alternative strategy to adding soil Se fertilizers or foliar application of Se fertilizers. The application of Se-enriched organic/green fertilizers has potential advantages over inorganic sources of Se, because Se can be gradually released from organic matter into soil solution. Interestingly, Bhatia et al. (2013) demonstrated the feasibility of producing Se-biofortified edible mushrooms grown in Se-rich agricultural by-products. Their bioaccessibility of Se may, however, be affected by the formation of indigestible Se-containing polysaccharides (Bhatia et al. 2013) and/or association of Se with chitin-containing structures in cell walls (Serafin Muñoz et al. 2006).

14.2.5 Natural Sources of Se

Another biofortification option is to exploit the possibility of producing Se-enriched food and feed products grown in soils naturally abundant in Se, such as in Enshi and Ziyang, China (Zhu et al. 2008); South Dakota, USA (Gerla et al. 2011), and Punjab, India (Dhillon and Dhillon 2009; Sharma et al. 2009). In China, Brazil, and in California, food products grown in different Se-rich regions produce food products with higher Se concentrations (Bañuelos et al. 2015b; da Silva et al. 2013). Similarly, Bañuelos (2002) reported higher Se concentrations in broccoli and other food crops irrigated with water loaded with naturally-occurring Se originated from soil drainage (Fig. 14.1).

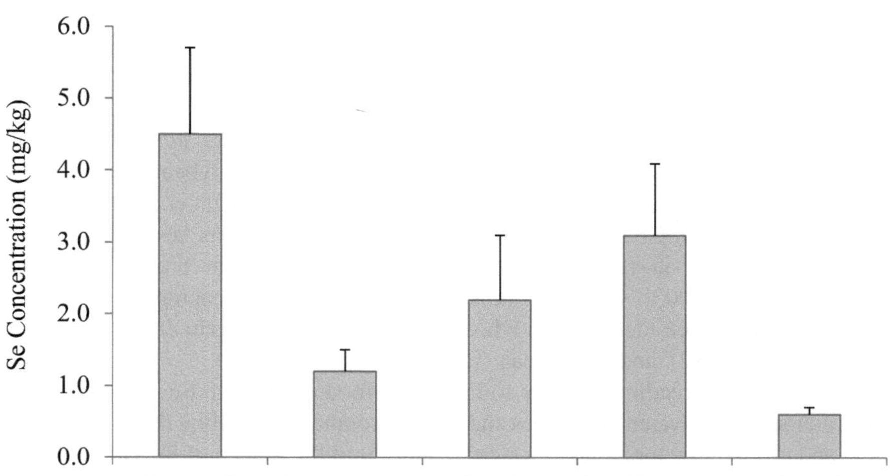

Fig.14.1 Examples of selenium biofortification in field-grown edible products using saline water (4–7 dS/m) naturally enriched with Se (150–250 μg/L). Values represent the mean with standard deviation

14.2.6 Microbial Assistance

Earlier work by de Souza et al. (1999) showed that rhizosphere bacteria from a seleniferous area enhanced root hair formation and consequently plant uptake of Se as selenate in Indian mustard (*Brassica juncea*). Furthermore, Yasin et al. (2015) showed that different bacterial consortia enhanced Se accumulation in Indian mustard grown in seleniferous soil. Different bacteria can either affect plant growth and/or play a role in bioavailability of Se from the soil, and this may be Se-species dependent. Consideration of the microflora surrounding roots of different crops grown in a Se-enriched growing medium might therefore provide additional information that is useful for creating effective Se-biofortification approaches for specific soil types. Importantly, Chander and Joergensen (2007) reported that Se-enriched soils (i.e. 20 µg Se/g soil) had no effect on some key soil microbial indices, e.g. microbial biomass C, microbial biomass N, adenosine triphosphate (ATP), adenylate energy charge (AEC), ATP-to-microbial biomass C and metabolic quotient (qCO_2).

14.2.7 Conventional Plant Breeding

Plant breeders need to screen existing crop varieties and accessions to determine whether sufficient genetic variation exists to breed for specific traits, e.g. related to Se absorption, speciation or bioavailability. At present, it is still not entirely clear whether sufficient genetic variation exists within modern breeding lines of crops to develop effective breeding strategies for increasing Se composition of edible plant products. Previous research has shown genetic variation of grain Se concentrations in cereal crops, since cereals are a major source of Se intake in many countries, including durum wheat (*Triticum turgidum* L.) (Rodríguez et al. 2011), barley (*Hordeum spontaneum* K. Koch) (Jun et al. 2011), oat (*Avena sativa* L.) (Eurola et al. 2008), mung bean (*Vigna radiate L.*) (Nair et al. 2015), soybean (*Glycine max* (L.) Merr.) (Yang et al. 2003) and rice (*Orzya sativa* L.) (Norton et al. 2010, 2012; Zhang et al. 2006). Genetic variation in seed Se concentrations has been reported among genotypes in several legumes (White 2015) and in broccoli varieties (Bañuelos et al. 2003), while chromosomal loci (QTLs) influencing Se concentrations have also been identified in wheat (Pu et al. 2014; Rongzhi et al. 2013), rice (Norton et al. 2012), and in soybean (Ramamurthy et al. 2014).

A successful breeding strategy to biofortify food crops with Se will depend on the interactions between genotypes and the environment, including those soil chemical and physical factors that may significantly limit the uptake of Se. Breeding is a long-term process that requires long-term input of resources. When promising high-yielding and efficient Se accumulation lines emerge, they must be repeatedly tested at multiple sites. The breeding, testing and releasing can take years to be completed. As Cakmak (2008) discussed for Zn breeding, the following steps need to be considered; (1) identification of a useful genetic variation and the most promising parents, (2) long-term crossing and back-crossing activities, (3) stability of target traits (i.e. high Se concentration) across the varied soil and climatic conditions, and finally (4) adaptation of the newly-developed biofortified genotypes over a range of crop and soil management practices utilized in different countries. Moreover, breeding or selecting crops for their ability to accumulate greater Se concentrations should lead to increased Se intake in animals and humans (White 2015). Importantly, the introduction or promotion of superior Se-accumulating varieties of a few staple crops should not neglect the importance of also preserving crop biodiversity and dietary diversity for their considerable potential roles in contributing to increased Se intakes.

14.2.8 Molecular and Genetic Engineering

Genetic engineering to deliver higher levels of Se accumulation in plants has been reviewed by others (Pilon-Smits and LeDuc 2009; Terry et al. 2000). Transgenic plants have been engineered with greater Se tolerance, Se accumulation or Se volatilization than their non-transgenic counterparts (Pilon-Smits 2012; Pilon-Smits and LeDuc 2009; Terry et al. 2000). The manipulation of Se transport and biochemistry may benefit the development of crops with greater Se tolerance that can grow on soils with high soil Se concentration. Importantly, it may also benefit crop quality through Se biofortification by enabling crops to accumulate greater Se and seleno-amino acid concentrations in edible parts of the plant that can be beneficial for preserving human and animal health. Overexpressing genes encoding transporters for selenate, selenite or seleno-amino acids in the plasma membrane of particular cells can increase the capacity for Se uptake and transport within the plant (White 2015). In non-accumulator plants, the conversion of.selenate to selenite within plastids appears to be the rate-liming step in the assimilation of Se into organic compounds (Pilon-Smits et al. 1999). In Arabidopsis, the overexpression of *Arabidopsis thaliana* (At) adenosine triphosphate sulfurylase (ATPS1) transgene, *Pseudomonas aeruginosa* (Pa) adenosine 5′-phospho sulphate reductase (APR) or both AtATPS1 and PaAPR resulted in greater concentrations of organic Se in leaves but a decrease of total Se (Sors et al. 2005). In contrast, the overexpression of AtATPS1 in Indian mustard resulted in greater concentrations of Se and organic Se in leaves (Bañuelos et al. 2005; Pilon-Smits et al. 1999; van Huysen et al. 2004). The overexpression of selenocysteine methyltransferase (SMT), with or without the overexpression of ATPS1, resulted in greater total Se, Se-selenomethylselenocysteine (SeMeSeCys) and γ-glutamyl-SeMeSeCys concentrations in Indian mustard and Arabidopsis, compared with untransformed control plants (Bañuelos et al. 2007; Ellis et al. 2004; LeDuc et al. 2004, 2006). Other work with the overexpression of genes encoding SeCys lyases or the overexpression of At selenium building protein (SBP)1 and increasing a plant's tolerance to selenate or selenite has been referenced elsewhere (White 2015). Our knowledge of the genetics of Se accumulation will increase as more molecular techniques are utilized. For other micronutrients, e.g. Zn (Cakmak 2008), genetic biofortification may be the most cost effective approach for improving concentrations in grains.

14.3 Mechanisms

14.3.1 Assimilation as Selenoamino Acids

There are a number of studies in animals and humans that suggest that the metabolic fate and function of dietary Se is dependent upon species of Se, e.g. pure selenite, pure SeMet or food-derived Se (Lyons et al. 2007). Selenite and selenate, SeMet, SeCys, and Se compounds that contain amino groups show the best assimilation (Kieliszek et al. 2012; Hoefig et al. 2011). The assimilation of Se is increased from a diet rich in low-molecular-weight proteins and certain vitamins (Fairweather-Tait et al. 2010). There may be two pools of Se in the body (Daniels 1996); there is an active Se pool providing Se for synthesis of the primary functionally important selenocompounds, while the second pool consists of SeMet-containing proteins that may provide seleno-amino acids for selenoprotein synthesis (Lyons et al. 2007).

Selenium biofortification is dependent upon plant S/Se metabolism (Terry et al. 2000). Generally, intracellular selenate is reduced to selenite via activation by ATPS and reduction by APS reductase. Selenite may be further reduced to selenide enzymatically (via sulfite reductase) or non-enzymatically (via glutathione). Subsequently, selenide is incorporated into SeCys via coupling with O-acetylserine, a step that is catalyzed by cysteine synthase. In turn, SeCys may be further metabolized into SeMet via the methionine cycle, which includes enzymatic transformation of SeCys to Se-cystathionine (SeCyst), Se-homocysteine and finally SeMet (Terry et al. 2000). Selenium metabolism within Se-enriched plants may vary for different crops, i.e. monocots vs dicots, or among different organs, e.g. fruit vs leaves vs seed vs tuber vs root. Today, one of the most common dietary supplements of Se for humans is Se-enriched yeast, which contains Se primarily as SeMet. In milk containing Se, more Se is absorbed by humans from SeMet than from selenite (Moser-Veillon et al. 1992). In addition, Luo et al. (1985) showed that SeMet was more effective than selenite in raising plasma and erythrocyte Se in men. When Se deficiency is diagnosed based on clinical signs, however, selenite food supplementation might be the preparation of choice. Using selenite via feed, water, or injection will reduce the short-term or acute Se-deficient-related health problems. However, when the goal is to meet physiological requirements of animals and humans to maintain a high productive and reproductive performance, Se-enriched food or feed, e.g. yeast, more adequately supplies the tissue with Se. The natural form of Se as SeMet provides animals and humans a better chance to synthesize additional selenoproteins (Lyons et al. 2007).

14.3.2 *Cellular Bioaccessibility and Bioavailability*

Several pharmacological factors influence the bioavailability of Se from nutraceuticals, including interaction with other micronutrients in the supplement, the formulation under which the supplement is usually taken, effects derived from taking other medications, timing, dose and schedule of supplementation, and Se health status of the human (Navarro-Alarcon et al. 2002). Compared to inorganic forms of Se, absorption of dietary Se (predominately organic Se compounds) is generally believed to be relatively bioeffective (e.g. ~80%) (Reilly 1996). Hence, it is important to know the chemical form of Se consumed with regard to its bioavailability. For example, SeMet is more bioavailable than inorganic Se and it can be non-specifically incorporated in body proteins and serve as a pool of SeMet that can be drawn on at times of depletion or increasing need (Dumont et al. 2006c). The bioavailability and benefits to human and animal health of dietary Se will depend upon not only the amounts of Se but also the chemical forms of Se supplied (Combs 2001; Finley 2005; Rayman et al. 2008), and importantly on the Se speciation in the Se-enriched food product (Thiry et al. 2013).

A topic of major concern pertains to whether the Se species in food sources can be easily bioaccessible by the human body and if so, whether they are stable in the conditions prevailing during human digestion. This is especially important for Se clinical and nutritional studies, since it is the Se species at the time of gastrointestinal absorption. Although many food sources contain SeMet, it is necessary to know whether they are easily accessible to humans. Others have reported that the species SeMet, $Se(Cys)_2$, and SeMeSeCys remained stable under gastrointestinal digestion and that the ɣ-glu-SeMeSeCys lost its glutamine part (Dumont et al. 2006c) but the bioaccessibility can vary among different food sources (Dumont et al. 2004, 2006a, b; Lavu et al. 2016; Vonderheide et al. 2002).

14.4 Benefits of Se Biofortification

14.4.1 *Humans*

The production of 'functional food' enriched with Se has created quite bit of attention worldwide. It is possible to provide consumers with a wide range of Se-enriched products to improve the general diet and help maintain good health. For example, Se enrichment of eggs produced in more than 25 countries (Pappas et al. 2006), meat and milk is a valuable option to improve the Se status of a general population. A crucial factor that needs to be emphasized is that the additional Se intake via food biofortification may well benefit people with low Se status, but the effects of Se on human health are multiple and complex (Rayman 2008). People of adequate or high Se status could be affected adversely and probably should not consume excessive Se-biofortified food products. Thus, it is important to assess the Se nutritional status of a given population when supplementing their diet with Se-biofortified agricultural products. For example, the inclusion of Se-enriched Brazil nuts in school meals in Macapa (in the Brazilian Amazon) provided to children with already high dietary Se intakes may result in an increased risk for Se toxicity (Martens et al. 2015).

14.4.2 Plants

The predominant form of Se in plants, based on extractions with enzymatic hydrolysis, is generally organic selenomethionine (SeMet), methylselenocysteine (MeSeCys) and ɣ-glutamyl-Se-selencysteine (ɣ-Glu-MeSeCys) (Hart et al. 2011). However, much more research is needed to determine how the form of Se applied affects Se speciation and the proportion of organic Se species accumulated in edible plant tissues resulting from a biofortification practice.

As observed previously for field-grown wheat (Broadley et al. 2010), maize grain and corn stover, yields were unaffected by Se applications up to 100 g Se per ha. These observations are consistent with other field studies of wheat (Curtin et al. 2008; Ducsay and Lozek 2006; Grant et al. 2007), despite evidence that plant growth may be stimulated by increased Se supply in controlled environment conditions (Hartikainen and Xue 1999; Lyons et al. 2009; Ríos et al. 2009; Turakainen et al. 2004; White et al. 2004; Xue and Hartikainen 2008). Selenium-induced growth stimulation in plants has been attributed to increased resistance to oxidative stress and the stimulation of S transport and assimilation pathways. Further studies are needed to assess these phenomena in a wider field context.

Selenium applied at varying concentrations is reported to improve antioxidant activities in plants, mainly relating to improved glutathione peroxidase (GSH-Px) and superoxide dismutase (SOD) activities, as well as decreased lipid peroxidation in some plant species (Businelli et al. 2015; Djanaguiraman et al. 2005; Hartikainen et al. 2000; Hsu et al. 2011; Proietti et al. 2013) and thus may boost plant defenses against insects and other pests (Freeman et al. 2007; Hladun et al. 2013; Ríos et al. 2008). In addition, D'Amato et al. (2014) found the Se supplementation on olive trees had positive effects on some oil properties, e.g. color intensity, stability and sensory quality. In peach and pear, higher fruit Se concentrations resulting from Se biofortification, slowed down the rate of fruit softening and thus increased shelf-life of these fruits (Pezzarossa et al. 2012). Negative effects of higher Se concentrations applied to lettuce have also been observed; ~15 µM selenite decreased growth and resulted in an intensification of peroxidative processes (Hawrylak-Nowak 2013). Also, Se tissue concentrations between 0.05 and 0.1% in *Brassica juncea* (L.) decreased pollen germination (Prins et al. 2011).

In regards to other Brassica crops, e.g. broccoli (*B. oleracea)*, a possible 'antagonism' may exist between sulfate and selenate uptake and assimilation (Finley et al. 2005). As a result, some caution was expressed that high contents of glucoraphanin (a major aliphatic GSL) and higher plant Se metabolites may be difficult to simultaneously achieve due to metabolic interference inside the plant (Hsu et al. 2011), especially because competition between sulfate and selenate for uptake and assimilation has been demonstrated by Lyi et al. (2005) and Finley et al. (2005). Thus, under certain conditions, selenate fertilization may influence S-metabolic activity (Finley et al. 2005; Van Hoewyk et al. 2008) but the extent of this interference will depend on the relative availability of both sulfate and selenate in the growth medium (Hsu et al. 2011). In most plants, sulfate and selenate share the initial pathway for uptake, assimilation and incorporation into cysteine (Cys) and selenocysteine, respectively (Li et al. 2008; Pilon-Smits et al. 2002; Sors et al. 2005). While interference with Cys biosynthesis could result in a negative impact of selenate fertilization on glucosinolates (GSL), selenate has been shown to induce the expression of certain sulfate transporters in *A. thaliana* (El Kassis et al. 2007). In this regard, Hsu et al. (2011) demonstrated that selenate field applied once at either 25.3 or 253 μmol/broccoli plant did not affect plant growth, contents of cysteine, glutathione, total GSL, or glucoraphanin (a major GSL). Hsu et al. (2011) concluded that broccoli can be fertilized with Se without reduction in GSL content.

14.4.3 For Animals

In general, organic forms of Se are absorbed and retained more readily by ruminants than inorganic forms (Qin et al. 2007). Selenomethionine, the major dietary chemical form of Se, can have several metabolic fates. Cells do not distinguish between methionine and SeMet during protein synthesis, so SeMet is incorporated into general body proteins in place of methionine depending on the methionine concentration and the number of methionine residues in protein (Shiobara et al. 1998). On the other hand, inorganic selenite is rapidly taken up by red blood cells and then released into plasma after reduction to hydrogen selenide, which is the key central molecular form of Se in regulated Se-metabolic pathways (Fairweather-Tait et al. 2010). Organic SeMet also functions as a source of Se for the synthesis of selenoproteins and has a half-life that is longer than selenite (at least in humans) (Swanson et al. 1991). Selenium's role in animal performance is based upon the functions of selenoproteins. Hall et al. (2013) reported that biofortifying alfalfa with Se fertilizers is a potential management tool to improve Se-status and animal performance when fed to weaned beef calves produced in areas with low soil Se concentrations. At this time, Oregon is the only state in the USA that allows the addition of Se fertilizers. In the USA, the FDA (2015) has regulated Se supplementation to ruminant diets at a level of 0.3 mg/kg from either sodium selenate or selenite[1]. Feeding Se-enriched alfalfa to animals contributes to heathier animals, as well as edible meat that has sufficient Se to be considered a good source of Se for human intake. However, the

[1] The food additive selenium is a nutrient administered in animal feed as sodium selenite or sodium selenate or in a controlled-release sodium selenite bolus. In complete feed for chickens, swine, turkeys, sheep, cattle, and ducks at a level not to exceed 0.3 part per million.

form of Se used to produce the Se-biofortified alfalfa may determine the Se bioaccessibility and bioavailability to the animal and ultimately the Se status in humans after consumption of the meat. The Se concentration in beef can be directly related to the Se concentration of the feed on which the animals graze (Benes et al. 2015) or on the Se-rich soil in which the forage crop was grown (Hintze et al. 2002). Such meat is a good source of dietary Se intake (Hintze et al. 2002) and is highly bioavailable for protein synthesis, depending on dietary Se levels (Finley et al. 2004).

Fish and other marine organisms can take up Se from water, plants, or other marine organisms, including tuna, trout, krill, oyster and mussel (Quijano et al. 2000). Uptake of water-soluble Se can occur via gills, epidermis, or the gut. Uptake via the diet remains, however, the predominant pathway. Interesting relationships are reported to exist between accumulated Se and mercury absorption in fish, as described by Ralston et al. (2008). A higher molar ratio of Se to Hg accumulation in fish tissues reduces the Hg toxicity to the organism. This type of research is important for aquaculture systems with varied water quality, which are commonly used in different regions of the world to produce fish for consumption.

14.5 Economics

Biofortification using genetic strategies is potentially cost-effective for those who rely primarily on their own food for sustenance (White and Broadley 2009). Cost effectiveness of Se biofortification has not been calculated for countries that are susceptible to Se deficiencies. However, estimates of cost-to-benefit quotients for fertilization with Se suggest high returns on financial investments, such as those in Finland (Horton 2006; Lyons et al. 2005). Detailed efforts are currently being pursued in Malawi (Chilimba and Broadley; unpublished) on defining the economic costs associated with biofortification and, importantly, on developing strategies to lower costs for the grower and consumer. Likely, the costs associated with biofortification constitute only a fraction of the costs that result from protecting public heath on a sustainable basis with Se supplementation (Joy et al. 2015c). There are, however, costs associated with both long-term breeding or the development and regulatory approval of transgenic crops that are more efficient at accumulating Se. Food fortification programs will rely on widely distributed, freshly-grown Se enriched food products, or industrially processed food items (Mayer et al. 2008).

Selenium yeast is an attractive source of Se due to its low cost and its ability to act as a precursor for selenoprotein synthesis. Using selenized yeast instead of conventional yeast for bakery industry can be an effective and economical approach for increasing Se intake as SeMet worldwide. Another useful Se product is the Brazil nut (*Bertholletia excelsia*) known for its high concentrations of Se (as SeMet) (Martens et al. 2015), although it is not a commonly consumed food stuff worldwide. One single Brazil nut can exceed 55 μg Se (the US RDA value for Se) but the Se content will be dependent on the soil Se content within Brazil, Peru, Bolivia, Columbia, Venezuela and Ecuador (Dumont et al. 2006b). Food products from the

Allium genus (e.g. garlic, onion) and Brassica genus (e.g. broccoli, Brussel sprouts, cabbage, cauliflower, collards, kohlrabi, mustards and kale) are also plants that can accumulate high levels of Se and be viable biofortified products for increasing Se intake (Bañuelos and Meek 1990).

14.6 Acceptance of Biofortification and Products

To implement the Se biofortification strategy it will be imperative to adequately educate, instruct, and demonstrate Se fertilizer application techniques to growers. Additional support may be needed by local agricultural organizations to provide growers with needed technical support or training which will incur additional expenses. Growers will need financial incentives as to why they should apply these new Se fertilizers for growing these new food crops, either from being able to obtain a better price for their products or from a subsidy to reflect the likely wider public health benefits. Depending on the nature of the financial incentive, mechanisms to separate Se-enriched crops from non-Se enriched crops might need to be in place and organized, e.g. through grower co-operative systems, state or private extension services, or by food processors and sellers. There is a huge range of unknowns in this research area. For example, in the absence of direct or indirect subsidies, growers will tend to focus on new crops that can be afforded by the more affluent segments of the population. Consumer acceptance of a more expensive Se-biofortified product may become a matter of concern when one cannot see the invisible trait of higher Se concentration. In addition, it is important to know if the Se-biofortified product has been altered in its cooking, storage sensory (i.e. taste, odor, color, texture) quality, all of which affect consumer acceptance of the new food products. A subsidy-based approach might alleviate some of these issues and (in theory) it could be more equitable. However, subsidies could create undesirable market distortions. It is clear that discussions regarding economics, subsidies and government support will be key in successfully executing biofortification strategies. Each geographical region is unique and so overall biofortification strategies will need to be designed to be crop, site or community specific. As a template to partially emulate, Se biofortification strategies should study the Harvest Plus approach of developing a global interdisciplinary alliance of research on biofortified crops (Saltzman et al. 2013). Countries, like Brazil, China, India, and parts of Africa are already actively involved with Se biofortification of a wide array of staple food crops (Saltzman et al. 2013).

14.7 Future and Considerations of Se Biofortification

Food-chain based approaches of Se biofortification are designed to increase Se intake through the diet and may represent the most desirable and sustainable method of reducing Se deficiencies worldwide. They are likely to be especially important for satisfying the nutritional needs in developing countries. Selenium biofortification is not expected to eliminate Se deficiencies or Se-related diseases but it will provide a practical and cost-effective way to increase Se intake for vulnerable people. Research programs need to be continued with focus on the most efficient Se application methods with inorganic and organic sources of Se for maximizing Se accumulation in food products. Bioavailability of the Se species within the raw or processed food product must also be considered in Se biofortification. Effective interactions and discussions need to take place among the researchers, agronomists and growers who study and develop strategies for micronutrient malnutrition across the world. Specific activities related to biofortification strategies developed for other micronutrients, e.g. Zn, Fe and I, can be useful after modification in developing more effective and flexible Se biofortification tools for many different soil conditions. Biofortification strategies will vary depending on the location, the types of crops consumed in the respective communities, which crops will be biofortified, the bioavailability of Se following processing and cooking, the acceptance of Se-biofortified food or new transgenic Se-enriched food products, the economic status of the consumers, and the cost of the Se-biofortified products. Bouis et al. (2003) reported that consumers in both developed and developing countries will accept food prepared from biofortified crops provided that they are not appreciably more expensive than the alternatives and that the biofortification strategy did not alter the food quality.

In conclusion, biofortification of Se based firstly upon the identification and planting of more efficient Se absorbing crop cultivars/varieties consumed in a target region, timely applications of Se fertilizers via roots or foliar surface, long term crop breeding, and possible genetic manipulation, all contribute to increasing Se intake and bioaccessibility of Se in humans and animals in Se-deficient regions. Lastly, another important objective for the scientific community is to develop sophisticated analytical techniques for Se such as X-ray microprobe analysis, as used in the Pickering Laboratory at the University of Saskatchewan in Canada (Yang et al. 2016) and provide food safety assessment programs for those Se-enriched products and validate the quality through reliable laboratory monitoring.

Acknowledgements The senior author acknowledges the financial support from California State University Fresno Agriculture Research Initiative to conduct the multitude of studies related to selenium-enriched agricultural products.

References

Abrams M, Shennan C, Zasoski R, Burau R (1990) Selenomethionine uptake by wheat seedlings. Agron J 82:1127–1130

Ajwa HA, Bañuelos GS, Mayland HF (1998) Selenium uptake by plants from soils amended with inorganic and organic materials. J Environ Qual 27:1218–1227

Alfthan G, Aspila P, Ekholm P, Eurola M, Hartikainen H, Hero H, Hietaniemi V, Root T, Salminen P, Venäläinen E-R, Aro A (2011) Nationwide supplementation of sodium selenate to commercial fertilizers: history and 25-year results from the Finnish selenium monitoring programme. CAB International and Food and Agriculture Organization of the United Nations (FAO), Rome

Bañuelos GS (2002) Irrigation of broccoli and canola with boron and selenium-laden effluent. J Environ Qual 31:1802–1808

Bañuelos GS, Meek DW (1990) Accumulation of selenium in plants grown on selenium-treated soil. J Environ Qual 19:772–777

Bañuelos G, Mead R, Wu L, Beuselinck P, Akohoue S (1992) Differential selenium accumulation among forage plant species grown in soils amended with selenium-enriched plant tissue. J Soil Water Conserv 47:338–342

Bañuelos GS, Pasakdee S, Finley JW (2003) Growth response and selenium and boron distribution in broccoli varieties irrigated with poor quality water. J Plant Nutr 26:2537–2549

Bañuelos GS, Terry N, Leduc DL, Pilon-Smits EAH, Mackey B (2005) Field trial of transgenic Indian mustard plants shows enhanced phytoremediation of selenium-contaminated sediment. Environ Sci Technol 39:1771–1777

Bañuelos GS, Leduc DL, Pilon-Smits EAH, Terry N (2007) Transgenic Indian mustard overexpressing selenocysteine lyase or selenocysteine methyltransferase exhibit enhanced potential for selenium phytoremediation under field conditions. Environ Sci Technol 41:599–605

Bañuelos GS, Arroyo I, Pickering IJ, Yang SI, Freeman JL (2015a) Selenium biofortification of broccoli and carrots grown in soil amended with Se-enriched hyperaccumulator Stanleya pinnata. Food Chem 166:603–608.

Bañuelos GS, Tang J, Hou Y, Yin X, Yuan L (2015b) Environmental pathways and dietary intake of selenium in a selenium rich rural community in China: a natural biofortification case study. Global advances in selenium research from theory to application: proceedings of the 4th international conference on selenium in the environment and human health 2015. CRC Press Boca Raton

Benes SE, Robinson PH, Cun GS (2015) Depletion of selenium in blood, liver and muscle from beef heifers previously fed forages containing high levels of selenium. Sci Total Environ 536:603–608

Bhatia P, Aureli F, D'Amato M, Prakash R, Cameotra SS, Nagaraja TP, Cubadda F (2013) Selenium bioaccessibility and speciation in biofortified Pleurotus mushrooms grown on selenium-rich agricultural residues. Food Chem 140:225–230

Blazina T, Sun Y, Voegelin A, Lenz M, Berg M, Winkel LHE (2014) Terrestrial selenium distribution in China is potentially linked to monsoonal climate. Nat Commun 5:4717

Bouis HE, Welch RM (2010) Biofortification—a sustainable agricultural strategy for reducing micronutrient malnutrition in the global south. Crop Sci 50:S-20–S-32

Bouis HE, Chassy BM, Ochanda JO (2003) Genetically modified food crops and their contribution to human nutrition and food quality. Trends Food Sci Technol 14:191–209

Broadley MR, White PJ, Bryson RJ, Meacham MC, Bowen HC, Johnson SE, Hawkesford MJ, McGrath SP, Zhao F-J, Breward N (2006) Biofortification of UK food crops with selenium. Proc Nutr Soc 65:169–181

Broadley MR, Alcock J, Alford J, Cartwright P, Foot I, Fairweather-Tait SJ, Hart DJ, Hurst R, Knott P, McGrath SP (2010) Selenium biofortification of high-yielding winter wheat (*Triticum aestivum* L.) by liquid or granular Se fertilisation. Plant Soil 332:5–18

Businelli D, D'Amato R, Onofri A, Tedeschini E, Tei F (2015) Se-enrichment of cucumber (*Cucumis sativus* L.), lettuce (*Lactuca sativa* L.) and tomato (*Solanum lycopersicum* L. Karst) through fortification in pre-transplanting. Sci Hortic 197:697–704

Cakmak I (2008) Enrichment of cereal grains with zinc: agronomic or genetic biofortification? Plant Soil 302:1–17

Campbell PG, Twiss MR, Wilkinson KJ (1997) Accumulation of natural organic matter on the surfaces of living cells: implications for the interaction of toxic solutes with aquatic biota. Can J Fish Aquat Sci 54:2543–2554

Carvalho SM, Vasconcelos MW (2013) Producing more with less: strategies and novel technologies for plant-based food biofortification. Food Res Int 54:961–971
Chander K, Joergensen RG (2007) Microbial biomass and activity indices after organic substrate addition to a selenium-contaminated soil. Biol Fertil Soils 44:241–244.
Chilimba AD, Young SD, Black CR, Rogerson KB, Ander EL, Watts MJ, Lammel J, Broadley MR (2011) Maize grain and soil surveys reveal suboptimal dietary selenium intake is widespread in Malawi. Sci Rep 1:72
Chilimba AD, Young SD, Black CR, Meacham MC, Lammel J, Broadley MR (2012a) Agronomic biofortification of maize with selenium (Se) in Malawi. Field Crop Res 125:118–128
Chilimba AD, Young SD, Black CR, Meacham MC, Lammel J, Broadley MR (2012b) Assessing residual availability of selenium applied to maize crops in Malawi. Field Crop Res 134:11–18.
Combs GF Jr (2001) Selenium in global food systems. Br J Nutr 85:517–547
Combs GF Jr, Combs SB (1984) The nutritional biochemistry of selenium. Annu Rev Nutr 4:257–280.
Curtin D, Hanson R, Lindley TN, Butler RC (2006) Selenium concentration in wheat (Triticum aestivum) grain as influenced by method, rate, and timing of sodium selenate application. N Z J Crop Hortic Sci 34:329–339
Curtin D, Hanson R, Van Der Weerden T (2008) Effect of selenium fertiliser formulation and rate of application on selenium concentrations in irrigated and dryland wheat (Triticum aestivum). N Z J Crop Hortic Sci 36:1–7
D'Amato R, Proietti P, Nasini L, Del Buono D, Tedeschini E, Businelli D (2014) Increase in the selenium content of extra virgin olive oil: quantitative and qualitative implications. Grasas Aceites 65:e025
da Silva EG, Mataveli LRV, Arruda MAZ (2013) Speciation analysis of selenium in plankton, Brazil nut and human urine samples by HPLC–ICP-MS. Talanta 110:53–57
Daniels LA (1996) Selenium metabolism and bioavailability. Biol Trace Elem Res 54:185–199.
de Souza MP, Chu D, Zhao M, Zayed AM, Ruzin SE, Schichnes D, Terry N (1999) Rhizosphere bacteria enhance selenium accumulation and volatilization by Indian mustard. Plant Physiol 119:565–574
Dhillon K, Dhillon S (2003) Distribution and management of seleniferous soils. Adv Agron 79:119–184
Dhillon KS, Dhillon SK (2009) Selenium concentrations of common weeds and agricultural crops grown in the seleniferous soils of northwestern India. Sci Total Environ 407:6150–6156
Diaz-Alarcon JP, Navarro-Alarcon M, Lopez-Garcia de la Serrana H, Lopez-Martinez MC (1994) Determination of selenium levels in vegetables and fruits by hydride generation atomic absorption spectrometry. J Agric Food Chem 42:2848–2851
Djanaguiraman M, Devi DD, Shanker AK, Sheeba JA, Bangarusamy U (2005) Selenium–an antioxidative protectant in soybean during senescence. Plant Soil 272:77–86
Duc M, Lefevre G, Fédoroff M (2006) Sorption of selenite ions on hematite. J Colloid Interface Sci 298:556–563
Ducsay L, Lozek O (2006) Effect of selenium foliar application on its content in winter wheat grain. Plant Soil Environ 52:78
Dumont E, Vanhaecke F, Cornelis R (2004) Hyphenated techniques for speciation of Se in in vitro gastrointestinal digests of Saccharomyces cerevisiae. Anal Bioanal Chem 379:504–511
Dumont E, De Pauw L, Vanhaecke F, Cornelis R (2006a) Speciation of Se in *Bertholletia excelsa* (Brazil nut): a hard nut to crack? Food Chem 95:684–692
Dumont E, Ogra Y, Vanhaecke F, Suzuki KT, Cornelis R (2006b) Liquid chromatography–mass spectrometry (LC–MS): a powerful combination for selenium speciation in garlic (*Allium sativum*). Anal Bioanal Chem 384:1196–1206
Dumont E, Vanhaecke F, Cornelis R (2006c) Selenium speciation from food source to metabolites: A critical review. Anal Bioanal Chem 385:1304–1323

El Kassis E, Cathala N, Rouached H, Fourcroy P, Berthomieu P, Terry N, Davidian J-C (2007) Characterization of a selenate-resistant Arabidopsis mutant. Root growth as a potential target for selenate toxicity. Plant Physiol 143:1231–1241

Ellis DR, Sors TG, Brunk DG, Albrecht C, Orser C, Lahner B, Wood KV, Harris HH, Pickering IJ, Salt DE (2004) Production of Se-methylselenocysteine in transgenic plants expressing selenocysteine methyltransferase. BMC Plant Biol 4:1

Eurola M, Hietaniemi V, Kontturi M (2008) Selenium content of Finnish oats in 1997-1999: effect of cultivars and cultivation techniques. Agric Food Sci 13:46–53

Fairweather-Tait SJ, Collings R, Hurst R (2010) Selenium bioavailability: current knowledge and future research requirements. Am J Clin Nutr 91:1484S–1491S

Fairweather-Tait SJ, Bao Y, Broadley MR, Collings R, Ford D, Hesketh JE, Hurst R (2011) Selenium in human health and disease. Antioxid Redox Signal 14:1337–1383

FDA (2015) Code of federal regulations Title 21 food and drugs. In: DHHS (ed) Part 573 – food additives permitted in feed and drinking water of animals

Fernández-Martínez A, Charlet L (2009) Selenium environmental cycling and bioavailability: a structural chemist point of view. Rev Environ Sci Biotechnol 8:81–110

Finley JW (2005) Proposed criteria for assessing the efficacy of cancer reduction by plant foods enriched in carotenoids, glucosinolates, polyphenols and selenocompounds. Ann Bot 95:1075–1096

Finley JW, Grusak MA, Keck A-S, Gregoire BR (2004) Bioavailability of selenium from meat and broccoli as determined by retention and distribution of 75Se. Biol Trace Elem Res 99:191–209

Finley JW, Sigrid-Keck A, Robbins RJ, Hintze KJ (2005) Selenium enrichment of broccoli: interactions between selenium and secondary plant compounds. J Nutr 135:1236–1238

Floor GH, Román-Ross G (2012) Selenium in volcanic environments: a review. Appl Geochem 27:517–531

Fordyce FM (2013) Selenium deficiency and toxicity in the environment. In: Selinus O (ed) Essentials of medical geology, Rev edn. Springer Netherlands, Dordrecht

Freeman JL, Bañuelos GS (2011) Selection of salt and boron tolerant selenium hyperaccumulator Stanleya pinnata genotypes and characterization of Se phytoremediation from agricultural drainage sediments. Environ Sci Technol 45:9703–9710

Freeman JL, Lindblom SD, Quinn CF, Fakra S, Marcus MA, Pilon-Smits EA (2007) Selenium accumulation protects plants from herbivory by Orthoptera via toxicity and deterrence. New Phytol 175:490–500

Gerla P, Sharif M, Korom S (2011) Geochemical processes controlling the spatial distribution of selenium in soil and water, west central South Dakota, USA. Environ Earth Sci 62:1551–1560

Grant CA, Buckley WT, Wu R (2007) Effect of selenium fertilizer source and rate on grain yield and selenium and cadmium concentration of durum wheat. Can J Plant Sci 87:703–708

Hall JA, Bobe G, Hunter JK, Vorachek WR, Stewart WC, Vanegas JA, Estill CT, Mosher WD, Pirelli GJ (2013) Effect of feeding selenium-fertilized alfalfa hay on performance of weaned beef calves. PLoS One 8:e58188

Hart D, Fairweather-Tait S, Broadley M, Dickinson S, Foot I, Knott P, McGrath S, Mowat H, Norman K, Scott P (2011) Selenium concentration and speciation in biofortified flour and bread: retention of selenium during grain biofortification, processing and production of Se-enriched food. Food Chem 126:1771–1778

Hartfiel W, Bahners N (1988) Selenium deficiency in the Federal Republic of Germany. Biol Trace Elem Res 15:1–12

Hartikainen H (2005) Biogeochemistry of selenium and its impact on food chain quality and human health. J Trace Elem Med Biol 18:309–318

Hartikainen H, Xue T (1999) The promotive effect of selenium on plant growth as triggered by ultraviolet irradiation. J Environ Qual 28:1372–1375

Hartikainen H, Xue T, Piironen V (2000) Selenium as an anti-oxidant and pro-oxidant in ryegrass. Plant Soil 225:193–200

Hawkesford MJ, Davidian J-C, Grignon C (1993) Sulphate/proton cotransport in plasma-membrane vesicles isolated from roots of Brassica napus L.: increased transport in membranes isolated from sulphur-starved plants. Planta 190:297–304

Hawrylak-Nowak B (2013) Comparative effects of selenite and selenate on growth and selenium accumulation in lettuce plants under hydroponic conditions. Plant Growth Regul 70:149–157

Hintze KJ, Lardy GP, Marchello MJ, Finley JW (2002) Selenium accumulation in beef: effect of dietary selenium and geographical area of animal origin. J Agric Food Chem 50:3938–3942

Hirschi KD (2009) Nutrient biofortification of food crops. Annu Rev Nutr 29:401–421

Hladun KR, Parker DR, Tran KD, Trumble JT (2013) Effects of selenium accumulation on phytotoxicity, herbivory, and pollination ecology in radish (*Raphanus sativus* L.) Environ Pollut 172:70–75

Hoefig CS, Renko K, Köhrle J, Birringer M, Schomburg L (2011) Comparison of different selenocompounds with respect to nutritional value vs. toxicity using liver cells in culture. J Nutr Biochem 22:945–955

Hoffmann FW, Hashimoto AC, Shafer LA, Dow S, Berry MJ, Hoffmann PR (2010) Dietary selenium modulates activation and differentiation of CD4+ T cells in mice through a mechanism involving cellular free thiols. J Nutr 140:1155–1161

Horton S (2006) The economics of food fortification. J Nutr 136:1068–1071

Hsu FC, Wirtz M, Heppel SC, Bogs J, Krämer U, Khan MS, Bub A, Hell R, Rausch T (2011) Generation of Se-fortified broccoli as functional food: impact of Se fertilization on S metabolism. Plant Cell Environ 34:192–207

Hu Q, Chen L, Xu J, Zhang Y, Pan G (2002) Determination of selenium concentration in rice and the effect of foliar application of Se-enriched fertiliser or sodium selenite on the selenium content of rice. J Sci Food Agric 82:869–872

Huysen T, Terry N, Pilon-Smits E (2004) Exploring the selenium phytoremediation potential of transgenic Indian mustard overexpressing ATP sulfurylase or cystathionine-γ-synthase. Int J Phytoremediation 6:111–118

Iwashita Y, Nishi K (2004) Cultivation of selenium-enriched vegetables in large scale. Biomed Res Trace Elem 15:72–75

Jardine TD, Kidd KA (2011) Low concentrations of selenium in stream food webs of eastern Canada. Sci Total Environ 409:785–791

Johns T, Eyzaguirre PB (2007) Biofortification, biodiversity and diet: a search for complementary applications against poverty and malnutrition. Food Policy 32:1–24

Joy EJM, Ander EL, Young SD, Black CR, Watts MJ, Chilimba ADC, Chilima B, Siyame EWP, Kalimbira AA, Hurst R, Fairweather-Tait SJ, Stein AJ, Gibson RS, White PJ, Broadley MR (2014) Dietary mineral supplies in Africa. Physiol Plant 151:208–229

Joy EJ, Broadley MR, Young SD, Black CR, Chilimba AD, Ander EL, Barlow TS, Watts MJ (2015a) Soil type influences crop mineral composition in Malawi. Sci Total Environ 505:587–595

Joy EJ, Kumssa DB, Broadley MR, Watts MJ, Young SD, Chilimba AD, Ander EL (2015b) Dietary mineral supplies in Malawi: spatial and socioeconomic assessment. BMC Nutr 1:42

Joy EJ, Stein AJ, Young SD, Ander EL, Watts MJ, Broadley MR (2015c) Zinc-enriched fertilisers as a potential public health intervention in Africa. Plant Soil 389:1–24

Jun Y, Fang W, Haibo Q, Guoxiong C, Eviatar N, Fahima T, Jianping C (2011) Natural variation in grain selenium concentration of wild barley, *Hordeum spontaneum*, populations from Israel. Biol Trace Elem Res 142:773–786

Kápolna E, Gergely V, Dernovics M, Illés A, Fodor P (2007) Fate of selenium species in sesame seeds during simulated bakery process. J Food Eng 79:494–501

Kápolna E, Hillestrøm PR, Laursen KH, Husted S, Larsen EH (2009) Effect of foliar application of selenium on its uptake and speciation in carrot. Food Chem 115:1357–1363

Kieliszek M, Błazejak S, Jedrzejczak R (2012) The capacity for binding selenium by fodder yeast strain Candida utilis ATCC 9950. Bromatol Chem Toksykol 45:628–633
Kikkert J, Berkelaar E (2013) Plant uptake and translocation of inorganic and organic forms of selenium. Arch Environ Contam Toxicol 65:458–465.
Kikkert J, Hale B, Berkelaar E (2013) Selenium accumulation in durum wheat and spring canola as a function of amending soils with selenite, selenate and or sulphate. Plant Soil 372:629–641.
Kulp TR, Pratt LM (2004) Speciation and weathering of selenium in Upper Cretaceous chalk and shale from South Dakota and Wyoming, USA. Geochim Cosmochim Acta 68:3687–3701.
Lavu RVS, Van De Wiele T, Pratti VL, Tack F, Du Laing G (2016) Selenium bioaccessibility in stomach, small intestine and colon: comparison between pure Se compounds, Se-enriched food crops and food supplements. Food Chem 197:382–387
LeDuc DL, Tarun AS, Montes-Bayon M, Meija J, Malit MF, Wu CP, Abdel Samie M, Chiang C-Y, Tagmount A, deSouza M (2004) Overexpression of selenocysteine methyltransferase in Arabidopsis and Indian mustard increases selenium tolerance and accumulation. Plant Physiol 135:377–383
LeDuc DL, Abdel Samie M, Móntes-Bayon M, Wu CP, Reisinger SJ, Terry N (2006) Overexpressing both ATP sulfurylase and selenocysteine methyltransferase enhances selenium phytoremediation traits in Indian mustard. Environ Pollut 144:70–76
Li HF, McGrath SP, Zhao FJ (2008) Selenium uptake, translocation and speciation in wheat supplied with selenate or selenite. New Phytol 178:92–102
Luo XM, Wei HJ, Yang CL, Xing J, Liu X, Qiao CH, Feng YM, Liu J, Liu YX, Wu Q et al (1985) Bioavailability of selenium to residents in a low-selenium area of China. Am J Clin Nutr 42:439–448
Lyi SM, Heller LI, Rutzke M, Welch RM, Kochian LV, Li L (2005) Molecular and biochemical characterization of the selenocysteine Se-methyltransferase gene and Se-methylselenocysteine synthesis in broccoli. Plant Physiol 138:409–420
Lyons G (2010) Selenium in cereals: improving the efficiency of agronomic biofortification in the UK. Plant Soil 332:1–4
Lyons GH, Judson GJ, Ortiz-Monasterio I, Genc Y, Stangoulis JC, Graham RD (2005) Selenium in Australia: selenium status and biofortification of wheat for better health. J Trace Elem Med Biol 19:75–82
Lyons MP, Papazyan TT, Surai PF (2007) Selenium in food chain and animal nutrition: lessons from nature – review. Asian Australas J Anim Sci 20:1135–1155.
Lyons GH, Genc Y, Soole K, Stangoulis J, Liu F, Graham R (2009) Selenium increases seed production in Brassica. Plant Soil 318:73–80
Martens IB, Cardoso BR, Hare DJ, Niedzwiecki MM, Lajolo FM, Martens A, Cozzolino SM (2015) Selenium status in preschool children receiving a Brazil nut–enriched diet. Nutrition 31:1339–1343
Masscheleyn PH, Delaune RD, Patrick WH Jr (1990) Transformations of selenium as affected by sediment oxidation-reduction potential and pH. Environ Sci Technol 24:91–96
Mayer JE, Pfeiffer WH, Beyer P (2008) Biofortified crops to alleviate micronutrient malnutrition. Curr Opin Plant Biol 11:166–170
Moser-Veillon PB, Mangels AR, Patterson KY, Veillon C (1992) Utilization of two different chemical forms of selenium during lactation using stable isotope tracers: an example of speciation in nutrition. Analyst 117:559–562
Nair RM, Thavarajah P, Giri RR, Ledesma D, Yang R-Y, Hanson P, Easdown W, Hughes JA (2015) Mineral and phenolic concentrations of mungbean [*Vigna radiata* (L.) R. Wilczek var. radiata] grown in semi-arid tropical India. J Food Compos Anal 39:23–32
Navarro-Alarcon M, Cabrera-Vique C (2008) Selenium in food and the human body: a review. Sci Total Environ 400:115–141

Navarro-Alarcon M, de la Serrana HL-G, Perez-Valero V, López-Martínez M (2002) Selenium concentrations in serum of individuals with liver diseases (cirrhosis or hepatitis): relationship with some nutritional and biochemical markers. Sci Total Environ 291:135–141

Nazıroğlu M (2009) Role of selenium on calcium signaling and oxidative stress-induced molecular pathways in epilepsy. Neurochem Res 34:2181–2191

Norton GJ, Deacon CM, Xiong L, Huang S, Meharg AA, Price AH (2010) Genetic mapping of the rice ionome in leaves and grain: identification of QTLs for 17 elements including arsenic, cadmium, iron and selenium. Plant Soil 329:139–153

Norton G, Duan G-L, Lei M, Zhu YG, Meharg A, Price A (2012) Identification of quantitative trait loci for rice grain element composition on an arsenic impacted soil: influence of flowering time on genetic loci. Ann Appl Biol 161:46–56

ODS (Office of Dietary Supplements) (2016) Selenium: Dietary supplement fact sheet. Health Information. US Department of Health and Human Services, National Institutes of Heath, Office of Dietary Supplements, Washington, DC

Pappas AC, Karadas F, Surai PF, Wood NA, Cassey P, Bortolotti GR, Speake BK (2006) Interspecies variation in yolk selenium concentrations among eggs of free-living birds: the effect of phylogeny. J Trace Elem Med Biol 20:155–160

Peak D, Sparks D (2002) Mechanisms of selenate adsorption on iron oxides and hydroxides. Environ Sci Technol 36:1460–1466

Pezzarossa B, Remorini D, Gentile ML, Massai R (2012) Effects of foliar and fruit addition of sodium selenate on selenium accumulation and fruit quality. J Sci Food Agric 92:781–786

Pilon-Smits EA (2012) Plant selenium metabolism—Genetic manipulation, phytotechnological applications, and ecological implications. In: Wong MH (ed) Environmental contamination: health risks and ecological restoration. CRC Press, Boca Raton

Pilon-Smits EAH (2015) Selenium in plants. In: Lüttge U, Beyschlag W (eds) Progress in botany. Springer International Publishing, Cham

Pilon-Smits EA, Hwang S, Lytle CM, Zhu Y, Tai JC, Bravo RC, Chen Y, Leustek T, Terry N (1999) Overexpression of ATP sulfurylase in Indian mustard leads to increased selenate uptake, reduction, and tolerance. Plant Physiol 119:123–132

Pilon-Smits EA, Garifullina GF, Abdel-Ghany S, Kato S-I, Mihara H, Hale KL, Burkhead JL, Esaki N, Kurihara T, Pilon M (2002) Characterization of a NifS-like chloroplast protein from Arabidopsis. Implications for its role in sulfur and selenium metabolism. Plant Physiol 130:1309–1318

Pilon-Smits EA, Quinn CF, Tapken W, Malagoli M, Schiavon M (2009) Physiological functions of beneficial elements. Curr Opin Plant Biol 12:267–274

Plant J, Kinniburgh D, Smedley P, Fordyce F, Klinck B (2004) Arsenic and selenium. In: Lollar BS, Holland HD, Turekian KK (eds) Treatise on geochemistry 9. Elsevier Ltd., San Diego

Poblaciones MJ, Rodrigo SM, Santamaría O (2013) Evaluation of the potential of peas (*Pisum sativum* L.) to be used in selenium biofortification programs under Mediterranean conditions. Biol Trace Elem Res 151:132–137

Poblaciones M, Rodrigo S, Santamaría O, Chen Y, McGrath S (2014) Agronomic selenium biofortification in Triticum durum under Mediterranean conditions: from grain to cooked pasta. Food Chem 146:378–384

Poggi V, Arcioni A, Filippini P, Pifferi PG (2000) Foliar application of selenite and selenate to potato (Solanum tuberosum): effect of a ligand agent on selenium content of tubers. J Agric Food Chem 48:4749–4751

Presser TS, Sylvester MA, Low WH (1994) Bioaccumulation of selenium from natural geologic sources in western states and its potential consequences. Environ Manag 18:423–436

Proietti P, Nasini L, Del Buono D, D'Amato R, Tedeschini E, Businelli D (2013) Selenium protects olive (Olea europaea L.) from drought stress. Sci Hortic 164:165–171

Prins CN, Hantzis LJ, Quinn CF, Pilon-Smits EA (2011) Effects of selenium accumulation on reproductive functions in *Brassica juncea and Stanleya pinnata*. J Exp Bot 15:5633–5640

Pu Z, Ma Y, He Q, Chen G, Wang J, Liu Y, Jiang Q, Wei L, Dai S, Wei Y, Zheng Y (2014) Quantitative trait loci associated with micronutrient concentrations in two recombinant inbred wheat lines. J Integr Agric 13:2322–2329

Qin S, Gao J, Huang K (2007) Effects of different selenium sources on tissue selenium concentrations, blood GSH-Px activities and plasma interleukin levels in finishing lambs. Biol Trace Elem Res 116:91–102

Quijano M, Moreno P, Gutiérrez AM, Pérez-Conde MC, Camara C (2000) Selenium speciation in animal tissues after enzymatic digestion by high-performance liquid chromatography coupled to inductively coupled plasma mass spectrometry. J Mass Spectrom 35:878–884

Ralston NV, Ralston CR, Blackwell JL, Raymond LJ (2008) Dietary and tissue selenium in relation to methylmercury toxicity. Neurotoxicology 29:802–811

Ramamurthy RK, Jedlicka J, Graef GL, Waters BM (2014) Identification of new QTLs for seed mineral, cysteine, and methionine concentrations in soybean [*Glycine max* (L.) Merr.] Mol Breed 34:431–445

Rayman MP (2000) The importance of selenium to human health. Lancet 356:233–241

Rayman MP (2008) Food-chain selenium and human health: emphasis on intake. Br J Nutr 100:254–268

Rayman MP (2012) Selenium and human health. Lancet 379:1256–1268

Rayman MP, Infante HG, Sargent M (2008) Food-chain selenium and human health: spotlight on speciation. Br J Nutr 100:238–253

Reilly C (1996) Selenium in food and health. Blackie Academic & Professional, London

Ríos J, Rosales M, Blasco B, Cervilla L, Romero L, Ruiz J (2008) Biofortification of Se and induction of the antioxidant capacity in lettuce plants. Sci Hortic 116:248–255

Ríos J, Blasco B, Cervilla L, Rosales M, Sanchez-Rodriguez E, Romero L, Ruiz J (2009) Production and detoxification of H2O2 in lettuce plants exposed to selenium. Ann Appl Biol 154:107–116

Rodríguez LH, Morales DA, Rodríguez ER, Romero CD (2011) Minerals and trace elements in a collection of wheat landraces from the Canary Islands. J Food Compos Anal 24:1081–1090

Rongzhi Y, Ru W, Wentao X, Jun Y, Gang Z, Fahima T, Jianping C (2013) QTL location and analysis of selenium content in tetraploid wheat grain. Guizhou Agric Sci 10:1–4

Ros G, van Rotterdam A, Bussink D, Bindraban P (2016) Selenium fertilization strategies for biofortification of food: an agro-ecosystem approach. Plant Soil 1:99–112

Saltzman A, Birol E, Bouis HE, Boy E, De Moura FF, Islam Y, Pfeiffer WH (2013) Biofortification: progress toward a more nourishing future. Global Food Secur 2:9–17.

Schomburg L, Köhrle J (2008) On the importance of selenium and iodine metabolism for thyroid hormone biosynthesis and human health. Mol Nutr Food Res 52:1235–1246

Seby F, Potin-Gautier M, Giffaut E, Borge G, Donard O (2001) A critical review of thermodynamic data for selenium species at 25 °C. Chem Geol 171:173–194

Seiler RL, Skorupa JP, Peltz LA (1999) Areas susceptible to irrigation-induced selenium contamination of water and biota in the Western United States. US Geol Surv Circ 1180:1–44

Serafin Muñoz AH, Kubachka K, Wrobel K, Gutierrez Corona JF, Yathavakilla SK, Caruso JA, Wrobel K (2006) Se-enriched mycelia of Pleurotus ostreatus: distribution of selenium in cell walls and cell membranes/cytosol. J Agric Food Chem 54:3440–3444

Shao S, Zheng B (2008) The biogeochemistry of selenium in Sunan grassland, Gansu, Northwest China, casts doubt on the belief that Marco Polo reported selenosis for the first time in history. Environ Geochem Health 30:307–314

Sharma N, Prakash R, Srivastava A, Sadana U, Acharya R, Prakash NT, Reddy AR (2009) Profile of selenium in soil and crops in seleniferous area of Punjab, India by neutron activation analysis. J Radioanal Nucl Chem 281:59–62

Shiobara Y, Yoshida T, Suzuki KT (1998) Effects of dietary selenium species on Se concentrations in hair, blood, and urine. Toxicol Appl Pharmacol 152:309–314

Shrestha B, Lipe S, Johnson K, Zhang T, Retzlaff W, Lin Z-Q (2006) Soil hydraulic manipulation and organic amendment for the enhancement of selenium volatilization in a soil–pickleweed system. Plant Soil 288:189–196
Shrift A, Ulrich JM (1969) Transport of selenate and selenite into Astragalus roots. Plant Physiol 44:893–896
Slekovec M, Goessler W (2005) Accumulation of selenium in natural plants and selenium supplemented vegetable and selenium speciation by HPLC-ICPMS. Chem Speciat Bioavailab 17:63–73
Smrkolj P, Stibilj V, Kreft I, Germ M (2006) Selenium species in buckwheat cultivated with foliar addition of Se (VI) and various levels of UV-B radiation. Food Chem 96:675–681
Sors TG, Ellis DR, Na GN, Lahner B, Lee S, Leustek T, Pickering IJ, Salt DE (2005) Analysis of sulfur and selenium assimilation in Astragalus plants with varying capacities to accumulate selenium. Plant J 42:785–797
Steinnes E (2009) Soils and geomedicine. Environ Geochem Health 31:523–535
Stroud J, Broadley M, Foot I, Fairweather-Tait S, Hart D, Hurst R, Knott P, Mowat H, Norman K, Scott P (2010a) Soil factors affecting selenium concentration in wheat grain and the fate and speciation of Se fertilisers applied to soil. Plant Soil 332:19–30
Stroud J, Li H, Lopez-Bellido F, Broadley M, Foot I, Fairweather-Tait S, Hart D, Hurst R, Knott P, Mowat H (2010b) Impact of sulphur fertilisation on crop response to selenium fertilisation. Plant Soil 332:31–40
Swanson C, Patterson B, Levander O, Veillon C, Taylor P, Helzlsouer K, McAdam P, Zech L (1991) Human [74Se] selenomethionine metabolism: a kinetic model. Am J Clin Nutr 54:917–926
Terry N, Zayed A, De Souza M, Tarun A (2000) Selenium in higher plants. Annu Rev Plant Biol 51:401–432
Thiry C, Schneider Y-J, Pussemier L, De Temmerman L, Ruttens A (2013) Selenium bioaccessibility and bioavailability in Se-enriched food supplements. Biol Trace Elem Res 152:152–160
Toler HD, Charron CS, Sams CE, Randle WR (2007) Selenium increases sulfur uptake and regulates glucosinolate metabolism in rapid-cycling Brassica oleracea. J Am Soc Hortic Sci 132:14–19
Turakainen M, Hartikainen H, Seppänen MM (2004) Effects of selenium treatments on potato (Solanum tuberosum L.) growth and concentrations of soluble sugars and starch. J Agric Food Chem 52:5378–5382
Ulrich JM, Shrift A (1968) Selenium absorption by excised Astragalus roots. Plant Physiol 43:14–20
Van Hoewyk D, Takahashi H, Inoue E, Hess A, Tamaoki M, Pilon-Smits EA (2008) Transcriptome analyses give insights into selenium-stress responses and selenium tolerance mechanisms in Arabidopsis. Physiol Plant 132:236–253
Van Metre DC, Callan RJ (2001) Selenium and vitamin E. Vet Clin N Am Food Anim Pract 17: 373–402, vii–viii
Vonderheide AP, Wrobel K, Kannamkumarath SS, B'Hymer C, Montes-Bayón M, Ponce de León C, Caruso JA (2002) Characterization of selenium species in Brazil nuts by HPLC-ICP-MS and ES-MS. J Agric Food Chem 50:5722–5728
Vuori E, Vääriskoski J, Hartikainen H, Vakkilainen P, Kumpulainen J, Niinivaara K (1989) Sorption of selenate by Finnish agricultural soils. Agric Ecosyst Environ 25:111–118
Wang Z, Gao Y (2001) Biogeochemical cycling of selenium in Chinese environments. Appl Geochem 16:1345–1351
Wen H, Carignan J (2007) Reviews on atmospheric selenium: emissions, speciation and fate. Atmos Environ 41:7151–7165
White PJ (2015) Selenium accumulation by plants. Ann Bot 117:217–235.
White PJ, Broadley MR (2009) Biofortification of crops with seven mineral elements often lacking in human diets – iron, zinc, copper, calcium, magnesium, selenium and iodine. New Phytol 182:49–84

White P, Bowen H, Parmaguru P, Fritz M, Spracklen W, Spiby R, Meacham M, Mead A, Harriman M, Trueman L (2004) Interactions between selenium and sulphur nutrition in Arabidopsis thaliana. J Exp Bot 55:1927–1937

Winkel LH, Trang PTK, Lan VM, Stengel C, Amini M, Ha NT, Viet PH, Berg M (2011) Arsenic pollution of groundwater in Vietnam exacerbated by deep aquifer exploitation for more than a century. Proc Natl Acad Sci 108:1246–1251

Winkel LH, Vriens B, Jones GD, Schneider LS, Pilon-Smits E, Bañuelos GS (2015) Selenium cycling across soil-plant-atmosphere interfaces: a critical review. Nutrients 7:4199–4239.

Winkler J (2011) Biofortification: improving the nutritional quality of staple crops. In: Pasternak C (ed) Access not excess. Smith-Gordon Publishing, London

Wood SM, Beckham C, Yosioka A, Darban H, Watson RR (2000) β-Carotene and selenium supplementation enhances immune response in aged humans. Intern Med 2:85–92

Wu Z, Bañuelos GS, Lin Z-Q, Liu Y, Yuan L, Yin X, Li M (2015) Biofortification and phytoremediation of selenium in China. Front Plant Sci 6:136

Xue T, Hartikainen H (2008) Association of antioxidative enzymes with the synergistic effect of selenium and UV irradiation in enhancing plant growth. Agric Food Sci 9:177–186

Yang F, Chen L, Hu Q, Pan G (2003) Effect of the application of selenium on selenium content of soybean and its products. Biol Trace Elem Res 93:249–256

Yang SI, George GN, Lawrence JR, Kaminskyj SG, Dynes JJ, Lai B, Pickering IJ (2016) Multispecies biofilms transform selenium oxyanions into elemental selenium particles: studies using combined synchrotron X-ray fluorescence imaging and scanning transmission X-ray microscopy. Environ Sci Technol 50(19):10343–10350

Yasin M, El Mehdawi AF, Jahn CE, Anwar A, Turner MF, Faisal M, Pilon-Smits EA (2015) Seleniferous soils as a source for production of selenium-enriched foods and potential of bacteria to enhance plant selenium uptake. Plant Soil 386:385–394

Yonghua L, Wuyi W, Kunli L, Hairong L (2008) Environmental behaviors of selenium in soil of typical selenosis area, China. J Environ Sci 20:859–864

Zeng H, Combs GF (2008) Selenium as an anticancer nutrient: Roles in cell proliferation and tumor cell invasion. J Nutr Biochem 19:1–7

Zhang L, Shi W, Wang X, Zhou X (2006) Genotypic differences in selenium accumulation in rice seedlings at early growth stage and analysis of dominant factors influencing selenium content in rice seeds. J Plant Nutr 29:1601–1618

Zhu J, Wang N, Li S, Li L, Su H, Liu C (2008) Distribution and transport of selenium in Yutangba, China: Impact of human activities. Sci Total Environ 392:252–261

Chapter 15
Effects of Selenium on Plant Metabolism and Implications for Crops and Consumers

Michela Schiavon, Leonardo Warzea Lima, Ying Jiang, and Malcolm J. Hawkesford

Abstract Selenium (Se) is an essential trace element for many organisms including humans, while in plants it can trigger a variety of beneficial effects. Plants absorb Se mainly in the form of selenate using high affinity root sulfate transporters. Consequently, availability of sulfur (S) has a major impact on Se accumulation due to competition effects of the two oxyanions. In addition, Se has an impact on S uptake through interference with intrinsic regulatory mechanisms. Inside cells, selenate can access the sulfate assimilation pathway and influence the production of S-organic compounds that are of vital importance in plant responses to biotic and abiotic stress conditions. Selenium has been reported to mitigate stress in plants because of its capacity to induce the synthesis of S- and nitrogen (N) compounds, in addition to stimulating the activity of antioxidant enzymes and metabolites. Selenium can also alter the uptake of certain microelements like molybdenum, which functions as a cofactor for the enzyme nitrate reductase. Therefore, Se at high doses may interfere with N assimilation, causing a decrease in the level of N-compounds with structural and/or regulatory functions. Selenium interactions with multiple metabolic pathways in plants have relevant implications for plants and consumers that feed on them. Managing such interactions are useful to biofortify crops with organic forms of Se endowed with beneficial properties (selenomethionine and methylselenocysteine) and in other nutraceuticals like glucosinolates and antioxidants. Furthermore, Se at low doses may improve plant productivity or

M. Schiavon (✉) • L.W. Lima
Department of Biology, Colorado State University, Fort Collins, CO 80523, USA
e-mail: michela.schiavon@colostate.edu; leolima@rams.colostate.edu

Y. Jiang
Department of Biology, Colorado State University, Fort Collins, CO 80523, USA
e-mail: jiangying@cau.edu.cn

College of Agronomy and Biotechnology, China Agricultural University, Beijing 100193, China

M.J. Hawkesford
Rothamsted Research, Harpenden AL5 2JQ, UK
e-mail: malcolm.hawkesford@rothamsted.ac.uk

E.A.H. Pilon-Smits et al. (eds.), *Selenium in plants*, Plant Ecophysiology 11,
DOI 10.1007/978-3-319-56249-0_15

phytoremediation potential by enhancing photosynthesis and increasing the capacity of plants to tolerate stress.

Keywords Metals • Oxidative stress • Nutraceuticals • Biofortification

15.1 Selenium Metabolism and Its Close Relationship with Sulfur

Selenium (Se) is an important trace element for humans and many animals, as a component of selenoenzymes that display pivotal roles in cell metabolism by functioning as protectors from oxidative stress and controllers of cell redox status (Rayman 2012; Roman et al. 2014). Some organic forms of Se, like selenomethionine (SeMet), methylselenol and Se-methylselenocysteine (MetSeCys), have recognized anticarcinogenic properties (Combs 2005; Jackson and Comb 2008; Zeng and Combs 2008; Fernandes and Gandin 2015). MetSeCys for instance, has been reported to inhibit 7,12-dimethylbenz (a) anthracene (DMBA)-induced mammary tumors and act as a chemopreventative agent that blocks cell cycle progression and proliferation of premalignant mammary lesions (El-Bayoumy and Sinha 2004).

Despite the essentiality of Se for humans and animals, human Se intake in the diet is often lower than the recommended daily dose of 50–70 μg, which is required for full expression of protective selenoproteins (Brown and Arthur 2001). As a result, Se deficiency is a issue for concern in many countries worldwide (Combs 2001; Rayman 2002), being associated with a variety of diseases, such as reduced immune and thyroid function (Rayman 2012; Roman et al. 2014).

15.1.1 Uptake and Transport of Se

Plants represent one of the main dietary sources of Se for humans and animals. Depending on soil chemical properties, Se is available to plants mainly as either selenate or selenite. Selenate is usually the main soluble form of Se in soil. It is absorbed by plant cells via plasma membrane sulfate transporters and can be assimilated through the sulfur (S) assimilation pathway into Se-amino acids (Sors et al. 2005). This is because selenate and sulfate share high chemical similarity. In this respect, Se can interfere with S transport and assimilation in plants depending on Se/S ratio in the growth medium and/or in the plant (White et al. 2004; Schiavon et al. 2012). In Se non-hyperaccumulator plant species, selenate often induces a S deficiency response, which generally involves the up-regulation of genes coding for sulfate transporters and sulfate assimilation enzymes (Van et al. 2008; Harris et al. 2014; Schiavon et al. 2015). At low doses, Se can therefore cause an increase of sulfate uptake rates in these species, while at high concentration it will reduce S entry into root cells via competition for transporters (White et al. 2004; Schiavon et al. 2012). The high levels of Se may lead to replacement of S-containing amino acids with Se-containing equivalents which then triggers the S-starvation responses directly through relief of feedback inhibition of gene expression, even in the presence of S.

Under S-deficiency and even in the presence of modest levels of Se, high levels of selenium will accumulate in plant tissues (Stroud et al. 2010b; Shinmachi et al. 2010). The S-deficiency results in an induction of sulfate transporters due to de-repression of gene expression (Smith et al. 1997); additionally S-deficient soils may aid in selenate uptake by reduced competition, resulting in substantially increased vegetative and grain tissue selenium (Stroud et al. 2010b; Shinmachi et al. 2010). Due to the positive health benefits of Se in human and animal diets considerable attention has been paid to enhancing content in food crops such as cereals (reviewed in Hawkesford and Zhao 2007). As already stated, S-fertilization as well as Se availability in soils (Fan et al. 2008; Stroud et al. 2010a) will strongly influence the total Se-accumulation, the chemical form of Se and its tissue cellular and subcellular localization. For example, Se usually accumulates in seed tissues in parallel with S (reflecting the replacement of S in S-containing amino acids) but localized hotspots are also apparent, perhaps indicating vacuolar sequestration (Moore et al. 2009).

In Se hyperaccumulators, Se tends to reduce S levels in tissues due to competition, but in these species there typically is no S deficiency response because the sulfate transporters and assimilatory enzymes are constitutively up-regulated (Schiavon et al. 2015).

15.1.2 Chemical Fate of Se Within the Plant

The two Se-amino acids produced in the S assimilation pathway are selenocysteine (SeCys) and selenomethionine (SeMet), which are analogues of the S-amino acids cysteine (Cys) and methionine (Met) (Fig. 15.1). In addition to being protein subunits, Cys and Met play several functions in cells. Cysteine is a component of glutathione (GSH), a pivotal molecule in plant responses to multiple types of stress, while Met is a precursor of aliphatic glucosinolates, which are compounds involved in plant-pathogen/herbivore interactions (Mithöfer and Boland 2012). Therefore, Se interactions with S metabolism at different levels may affect the capacity of plants to cope with stress. Also, a secondary effect of such interactions is the capacity of Se to interfere with N metabolism, given that the S and N pathways come together at the level of cys synthesis. Cys is at a key regulatory point and may influence flux through both the N-assimilatory pathway, particularly regulating the provision of the cysteine precursor, O-acetylserine, and in the synthesis of glutathione, both suggested regulatory molecules for S-assimilation (Leustek and Saito 1999).

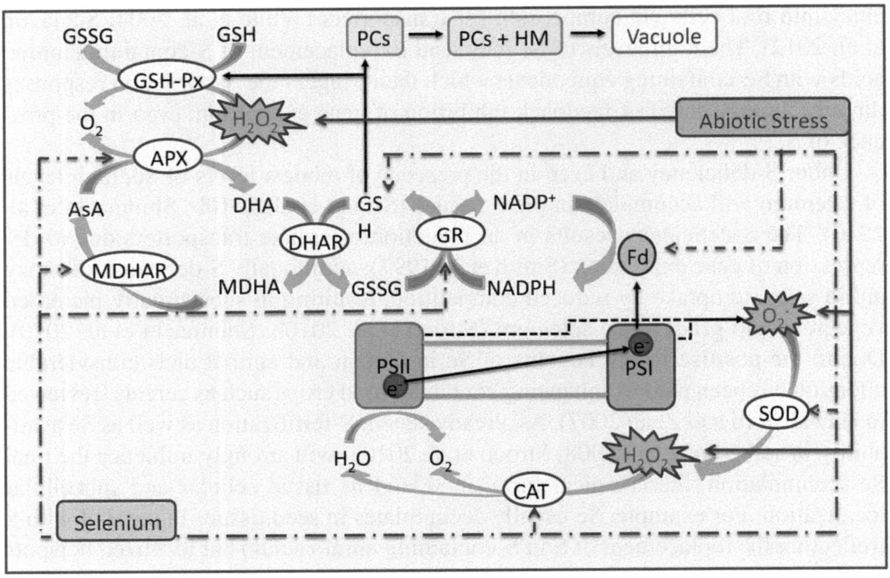

Fig. 15.1 Selenium induces enhanced antioxidant activity which stimulates plant productivity and resistance to oxidative stress. The image illustrates the pathway in photosynthetic tissues. *Enzymes*: Ascorbate peroxidase (APX), Catalase (CAT), Dehydroascorbate reductase (DHAR), Glutathione peroxidase (GSH-Px) Glutathione reductase (GR), Monodehydroascorbate reductase (MDHAR), SOD (Superoxide dismutase). *Metabolites*: AsA (Ascorbate), GSH (reduced glutathione), PCs (Phytochelatins). *Reactive Oxygen Species*: Superoxide radical ($O_2^{\cdot-}$); Hydrogen peroxide (H_2O_2)

15.2 Beneficial Effects of Se-induced Antioxidants (Enzymes and Metabolites) on Plant Productivity and Oxidative Stress Resistance

Plants can be faced with different environmental conditions that generate oxidative stress via production of Reactive Oxygen Species (ROS), and must activate different strategies to overcome it. ROS are the unstable and partially reduced forms of atmospheric oxygen (O_2), which show a great capacity to oxidize other cell compounds. These molecules are formed from the transfer of one, two or three electrons to the O_2 molecule, thus forming the superoxide radical ($O_2^{-\cdot}$), hydrogen peroxide (H_2O_2) and hydroxyl radical ($OH^{\cdot}$), respectively. This is particularly prone to happen in electron transfer processes in mitochondria, chloroplasts and peroxisomes (Shieber and Chandel 2014).

Various cellular defense responses are important for maintaining low concentrations of ROS, and involve both enzymatic and non-enzymatic antioxidant mechanisms (Fig. 15.2). Superoxide dismutases (SOD) constitute the first enzymatic barrier against oxidative stress by the dismutation reaction of $O_2^{-\cdot}$ in order to form O_2 and H_2O_2 (Shieber and Chandel 2014). Subsequently, H_2O_2 can be quickly con-

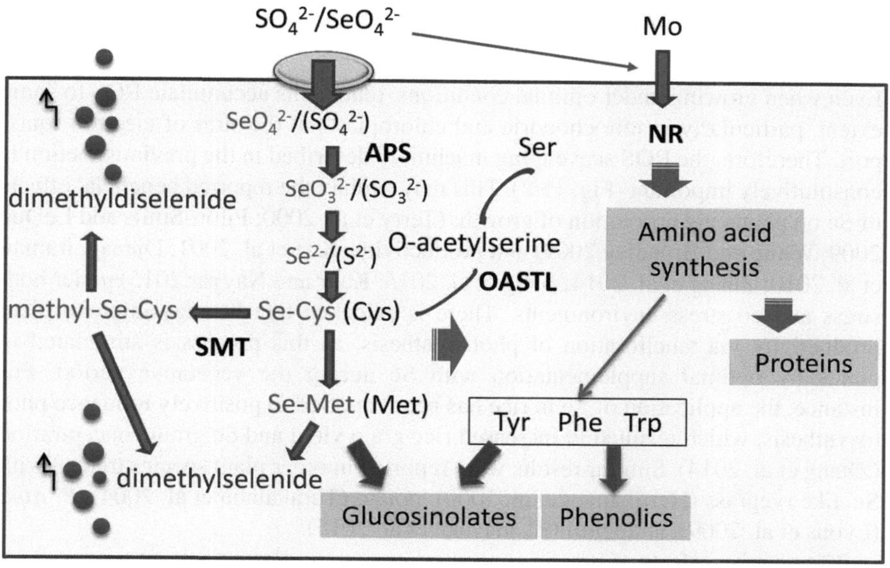

Fig. 15.2 Interaction of Se assimilation with other metabolic pathways. Selenium can influence the synthesis of glucosinolates by altering the content of precursor amino acids. Selenium can also alter the uptake of molybdenum (Mo), which is a cofactor of the enzyme nitrate reductase (NR), thus exerting an effect on nitrogen (N) assimilation into amino acids, proteins and phenolics

verted into H_2O and O_2 by specific peroxidases (POX), enzymes such as catalase (CAT) and glutathione peroxidase (GSH-Px) (Roychoudhury et al. 2012). High concentration of H_2O_2 in the cellular environment as a response to a stressful condition or SOD activity can cause oxidative damage. Non-enzymatic molecules implied in ROS detoxification are also important to preserve the cellular redox state, and mainly include the reduced form of glutathione (GSH), ascorbate, phytochelatins (PCs), proline, flavonoids, alkaloids and carotenoids (Foyer and Noctor 2012).

Selenium has been reported to help plants cope with stress by stimulating the plant cell antioxidant capacity through the enhancement of the activity of antioxidant enzymes (SOD, CAT and GSH-Px) and the synthesis of GSH, PCs, ascorbate, proline, flavonoids, alkaloids and carotenoids. Furthermore, Se may induce the spontaneous dismutation of the superoxide radical ($O_2^{-\bullet}$) into H_2O_2 (Feng et al. 2013). As a result of Se-increased antioxidant defense systems, lower levels of lipid peroxidation were observed under metal-induced oxidative stress conditions, because of reduced ROS accumulation (Feng and Wei 2012) (Fig. 15.2).

In addition to its function in mitigating heavy metal stress in plants, Se at low dosage has been shown to protect plants from a variety of other abiotic stresses including drought, cold, heat, salinity, and UV-B radiation, which also cause oxidative stress (Feng et al. 2013; Kaur et al. 2016).

15.2.1 Effects of Se on Plant Productivity

Even when growing under optimal conditions, plant cells accumulate ROS to some extent, particularly in mitochondria and chloroplasts at the sites of electron transport. Therefore, the ROS scavenging machinery described in the previous section is constitutively important (Fig. 15.2). This may explain the reported beneficial effects of Se on plants via promotion of growth, (Terry et al. 2000; Pilon-Smits and LeDuc 2009; White and Broadley 2009) and productivity (Xue et al. 2001; Djanaguiraman et al. 2010; Zhang et al. 2014; Jiang et al. 2015; Kaur and Nayyar 2015) under both stress and no stress environments. There is evidence that Se may improve plant productivity via amelioration of photosynthesis, as this process is stimulated in plants by optimal supplementation with Se during the vegetative period. For instance, the application of Se in rice has been reported to positively influence photosynthesis, which resulted in increased rice grain yield and Se grain concentration (Zhang et al. 2014). Similar results were reported in other plant species treated with Se, like ryegrass (Hartikainen et al. 2000), potato (Turakainen et al. 2004), *B. rapa* (Lyons et al. 2009), and lentil (Ekanayake et al. 2015).

The positive effects of low Se concentrations on the photosynthetic process may be explained via the enhancement of the antioxidant activity in cells at different levels (Fig. 15.2). Selenium can up-regulate the amount and activity of antioxidant enzymes (GSH-Px, GR, SOD, APX and CAT) and metabolites (GSH, ascorbate) resulting in higher ROS scavenging capacity of plants, as well-documented under stress conditions (Germ et al. 2007; Tadina et al. 2007; Djanaguiraman et al. 2010; Feng et al. 2013). In addition to this effect on the antioxidant machinery, appropriate Se concentrations could significantly improve photosynthesis by increasing the production of chlorophyll (Hawrylak-Nowak 2009; Yao et al. 2011; Liu et al. 2011; Zhang et al. 2014), stomatal conductance, intercellular CO_2 concentration, and transpiration efficiency (Germ et al. 2007; Djanaguiraman et al. 2010; Zhang et al. 2014).

In other photosynthetic organisms such as algae, no significant effect of Se on photosynthesis or modification of chloroplast ultrastructure were observed, with the exception of the increase in content of carotenoids, which are known to act as important intracellular antioxidants (Schiavon et al. 2012).

15.2.2 Heavy Metals

As mentioned in the previous sections, Se can stimulate the cell antioxidant capacity in plants that grow in the presence of heavy metals through the enhanced activity of antioxidant enzymes and the synthesis of non-enzymatic metabolites such as GSH and PCs, and may induce the spontaneous dismutation of the superoxide radical ($O_2^{-\bullet}$) into H_2O_2 (Feng et al. 2013). The lower concentration of ROS would result in reduced lipid peroxidation generally caused by metal-induced oxidative stress (Feng and Wei 2012). The interactions of Se with a number of toxic elements are highlighted below.

15.2.2.1 Cadmium (Cd)

Cadmium (Cd) is one of the most toxic among heavy metals. This metal can be complexed with the organic fraction of soil, and be released as Cd^{2+}, which is easily assimilated by plants through membrane transporters involved in the uptake of chemically similar nutrients, like Ca^{2+}, Fe^{2+}, Mg^{2+}, Cu^{2+} and Zn^{2+} (Qin et al. 2013). The presence of high concentrations of Cd in soil can cause a decrease of plant capacity to accumulate these and other nutrients and affect the synthesis of molecules such as chlorophylls, carotenoids and a broad spectrum of proteins, including antioxidant enzymes, which contain one or more of these nutrient metals in their active sites to function as catalysts (Cuypers et al. 2010; Hasanuzzaman et al. 2012). For instance, Cd can replace Zn, Cu or Fe in the active sites of antioxidant metalloenzymes, e.g. SOD and CAT, thus causing their inactivation (Cuypers et al. 2010).

Recent studies reported the positive effect of Se on the activity of antioxidant enzymes in response to Cd stress. Lin et al. (2012) showed that the application of 3 μM Se to rice (*Oryza sativa*) plants can increase the activity of SOD, peroxidase or guaiacol peroxidase (POD/GSH-Px) enzymes in roots and leaves. Treatment with 50 μM Se was shown to enhance the activity of CAT, GSH-Px, glutathione reductase (GR), ascorbate peroxidase (APX) and enzymes related to the ascorbate-glutathione cycle, like monodehydroascorbate reductase (MDHAR) and dehydroascorbate reductase (DHAR), as well as non-enzymatic compounds of this cycle like GSH, especially in the oxidized form (GSSG), in oilseed rape (*Brassica napus*) (Hasanuzzaman et al. 2012). Similarly, 5 and 10 μM of Se increased the activity of CAT, APX and GR in leaves of sunflower (*Helianthus annuus*) (Saidi et al. 2014a), and concomitantly decreased ROS production, lipid peroxidation, oxidative stress, and recovered the membrane physicochemical characteristics.

The non-enzymatic mechanism for cellular detoxification against Cd works in parallel to the enzymatic system to maintain proper cellular redox state. Phytochelatins represent one of the most important strategies used by plants to counteract Cd stress by complexing this element and storing it in the vacuole, as is the case for other heavy metals (Foyer and Noctor 2012). In this context, S as a component of the amino acid Cys plays an essential role in GSH and PCs synthesis (Roychoudhury et al. 2012). Plants absorb more S when they grow in the presence of Cd in order to synthetize more GSH and PCs that chelate Cd (Feng et al. 2013).

15.2.2.2 Arsenic (As)

Arsenic (As) is a metalloid that is mainly found in the forms of arsenate (AsO_4^{3-}) or arsenite (AsO_3^{-3}) in soils and waters. Arsenic contamination in soils is mainly due to anthropogenic activities like the application of pesticides and sewage sludge on crop fields, as well as to other activities not directly related to agriculture, such as mining and metal melting. The detrimental effect triggered by this metal in plants is related to a reduction of growth and development caused by photosynthesis inhibition, inefficient nutrition and oxidative stress (Malik et al. 2012; Han et al. 2015).

Similar to Cd^{2+}, As binds to the S presented in the thiol (SH) group of GSH. When Se is provided to plants, it may compete with As for the binding to thiol groups, thus actively reducing As absorption (Han et al. 2015). Likewise, 5 μM Se was reported to reduce As uptake in mung bean (*Phaseolus aureus* Roxb.), and alleviate oxidative stress by enhancing the activity of SOD, POD, APX enzymes and the synthesis of GSH and ascorbic acid (ASC) (Malik et al. 2012). Similar results were found in tobacco (*Nicotiana tabacum*) plants treated with 0.1 mg/L selenite (Han et al. 2015). Srivastava et al. (2009) found that 5 μM and 10 μM selenate decreased the lipid peroxidation process in *Pteris vittata* L., likely because of higher production of the non-protein thiol GSH.

15.2.2.3 Lead (Pb)

Lead is one of the most dangerous pollutants worldwide, with as main sources fertilizers, pesticides, mining, metal smelting, automobile fumes and industrial waste or discharge. This heavy metal is considered carcinogenic to humans, as it causes DNA damage and inhibition of DNA synthesis. In plants, Pb can disrupt membrane structure and permeability, causing dehydration and decreased electron transport in photosynthesis. Lead also binds to the thiol groups of amino acids, enzymes and proteins and, as a result, induces the overproduction of ROS and oxidative stress (Mroczek-Zdyrska and Wojcik 2012).

The beneficial effects of Se against Pb stress in plants have been described by several authors recently and are directly related to ROS scavenging in cells (Mroczek-Zdyrska and Wojcik 2012; Yuan et al. 2013; Hu et al. 2014). For instance, 1.5 μM selenite supply lowered the superoxide radical ($O_2^{-\bullet}$) production and concentration in the apical part of the root in *Vicia faba L. minor*, and increased the activity of POD/GSH-Px enzymes and non-protein thiol content (Mroczek-Zdyrska and Wojcik 2012). In addition, 1 μM selenite improved the leaf biomass of coleus (*Coleus blumei* Benth.) and decreased the rate of lipid peroxidation, likely because of the higher GSH level in roots (Yuan et al. 2013). Furthermore, Hu et al. (2014) demonstrated that 0.5 mg/kg selenite could reduce Pb accumulation in rice (*Oryza sativa*) shoot and husk tissues.

15.2.2.4 Other Heavy Metals

Excess of some metal micronutrients in plants can increase the production of ROS and cause decreased activities of antioxidant enzymes, via denaturation and inactivation. Manganese (Mn) is an important microelement for plants, but at high concentration it can be toxic. The toxicity is related to photosynthesis suppression, membrane integrity disruption, lower protein metabolism and oxidative stress. As shown by Saidi et al. (2014b), 5 μM of selenate can effectively counteract the detrimental effects of Mn in sunflower (*Helianthus annuus*) by improving CAT, APX and GSH-Px activities.

Chromium (Cr) has no biological function in plants and can be toxic at any concentration in soil, especially near areas with industrial activities. Selenium supply at 3 μM enhanced SOD activity in rice roots, alleviating the toxic effects of Cr on growth, and increased H^+ ATPase activity, thus protecting the plants from Cr-induced oxidative stress (Cao et al. 2013).

Mercury (Hg) is also a harmful environmental pollutant, and soil contamination by this metal comes from mining, metal smelting, and industrial activities. Its presence in plants causes growth inhibition, oxidative stress, lipid peroxidation, and reduced chlorophyll production and photosynthesis (Zhao et al. 2013). Selenite and selenate treatment improve growth of garlic (*Allium sativum*) under Hg stress and reduced Hg absorption, translocation and accumulation in roots and leaves, when applied at levels higher than 1 mg/L (Zhao et al. 2013).

The maintenance of cellular homeostasis under heavy metal contamination depends on several interlinked and complex mechanisms that together constitute the antioxidant defenses. The contributions of the various components may differ depending on various factors such as plant species, concentration, exposure time, nutrient concentration in soil, plant developmental stage, organs, and tissues analyzed. Thus, plant defense against heavy metals and other abiotic stresses is a dynamic and adaptive system. There is extensive evidence that different forms of Se can improve both enzymatic and non-enzymatic antioxidant responses, and thus counteract heavy metal induced stress.

15.2.3 Drought and Salt Stress

Water stress, particularly drought stress, causes the production of ROS in plants. The protective role of Se against this type of stress has been reported in various plant species and it occurs by quenching the accumulation of ROS via regulation of the level of enzymatic and non-enzymatic antioxidants (Pukacka et al. 2011; Yao et al. 2012; Durán et al. 2015). Yao et al. (2011) showed that optimum Se concentrations could help wheat seedlings maintain high growth performance under drought stress by significantly increasing the peroxidase and CAT activities that lower the level of ROS. In particular, Yao et al. (2012) observed that Se improved the recovery of wheat seedlings from drought stress after re-watering because Se turned the rate of $O_2^{\bullet-}$ production, MDA content, and CAT activity back to the control values. The reduction of ROS levels by Se in plants subjected to drought stress has also been observed in other plant species like rapeseed seedlings (Hasanuzzaman et al. 2010; Hasanuzzaman and Fujita 2011), *Trifolium repens* L. (Wang 2011) and wheat (Nawaz et al. 2013, 2015).

The effects of drought stress in plants widely overlap with those caused by salt stress, as they are both able to generate osmotic stress. The activity of the enzymes SOD and POD was increased by 10 μM Se in cowpea plants grown in the presence of 50 mM NaCl (Manaf 2016). In tomato (*Solanum lypersicon)*, Se was found to alleviate salt-induced oxidative stress by up- regulation of the antioxidant defense systems (Diao et al. 2014). In a previous study, Hawrylak-Nowak (2009) suggested that Se could enhance salt tolerance in plants by protecting the cell membranes against lipid peroxidation due to the antioxidative activity of Se at low concentration. Furthermore, the growth-promoting effect of Se under salt stress conditions could be due to the increased accumulation of proline accumulation and/or a decrease in the content of chloride ions (reduced salt uptake) in shoot tissues.

15.2.4 Extreme Temperatures

Similar to drought stress, high temperature and cold can increase the production of ROS in plants, particularly in species that possess low antioxidant capacity to detoxify ROS (Wang et al.2009; Djanaguiraman et al. 2010). Also under these types of stress, Se has been observed to protect plants from oxidative damage. In wheat, for instance, Se application was reported to ameliorate the symptoms related to cold stress by reducing MDA content and via enhanced production of antioxidant metabolites, such as anthocyanins, flavonoids, and phenolic compounds (Chu et al. 2010). Similar results were observed in potato (Seppänen et al. 2003), cucumber (Hawrylak-Nowak 2009), and sorghum plants (Abbas 2012, 2013) grown under low temperature and treated with Se.

With respect to heat stress, in a recent study Iqbal et al. (2015) found that exogenous application of Se reduced oxidative stress and induced heat tolerance in spring wheat, thus avoiding loss of grain yield. In these plants, Se-mediated the up-regulation of antioxidative systems, both enzymatic and non-enzymatic.

15.2.5 UV-B Stress

Increasing level of ultraviolet-B (UV-B) light because of thinning of the stratospheric ozone layer is one of the abiotic stress factors that can affect almost every aspect of plant productivity (Yao et al. 2013). Selenium can display a protective effect in plants against the harmful effects of UV-B radiation. Yao et al. (2010) showed that adequate Se supplementation (1.0 mg/kg) to wheat had a protective role in plants subjected to UV-B, via the decrease of oxidative stress-related damage to cellular components produced by high level of this type of radiation. Similar findings have been reported for lettuce and ryegrass (Xue and Hartikainen 2000).

In another study on wheat, Yao et al. (2010) reported that Se fertilization induced an evident increase in chlorophyll content, spike length, weight per spike, grain yield, protein content, N, Fe, Cu, and Se concentration under UV-B stress, leading to improved yield and quality of winter wheat to some extent. Other possible mechanisms by which Se may protect plants against UV-B stress are through increased levels of compounds that either absorb UV light (in the epidermis) or can reflect UV light (Golob et al. 2017).

15.3 Effects of Se on Plant Nutraceuticals

Given the importance of Se in human nutrition and Se deficiency-related issues existing in many areas worldwide, in recent years many efforts have been made to increase the concentration of Se in crops, especially when they are cultivated in soils that are low in this element. To date, the results achieved are promising, as several plant species have been successfully biofortified with organic forms of Se (Thavarajah et al. 2008; Brummell et al. 2011; Schiavon et al. 2013; Avila et al. 2014; Poblaciones et al. 2014; Rodrigo et al. 2014; Bañuelos et al. 2015; Bachiega et al. 2016). These plants represent high-nutrition value food that can be used to counteract the problem of Se deficiency where it occurs.

Biofortification is generally defined as the agricultural process aimed to improve the uptake and accumulation of specific phytochemicals in food derived-products by plant breeding, genetic engineering, and manipulation of agronomic practices (Rouached 2013; Wu et al. 2015). Despite the advantage they offer, biofortification technologies must be carefully performed in the case of Se because the concentration of this element in plant tissues should not exceed the threshold that is toxic for the plant and for the organisms that feed on it (Finley 2006). Selenium at high dosage may exert detrimental effects on human and animal metabolism due to Se replacement of S in proteins (Wilber 1980; Vinceti et al. 2001; Misra et al. 2015). Furthermore, Se biofortification may positively or negatively influence the synthesis of other health promoting compounds in plants (Robbins et al. 2005; Schiavon et al. 2013; Bachiega et al. 2016).

On this account, the challenge of Se biofortification is to produce plants enriched in organic Se forms without adversely impacting the synthesis of other nutraceuticals. Encouragingly, Se at low doses has been reported to enhance the levels of other beneficial health compounds in some studies (Schiavon et al. 2013, 2016; Avila et al. 2014; Tian et al. 2016). It is clear that Se biofortification programs, to be successful, should take into account the interactions of Se with the plant pathways that produce nutraceuticals, in addition to the concentration and the form of Se used to enrich plants in this element and/or the method employed for achieving Se enrichment. In the next section, the interactions of Se with plant metabolic processes involved in the synthesis of a number of (other) nutraceutical compounds are highlighted.

15.3.1 Glucosinolates

As already mentioned, Se can replace S in many S-containing compounds, including the Se-amino acids SeCys and SeMet. Selenomethionine and the methylated form of SeCys (SeMetCys) provide important beneficial properties to humans as their supplementation can alleviate thyroid disorders, prevent different types of cancer, treat male infertility, and enhance the immune system (Rayman 2012; Roman et al. 2014). These organic forms of Se can be produced by different plant species, either growing on Se-containing soil, or after Se fertilization (Sepúlveda et al. 2013).

In *Brassicaceae* spp. the S-amino acid methionine (Met), in addition to being an essential constituent of proteins, is a precursor of the anticarcinogenic aliphatic glucosinolates (GLSs). Therefore, as a consequence of Se interference with S assimilation in plants, Se fertilization may affect the levels of Met-derived GLSs in these plants. Contrasting results are reported in this respect. A weak decrease in aliphatic GLSs, especially glucoraphanin, was observed by Robbins et al. (2005) and Barickman et al. (2013) after supply of broccoli (*Brassica oleracea*) plants with high Se concentrations. The level of sulforaphane, a sulfur-containing aglycon produced during the GLS hydrolysis mediated by myrosinase, significantly decreased in response to Se application. In contrast, Sepúlveda et al. (2013) did not measure any variation in the content of GLSs and sulforaphane, nor in myrosinase activity in the same plant species treated with 100 μM selenate. However, when Se dosage applied to plants was lower than 0.8 mg/L (10 μM) or S concentration in the medium was increased, plants could maintain high levels of GLSs in their tissues (Barickman et al. 2013). This was likely because low Se concentration can stimulate S uptake in plants (Harris et al. 2014), thus promoting the synthesis of S-organic compounds.

In addition to this Se concentration-related effect on S assimilation, the chemical form of Se used in biofortification approaches and the method of supplementation must be considered. For instance, when Se in the form of selenium dioxide (SeO_2) was supplemented via root irrigation to *Brassica rapa* plants, an increase of several GLSs was observed, including the aliphatic GLSs glucobrassicanapin and glucoallysin (Thiruvengadam and Chung 2015).

A differential effect of Se on the levels of GLSs in broccoli plants was observed between plant organs (Avila et al. 2014). GLSs in the florets of broccoli treated with selenate were reduced, while GLS levels in the sprouts were not affected. Rather, sprouts were enhanced in the content of the potent anticarcinogenics glucoraphanin and SeMetCys, and therefore exhibited improved potential anticancer activity. Tian et al. (2016) also observed an increase of myrosinase activity and sulforaphane in broccoli sprouts treated with 100 μM selenite or selenate; meanwhile the amount of GLSs was unchanged. The same authors reported up-regulation of genes related to GLSs biosynthesis.

In recent years, Se-glucosinolates have also been identified in plants. Matich et al. (2012, 2015) in particular, showed that Brassicaceae spp. fertilized with Se contained (methylseleno) glucosinolates and their Se-containing aglycons. The major aliphatic Se-GLSs identified were glucoselenoraphanin and glucoselenoerucin in broccoli. In these species, Se-GLSs concentrations exceeded that reported for their S analogs. Results obtained in these studies have important implications for human health, because it seems the Se-containing isothiocyanates derived from Se-GLSs are more potent anticarcinogenic compounds than their S counterparts (Emmert et al. 2010).

15.3.2 Health Beneficial Nitrogen Containing-Compounds

The S and N metabolic pathways are strictly associated (Bielecka et al. 2015; Zhang et al. 2015) and a number of metabolites in plants contain both of these elements in their structure (e.g. cysteine, methionine, GSH, coenzyme A, GLSs). As a result of Se interaction with S assimilation, the N metabolic pathway may undergo changes in the synthesis of N compounds. Selenium can influence N metabolism also by interfering with the uptake of molybdenum (Mo) (Harris et al. 2014), which is a cofactor of nitrate reductase (NR), the enzyme that mediates the conversion of nitrate to nitrite in N assimilation. As a result of decreased nitrate reduction, the synthesis of all amino acids could be affected.

Some amino acids like methionine, tryptophan, phenylalanine and tyrosine, function as precursors for the synthesis of glucosinolates (Agerbirk and Olsen 2012). The same amino acids, with the exception of methionine, also function as precursors of other important metabolic compounds, including auxins, phenylpropanoids, tannins and alkaloids, synthesized through the shikimate pathway. Phenylpropanoids in particular, are reactive metabolites present in a wide range of plant-derived foods and display an important role in welfare and human health due to their antioxidant and antimicrobial properties (Ozcan et al. 2014). Among phenylpropanoids, phenolic acids and flavonoids have additional anti-carcinogenic and anti-mutagenic effects since they act as protective agents of DNA against free radicals, by inactivating carcinogens, inhibiting enzymes involved in pro-carcinogen activation and by activating xenobiotics detoxification enzymes (Ramos 2008).

In *Brassica rapa*, the application of SeO_2 caused the enhancement of phenolics and flavonoid accumulation, as well as the up-regulation of genes related to their biosynthesis (Thiruvengadam and Chung 2015). Similar results were obtained by Bachiega et al. (2016) in broccoli, especially at the stage of seedlings, as application of selenate increased their phenolic compounds content and antioxidant activity, and in tomato (*Solanum lycopersicum*), where the stimulation of flavonoids and phenolic acids was also observed after selenate supplementation, and fruits enriched in naringenin, chalcone and kaempferol were generated (Schiavon et al. 2013). However, Robbins et al. (2005) reported contrasting findings in broccoli, as Se fertilization in this case decreased the level of phenolics, without altering the profile distribution of specific compounds.

Given the role of phenolics as antioxidants in plants, it cannot be excluded that the capacity of Se to alleviate some types of stress in plants may be in part related to the stimulation of the shikimate pathway, in addition to its potential to elicit other antioxidant enzymatic and non-enzymatic systems.

15.4 General Conclusions and Further Prospects

Despite not being essential for plants, Se has been shown to exert beneficial effects on them depending on the chemical form supplied and the plant species. For instance, Se can improve plant defense systems by detoxifying intracellular free radicals directly, acting as antioxidant, and/or indirectly by increasing the activity of enzymatic (SOD, CAT, POX, GR) and non-enzymatic (GSH, proline, flavonoids, alkaloids carotenoids and PCs) antioxidants, which may help plants scavenge ROS and prevent oxidative stress. ROS may be produced by plant electron transport processes, even under optimal conditions, explaining the beneficial effect of Se on photosynthetic performance. ROS accumulation is particularly high during biotic and abiotic stress, which is when Se supplementation can be particularly beneficial for plants. In the case of heavy metals, Se may also reduce metal translocation from the root to the shoot by stimulating sulfate uptake and assimilation and the associated production of metal chelators GSH and PCs. All these processes result in better management of ROS production and concentrations in cells and, as a result, reduced oxidative stress-induced damage to cell membranes, proteins, DNA and other structures.

Low Se concentrations also have a beneficial effect on plants in terms of productivity and nutritional value. The enrichment of plants in organic forms of Se with recognized health properties, as well as in other precious nutraceuticals, through Se biofortification practices has significant implications in human and animal nutrition, especially in areas poor in Se where the local populations suffer of Se deficiency related-health issues.

Acknowledgments Michela Schiavon, Leonardo Warzea Lima and Ying Jiang were supported by National Science Foundation grant IOS-1456361. Rothamsted Research receives support from the Biotechnology and Biological Sciences Research Council (BBSRC) of the UK as part of the 20:20 Wheat project.

References

Abbas SM (2012) Effects of low temperature and selenium application on growth and the physiological changes in sorghum seedlings. J Stress Physiol Biochem 8:268–286

Abbas SM (2013) Low levels of selenium application attenuate low temperature stress in sorghum (*Sorghum bicolor* L.) seedlings. Pak J Bot 45:1597–1604

Agerbirk N, Olsen CE (2012) Glucosinolate structures in evolution. Phytochemistry 77:16–45

Avila FW, Yang Y, Faquin V, Ramos SJ, Guilherme LR, Thannhauser TW, Li L (2014) Impact of selenium supply on Se-methylselenocysteine and glucosinolate accumulation in selenium-biofortified *Brassica* sprouts. Food Chem 165:578–586

Bachiega P, Salgado JM, de Carvalho JE, Ruiz AL, Schwarz K, Tezotto T, Morzelle MC (2016) Antioxidant and antiproliferative activities in different maturation stages of broccoli (*Brassica oleracea* italica) biofortified with selenium. Food Chem 190:771–776

Bañuelos GS, Arroyo I, Pickering IJ, Yang SI, Freeman JL (2015) Selenium biofortification of broccoli and carrots grown in soil amended with Se-enriched hyperaccumulator *Stanleya pinnata*. Food Chem 166:603–608

Barickman TC, Kopsell DA, Sams CE (2013) Selenium influences glucosinolate and isothiocyanates and increases sulfur uptake in *Arabidopsis thaliana* and rapid-cycling *Brassica oleracea*. J Agric Food Chem 61:202–209

Bielecka M, Watanabe M, Morcuende R, Scheible WR, Hawkesford MJ, Hesse H, Hoefgen R (2015) Transcriptome and metabolome analysis of plant sulfate starvation and resupply provides novel information on transcriptional regulation of metabolism associated with sulfur, nitrogen and phosphorus nutritional responses in Arabidopsis. Front Plant Sci 5:805

Brown KM, Arthur JR (2001) Selenium, selenoproteins and human health: a review. Public Health Nutr 4:593–599

Brummell DA, Watson LM, Pathirana R, Joyce NI, West PJ, Hunter DA, McKenzie MJ (2011) Biofortification of tomato (*Solanum lycopersicum*) fruit with the anticancer compound methylselenocysteine using a selenocysteine methyltransferase from a selenium hyperaccumulator. J Agric Food Chem 59:10987–10994

Cao F, Wang N, Zhang M, Dai H, Dawood M, Zhang G, Wu F (2013) Comparative study of alleviating effects of GSH, Se and Zn under combined contamination of cadmium and chromium in rice (*Oryza sativa*). Biometals 26:297–308

Chu J, Yao X, Zhang Z (2010) Responses of wheat seedlings to exogenous selenium supply under cold stress. Biol Trace Elem Res 136:355–363

Combs GF Jr (2001) Selenium in global food systems. Braz J Nutr 85(5):517–547

Combs GF Jr (2005) Current evidence and research needs to support a health claim for selenium and cancer prevention. J Nutr 135:343–347

Cuypers A, Plusquin M, Remans T, Jozefczak M, Keunen E, Gielen H, Opdenakker K, Nair AR, Munters E, Artois TJ, Nawrot T, Vangronsveld J, Smeets K (2010) Cadmium stress: an oxidative challenge. Biometals 23:927–940

Diao M, Ma L, Jianwei W, Jinxia C, Aifei F, Huiying L (2014) Selenium promotes the growth and photosynthesis of tomato seedlings under salt stress by enhancing chloroplast antioxidant defense system. J Plant Growth Regul 33:671–682

Djanaguiraman M, Prasad PV, Seppanen M (2010) Selenium protects sorghum leaves from oxidative damage under high temperature stress by enhancing antioxidant defense system. Plant Physiol Biochem 48:999–1007

Durán P, Acuña JJ, Gianfreda L, Azcón R, Funes-Collado V, Mora ML (2015) Endophytic selenobacteria as new inocula for selenium biofortification. Appl Soil Ecol 96:319–326

Ekanayake LJ, Thavarajah D, Vial E, Schatz B, Mcgee R, Thavarajah P (2015) Selenium fertilization on lentil (*Lens culinaris*, Medikus) grain yield, seed selenium concentration, and antioxidant activity. Field Crop Res 177:9–14

El-Bayoumy K, Sinha R (2004) Mechanisms of mammary cancer chemoprevention by organoselenium compounds. Mutat Res Fundam Mol Mech of Mutagen 551:181–197

Emmert SW, Desai D, Amin S, Richie JP Jr (2010) Enhanced Nrf2-dependent induction of glutathione in mouse embryonic fibroblasts by isoselenocyanate analog of sulforaphane. Bioorg Med Chem Lett 20:2675–2679

Fan M-S, Zhao FJ, Poulton PR, McGrath SP (2008) Historical changes in the concentrations of selenium in soil and wheat grain from the Broadbalk experiment over the last 160 years. Sci Total Environ 389:532–538

Feng RW, Wei CY (2012) Antioxidative mechanisms on selenium accumulation in *Pteris vittata* L., a potential selenium phytoremediation plant. Plant Soil Environ 58:105–110

Feng R, Wei C, Tu S (2013) The roles of selenium in protecting plants against abiotic stresses. Environ Exp Bot 87:58–68

Fernandes AP, Gandin V (2015) Selenium compounds as therapeutic agents in cancer. Biochim Biophys Acta 1850:1642–1660

Finley JW (2006) Bioavailability of selenium from foods. Nutr Rev 64:146–151

Foyer CH, Noctor G (2012) Managing the cellular redox hub in photosynthetic organisms. Plant Cell Environ 35:199–201

Germ M, Kreft I, Stibilj V, Urbanc-Bercic O (2007) Combined effects of selenium and drought on photosynthesis and mitochondrial respiration in potato. Plant Physiol Biochem 45:162–167

Golob A, Kavčič J, Stibilj V, Gaberščik A, Vogel-Mikuš K, Germ M (2017) The effect of selenium and uv radiation on leaf traits and biomass production in *Triticum aestivum* L. Ecotoxicol Environ Saf 136:142–149

Han D, Xiong S, Tu S, Liu J, Chen C (2015) Interactive effects of selenium and arsenic on growth, antioxidant system, arsenic and selenium species of *Nicotiana tabacum* L. Environ Exp Bot 117:12–19

Harris J, Schneberg KA, Pilon-Smits EA (2014) Sulfur-selenium-molybdenum interactions distinguish selenium hyperaccumulator *Stanleya pinnata* from non-hyperaccumulator *Brassica juncea* (Brassicaceae). Planta 239:479–491

Hartikainen H, Xue T, Piironen V (2000) Selenium as an anti-oxidant and pro-oxidant in ryegrass. Plant Soil 43:193–200

Hasanuzzaman M, Fujita M (2011) Selenium pretreatment upregulates the antioxidant defense and methylglyoxal detoxification system and confers enhanced tolerance to drought stress in rapeseed seedlings. Biol Trace Elem Res 143:1758–1776

Hasanuzzaman M, Hossain MA, Fujita M (2010) Selenium in higher plants: physiological role, antioxidant metabolism and abiotic stress tolerance. J Plant Sci 5:354–375

Hasanuzzaman M, Hossain MA, Fujita M (2012) Exogenous selenium pretreatment protects rapeseed seedlings from cadmium-induced oxidative stress by upregulating antioxidant defense and methylglyoxal detoxification systems. Biol Trace Elem Res 149:248–261

Hawkesford MJ, Zhao FJ (2007) Strategies for increasing the selenium content of wheat. J Cereal Sci 46:282–292

Hawrylak-Nowak B (2009) Beneficial effects of exogenous selenium in cucumber seedlings subjected to salt stress. Biol Trace Elem Res 132:259–269

Hu Y, Norton GJ, Duan G, Huang Y, Liu Y (2014) Effect of selenium fertilization on the accumulation of cadmium and lead in rice plants. Plant Soil 384:131–140

Iqbal M, Hussain I, Liaqat H, Ashraf MA, Rasheed R, Rehman AU (2015) Exogenously applied selenium reduces oxidative stress and induces heat tolerance in spring wheat. Plant Physiol Biochem 94:95–103

Jackson MI, Combs GF Jr (2008) Selenium and anticarcinogenesis: underlying mechanisms. Curr Opin Clin Nutr Metab Care 11:718–726

Jiang Y, Zeng ZH, Bu Y, Ren CZ, Li JZ, Han JJ, Tao C, Zhang K, Wang XX, Li YJ, Lu GX, Hu YG (2015) Effects of selenium fertilizer on grain yield, se uptake and distribution in common buckwheat (*Fagopyrum esculentum* moench). Plant Soil Environ 61:371–377

Kaur S, Nayyar H (2015) Selenium fertilization to salt-stressed mungbean (*Vigna radiata* L. Wilczek) plants reduces sodium uptake, improves reproductive function, pod set and seed yield. Sci Hortic 197:304–317

Kaur S, Kaur N, Siddique KHM, Nayyar H (2016) Beneficial elements for agricultural crops and their functional relevance in defence against stresses. Arch Agron Soil Sci 62:905–920

Leustek T, Saito K (1999) Sulfate transport and assimilation in plants. Plant Physiol 120:637–643

Lin L, Zhou W, Dai H, Cao F, Zhang G, Wu F (2012) Selenium reduces cadmium uptake and mitigates cadmium toxicity in rice. J Hazard Mater 235-236:343–351

Liu K, ZhaoY CF, Gu Z, Bu G (2011) Enhanced glutathione peroxidases (GPx) activity in young barley seedlings enriched with selenium. Afr J Biotechnol 10:11483–11487
Lyons GH, Genc Y, Soole K, Stangoulis JCR, Liu F, Graham RD (2009) Selenium increases seed production in *Brassica*. Plant Soil 318:73–80
Malik JA, Goel S, Kaur N, Sharma S, Singh I, Nayyar H (2012) Selenium antagonises the toxic effects of arsenic on mungbean (*Phaseolus aureus* Roxb.) plants by restricting its uptake and enhancing the antioxidative and detoxification mechanisms. Environ Exp Bot 77:242–248
Manaf HH (2016) Beneficial effects of exogenous selenium, glycine betaine and seaweed extract on salt stressed cowpea plant. Ann Agric Sci 61:41–48
Matich AJ, McKenzie MJ, Lill RE, Brummell DA, McGhie TK, Chen RK, Rowan DD (2012) Selenoglucosinolates and their metabolites produced in *Brassica* spp. fertilized with sodium selenate. Phytochemistry 75:140–152
Matich AJ, McKenzie MJ, Lill RE, McGhie TK, Chen RK, Rowan DD (2015) Distribution of selenoglucosinolates and their metabolites in *Brassica* treated with sodium selenate. J Agric Food Chem 63:1896–1905
Misra S, Boylan M, Selvam A, Spallholz JE, Björnstedt M (2015) Redox-active selenium compounds–from toxicity and cell death to cancer treatment. Nutrition 7:3536–3556
Mithöfer A, Boland W (2012) Plant defense against herbivores: chemical aspects. Annu Rev Plant Biol 63:431–450
Moore KL, Schroder M, Lombi E, Zhao FJ, McGrath SP, Hawkesford MJ, Shewry PR, Grovenor CRM (2009) NanoSIMS analysis of arsenic and selenium in cereal grain. New Phytol 185:434–445
Mroczek-Zdyrska M, Wojcik M (2012) The influence of selenium on root growth and oxidative stress induced by lead in *Vicia faba L. minor* plants. Biol Trace Elem Res 147:320–328
Nawaz F, Ashraf MY, Ahmad R, Waraich EA (2013) Selenium (Se) seed priming induced growth and biochemical changes in wheat under water deficit conditions. Biol Trace Elem Res 151:284–293
Nawaz F, Ahmad R, Ashraf MY, Waraich EA, Khan SZ (2015) Effect of selenium foliar spray on physiological and biochemical processes and chemical constituents of wheat under drought stress. Ecotoxicol Environ Saf 113:191–200
Ozcan T, Akpinar-Bayizit A, Yilmaz-Ersan L, Delikanli B (2014) Phenolics in human health. Int J Chem Engin Appl 5:393–396
Pilon-Smits EA, LeDuc DL (2009) Phytoremediation of selenium using transgenic plants. Curr Opin Biotechnol 20:207–212
Poblaciones MJ, Rodrigo S, Santamaria O, Chen Y, McGrath SP (2014) Selenium accumulation and speciation in biofortified chickpea (*Cicer arietinum* L.) under Mediterranean conditions. J. Sci Food Agric 94:1101–1106
Pukacka S, Ratajczak E, Kalemba E (2011) The protective role of selenium in recalcitrant acer saccharium l. seeds subjected to desiccation. J Plant Physiol 168:220–225
Qin Q, Li X, Wu H, Zhang Y, Feng Q, Tai P (2013) Characterization of cadmium (108Cd) distribution and accumulation in *Tagetes erecta* L. seedlings: effect of split-root and of remove-xylem/ phloem. Chemosphere 93:2284–2288
Ramos S (2008) Cancer chemoprevention and chemotherapy: dietary polyphenols and signaling pathways. Mol Nutr Food Res 52:507–526
Rayman MP (2002) The argument for increasing selenium intake. Proc Nutr Soc 61:203–215
Rayman MP (2012) Selenium and human health. Lancet 379:1256–1268
Robbins RJ, Keck AS, Banuelos G, Finley JW (2005) Cultivation conditions and selenium fertilization alter the phenolic profile, glucosinolate, and sulforaphane content of broccoli. J Med Food 8:204–214
Rodrigo S, Santamaria O, Chen Y, McGrath SP, Poblaciones MJ (2014) Selenium speciation in malt, wort, and beer made from selenium-biofortified two-rowed barley grain. J Agric Food Chem 62:5948–5953
Roman M, Jitaru P, Barbante C (2014) Selenium biochemistry and its role for human health. Metallomics 6:25–54

Rouached H (2013) Recent developments in plant zinc homeostasis and the path toward improved biofortification and phytoremediation programs. Plant Signal Behav 8(1):e22681–e22154
Roychoudhury A, Basu S, Sengupta DN (2012) Antioxidants and stress-related metabolites in the seedlings of two indica rice varieties exposed to cadmium chloride toxicity. Acta Physiol Plant 34:835–847
Saidi I, Chtouron Y, Djebali W (2014a) Selenium alleviates cadmium toxicity by preventing oxidative stress in sunflower (*Helianthus annuus*) seedlings. J Plant Physiol 171:85–91
Saidi I, Nawel N, Djebali W (2014b) Role of selenium in preventing manganese toxicity in sunflower (*Helianthus annuus*) seedling. S Afr J Bot 94:88–94
Schiavon M, Pittarello M, Pilon-Smits EAH, Wirtz M, Hell R, Malagoli M (2012) Selenate and molybdate alter sulfate transport and assimilation in *Brassica juncea* L. Czern.: implications for phytoremediation. Environ Exp Bot 75:41–51
Schiavon M, Dall' Acqua S, Mietto A, EAH P-S, Sambo P, Masi A, Malagoli M (2013) Selenium fertilization alters the chemical composition and antioxidant constituents of tomato (*Solanum lycopersicon* L.) J Agric Food Chem 61:10542–10554
Schiavon M, Pilon M, Malagoli M, Pilon-Smits EA (2015) Exploring the importance of sulfate transporters and ATP sulphurylases for selenium hyperaccumulation-a comparison of *Stanleya pinnata* and *Brassica juncea* (Brassicaceae). Front Plant Sci 6:2
Schiavon M, Berto C, Malagoli M, Trentin A, Sambo P, Dall'Acqua S, Pilon-Smits EAH (2016) Selenium biofortification in radish enhances nutritional quality via accumulation of methyl-selenocysteine and promotion of transcripts and metabolites related to glucosinolates, phenolics, and amino acids. Front Plant Sci 7:1371
Seppänen M, Turakainen M, Hartikainen H (2003) Selenium effects on oxidative stress in potato. Plant Sci 165:311–319
Sepúlveda I, Barrientos H, Mahn A, Moenne A (2013) Changes in SeMSC, glucosinolates and sulforaphane levels, and in proteome profile in broccoli (*Brassica oleracea var. Italica*) fertilized with sodium selenate. Molecules 18:5221–5234
Shieber M, Chandel NS (2014) ROS function in redox signaling and oxidative stress. Curr Biol 24:453–462
Shinmachi F, Buchner P, Stroud JL, Parmar S, Zhao FJ, McGrath SP, Hawkesford MJ (2010) Influence of sulfur deficiency on the expression of specific sulfate transporters and the distribution of sulfur, selenium and molybdenum in wheat. Plant Physiol 153:327–336
Smith FW, Hawkesford MJ, Ealing PM, Clarkson DT, Vanden Berg PJ, Belcher AR, Warrilow AGS (1997) Regulation of expression of a cDNA from barley roots encoding a high affinity sulphate transporter. The Plant J 12:875–884
Sors TG, Ellis DR, Salt DE (2005) Selenium uptake, translocation, assimilation and metabolic fate in plants. Photosynth Res 86:373–389
Srivastava M, Ma LQ, Rathinasabapathi B, Srivastava P (2009) Effects of selenium on arsenic uptake in arsenic hyperaccumulator *Pteris vittata* L. Bioresour Technol. 100:1115–1121
Stroud J, Broadley M, Foot I, Fairweather-Tate S, Hart D, Hurst R, Knott P, Mowat H, Norman K, Scott P, Tucker M, White P, McGrath S, Zhao F (2010a) Soil factors affecting selenium concentration in wheat grain and the fate and speciation of Se fertilisers applied to soil. Plant Soil 332:19–30
Stroud J, Zhao FJ, Buchner P, Shinmachi F, McGrath SP, Abecassis J, Hawkesford MJ, Shewry PR (2010b) Impacts on sulphur nutrition on selenium and molybdenum concentrations in wheat grain. J Cereal Sci 52:111–113
Tadina N, Germ M, Kreft I, Breznik B, Gaberščik A (2007) Effects of water deficit and selenium on common buckwheat (*Fagopyrum esculentum*, moench.) plants. Photosynthetica 45:472–476
Terry N, Zayed AM, de Souza MP, Tarun AZ (2000) Selenium in higher plants. Annu Rev Plant Physiol Plant Mol Biol 51:401–432
Thavarajah D, Ruszkowski J, Vandenberg A (2008) High potential for selenium biofortification of lentils (*Lens culinaris* L.) J Agric Food Chem 56:10747–10753

Thiruvengadam M, Chung IM (2015) Selenium, putrescine, and cadmium influence health-promoting phytochemicals and molecular-level effects on turnip (*Brassica rapa ssp. rapa*). Food Chem 173:185–193

Tian M, Xu X, Liu Y, Xie L, Pan S (2016) Effect of Se treatment on glucosinolate metabolism and health-promoting compounds in the broccoli sprouts of three cultivars. Food Chem 190:374–380

Turakainen M, Hartikainen H, Seppänen MM (2004) Effects of selenium treatments on potato (*Solanum tuberosum* L.) growth and concentrations of soluble sugars and starch. J Agric Food Chem 52:5378–5382

Van HD, Takahashi H, Inoue E, Hess A, Tamaoki M, Pilon-Smits EA (2008) Transcriptome analyses give insights into selenium-stress responses and selenium tolerance mechanisms in arabidopsis. Physiol Plant 132:236–253

Vinceti M, Wei ET, Malagoli C, Bergomi M, Vivoli G (2001) Adverse health effects of selenium in humans. Rev Environ Health 16:233–251

Wang CQ (2011) Water-stress mitigation by selenium in *Trifolium repens* L. J Plant Nutr Soil Sci 174:276–282

Wang Y, Yang ZM, Zhang QF et al (2009) Enhanced chilling tolerance in *Zoysia matrella* by pre-treatment with salicylic acid, calcium chloride, hydrogen peroxide or 6-benzylaminopurine. Biol Plant 53:179–182

White PJ, Broadley MR (2009) Biofortification of crops with seven mineral elements often lacking in human diets – iron, zinc, copper, calcium, magnesium, selenium and iodine. New Phytol 182:49–84

White PJ, Bowen HC, Parmaguru P, Fritz M, Spracklen WP, Spiby RE, Meacham MC, Mead A, Harriman M, Trueman LJ, Smith BM, Thomas B, Broadley MR (2004) Interactions between selenium and sulphur nutrition in *Arabidopsis thaliana*. J Exp Bot 55:1927–1937

Wilber CG (1980) Toxicology of selenium: a review. Clin Toxicol 17:171–230

Wu Z, Bañuelos GS, Lin ZQ, Liu Y, Yuan L, Yin X, Li M (2015) Biofortification and phytoremediation of selenium in China. Front Plant Sci 6:136

Xue T, Hartikainen H, Vieno Piironen V (2001) Antioxidative and growth-promoting effect of selenium on senescing lettuce. Plant Soil 237:55–61

Yao XQ, Chu JZ, Ba CJ (2010) Antioxidant responses of wheat seedlings to exogenous selenium supply under enhanced ultraviolet-B. Biol Trace Elem Res 136:96–105

Yao X, Chu J, He X, Ba C (2011) Protective role of selenium in wheat seedlings subjected to enhanced UV-B radiation. Russ J Plant Physiol 58:283–289

Yao X, Chu J, Liang L, Geng W, Li J, Hou G (2012) Selenium improves recovery of wheat seedlings at rewatering after drought stress. Russ. J. Plant Physiol 59:701–707

Yao X, Jianzhou C, Xueli H, Binbin L, Jingmin L, Zhaowei Y (2013) Effects of selenium on agronomical characters of winter wheat exposed to enhanced ultraviolet-B. Ecotoxicol Environ Saf 92:320–326

Yuan J, Hu M, Zhou Z (2013) Selenium treatment mitigates the effect of lead exposure in *Coleus blumei* benth. Environ Anal Toxicol 3:191

Zeng H, Combs GF Jr (2008) Selenium as an anticancer nutrient: roles in cell proliferation and tumor cell invasion. J Nutr Biochem 19:1–7

Zhang M, Tang S, Huang X, Zhang F, Pang Y, Huang Q (2014) Selenium uptake, dynamic changes in selenium content and its influence on photosynthesis and chlorophyll fluorescence in rice (*Oryza sativa*, L.) Environ Exp Bot 107:39–45

Zhang Q, Lee BR, Park SH, Zaman R, Avice JC, Ourry A, Kim TH (2015) Sulfate resupply accentuates protein synthesis in coordination with nitrogen metabolism in sulfur deprived *Brassica napus*. Plant Physiol Biochem 87:1–8

Zhao J, Gao Y, Li YF, Hu Y, Peng X, Dong Y, Li B, Chen C, Chai Z (2013) Selenium inhibits the phytotoxicity of mercury in garlic (*Allium sativum*). Environ Res 125:75–81

Chapter 16
Overview and Prospects of Selenium Phytoremediation Approaches

Karaj S. Dhillon and Gary S. Bañuelos

Abstract Phytoremediation, a plant based technology, is perceived as a novel, low-cost, eco-friendly technology for *in-situ* management of Se-contaminated soil and water resources. Among the different phytoremediation mechanisms, phytoextraction, phytovolatilization, and rhizofiltration are primarily responsible for the management of Se in a contaminated environment. Selection of the best-suited plant and cultivation strategies are crucial for the success of phytoremediation technology at any given site. For example, *Brassica*-based cropping systems are about two times more efficient than agroforestry-based systems in removing Se from the contaminated sites. In addition, the potential of several transgenic approaches have been highlighted for further increasing Se accumulation, volatilization, and tolerance by plant species selected for phytoremediation. The accumulation of Se in plant tissues may also act as a deterrent for a number of herbivores like crickets, grasshoppers, prairie dogs, etc., while the entry of Se into the food chain can be minimized by growing non-food plants, e.g. flowers, in Se-contaminated soils. In this chapter, a number of alternatives for safe disposal and utilization of Se-rich biomass have also been discussed. These options will greatly help in promoting the adoption of phytoremediation as a vital tool in sustainable management of Se-contaminated soil and water resources.

Keywords Selenium • Phytoremediation • Transgenics • Ecological implications • Safe biomass disposal

K.S. Dhillon (✉)
Department of Soil Science, Punjab Agricultural University, Ludhiana, Punjab, India
e-mail: dhillon_karaj@yahoo.com

G.S. Bañuelos
USDA, Agricultural Research Service, San Joaquin Valley Agricultural Sciences Center, 9611 South Riverbend Avenue, Parlier, CA 93648-9757, USA
e-mail: gary.banuelos@ars.usda.gov

E.A.H. Pilon-Smits et al. (eds.), *Selenium in plants*, Plant Ecophysiology 11,
DOI 10.1007/978-3-319-56249-0_16

16.1 Introduction

Selenium (Se) – a naturally occurring element, was discovered by the Swedish chemist Jons Jakob Berzelius in 1817 and named after the Greek moon goddess 'Selene'. Selenium belongs to the Group 16 (previously Group VIa) in the Periodic Table; the group that also contains oxygen (O), sulphur (S) and tellurium (Te). It is a chalcophile (S-loving) and replaces S in common sulfide minerals such as pyrite, chalcopyrite, pyrrhotite and sphalerite. It is a constituent of several rare minerals including crookesite [$(Cu, Tl, Ag)_2Se$], berzelianite (Cu_2Se) and tiemannite (HgSe) (Fordyce 2013).

Parent material constitutes an important factor controlling the level of Se in geo-ecosystems (Rosenfeld and Beath 1964). Anthropogenic activities such as on-land disposal of coal generated fly ash, mine tailings and use of agricultural drainage and underground water for crop production can lead to Se accumulation in toxic levels (Dhillon and Dhillon 2003a). The Se content of most soils is very low, ranging from 0.01 to 2 mg/kg (world mean is 0.4 mg/kg), but high concentrations up to 1200 mg/kg have been reported for some regions (Fleming 1980; Neal 1995). Soils can be classified as seleniferous or non-seleniferous depending upon the Se level of cultivated agricultural crops grown on that soil. For example, soils containing 0.1–0.5 mg/kg Se may be considered as seleniferous (Ravikovitch and Margolin 1957; Dhillon et al. 1992) because forages grown on such soils can accumulate >5 mg Se/kg – the maximum permissible level for animal consumption. Soils with excess, adequate or deficient Se levels may exist side by side because Se content of the soil is influenced by leaching and hydrological transport processes (Wang and Gao 2001; Dhillon et al. 2008b). Some areas from where leaching has taken place, have developed into deficient regions, while Se-enriched regions can arise where the leachate has been deposited. Depending upon the redox conditions of the soil environment, Se may exist in four oxidation states: Se^{2-} (selenide), Se^0 (elemental Se), Se^{4+} (selenite) and Se^{6+} (selenate). Selenium is highly mobile under oxidizing conditions and its mobility decreases with decreasing pH (Gondi et al. 1992). Among the Se species in soil solution, selenate predominates at high redox (pE + pH >15), selenite in the medium redox range (pE + pH 7.5–15) and selenide species only at low redox (pE + pH <7.5) (Elrashidi et al. 1987, 1989). Presence of Se-contaminated soils and cases of Se poisoning in animals and humans have been reported from several countries including Ireland, Australia, United States, China and India (Dhillon and Dhillon 2003b).

The discovery of Se as an essential element for animals in the 1950s (Schwarz and Foltz 1957) prompted researchers to examine Se levels in all components of the environment and monitor its entry into the food chain. Large variations in dietary Se intake (ranging from 3 to 7000 μg/d in adults may be attributed to the wide range of Se content in the environment on the global scale (Fordyce 2013). For example, the range of mean Se content in cereal grains varies from 142 to 970 μg/kg for regions with high soil Se levels and from 14 to 90 μg/kg for regions with low soil Se levels. On the other hand, most forage and crop plants as well as grasses, contain Se less than 25 mg/kg dry matter (DM) and do not accumulate Se more than 100 mg/kg DM even when grown on seleniferous soils. Dietary Se intake in the endemic area of northwest India was 475±53 μg/d in women and 632±31 μg/d in men as compared to 52±1 and 65±2 μg/d, respectively, in the nonendemic region (Hira et al. 2004).

Evidence is lacking on whether Se is essential for vegetation growth, but plants can absorb, assimilate, and accumulate Se in leaves and roots. The capability of plants to take up substantial amount of Se is now being utilized to remove excess Se from contaminated areas. This process has been termed as 'phytoremediation'. The term phytoremediation is a combination of the Greek word 'phyto,' meaning plant, and the Latin word 'remedium,' for restoration. Although phytoremediation has been recognized and documented for more than 300 years, its use on scientific lines was not started until the early 1980s (Lasat 2000). The idea of using plants that accumulate metals to selectively remove and recycle excessive soil metals was introduced in 1983 (Chaney 1983). Since that time, its popularity has been increasing as a potential practical and cost-effective technology for remediation of contaminated environments (Cunningham et al. 1995; Chaney et al. 1995; Salt et al. 1995). Phytoremediation of Se-contaminated soils can be a non-polluting and cost effective way to remove or stabilize Se over time that might otherwise be leached out of the soil by excessive irrigation or rain water to contaminate groundwater, surface waters, or drainage waters. However, due to some inherent limitations, phytoremediation may not be fully effective in all types of contaminated sites, especially when the contamination runs too deep or the concentration is too high (Cunningham et al. 1995). Importantly, the green process takes time. The efficiency of phytoremediation may be greatly increased through the application of recent technological advances in plant breeding, genetic engineering, and by manipulation of agronomic practices. In this chapter we have attempted to review phytoremediation strategies at field scale that have helped to reduce the impact of high Se levels in the soil.

16.2 Phytoremediation Approaches

There are several reports emanating from Australia, China, Ireland, India and the United States showing that entry of excess Se from contaminated soils to the food chain results in serious health hazards in animals and humans (Fordyce 2013). Besides essential nutrients, plants are also able to absorb and accumulate potentially toxic metals like Se. Selecting specific plants having high Se absorbing capacity and growing these on contaminated soils for removal of Se in biomass has been termed as 'phytoremediation'. Conventional decontamination techniques like soil washing and soil replacement appear to be too costly, environmentally destructive and are *ex-situ* approaches (Cunningham and Ow 1996). On the other hand, phytoremediation is considered to be environment friendly and highly cost effective as an *in-situ* approach (Salt et al. 1995; Chaney et al. 1997). Among different phytoremediation strategies, brief procedural details are discussed below for (1) phytoextraction, (2) phytovolatilization and (3) rhizofiltration, all of which are considered more suitable for remediation of Se-contaminated soils and waters.

16.2.1 Phytoextraction

Phytoextraction involves the cultivation of higher plants for removal of contaminants from affected soils. After sufficient plant growth and metal accumulation, the above-ground plant parts are harvested and suitably disposedoff (Kumar et al. 1995). The success of phytoextraction is determined by the successful cultivation of crops and the plant characteristics related to biomass generation and uptake capacity.

Some plant species growing on seleniferous soils are considered Se-hyperaccumulators, as they are able to accumulate Se up to 15,000 mg/kg DM. Unfortunately, the majority of Se- hyperaccumulators have slow growth and limited biomass production, thereby leading to insufficient Se removal from contaminated soils and are wild species with no economic value (Chaney 1983). Depending on Se accumulation capacity, *Brassica* oilseed species like canola (*Brassica napus*) and Indian mustard (*Brassica juncea*) were found to be especially suitable for phytoextraction of Se compared to other cultivated crops (Bañuelos et al. 1997; Ajwa et al. 1998).

Chelating chemicals like EDTA, DTPA or organic manures are known to increase the availability of heavy metals to plant roots, thereby increasing the efficiency of phytoextraction (Sims and Johnson 1991). The effectiveness of different chelating agents, however, varies according to the plant and the heavy metals under investigation (Evangelou et al. 2007). Surprisingly, use of chelating agents for enhancing mobility of Se in soil has attracted the least attention of selenophiles. In a review dealing with the role of chelating agents in mobilizing metal contaminants in soil, only one publication pertained to Se among 30 papers published in a decade (Bolan et al. 2014). Esringu and Turan (2012) showed 12–20 time increase in Se removal by applying 7.5 mmol/kg EDDS and 1.0 mmol/kg DTPA in a greenhouse experiment. On the other hand, addition of organic amendments like poultry manure and press mud decreased Se uptake by 44–97% in different crops both under greenhouse and field conditions (Dhillon et al. 2010). Selenium concentration in wheat grains was reduced from 1350 to 160 μg/kg when the organic matter content in the plough layer increased from 1.4% to 39% (Johnsson 1991). There also exist reports that chelate-assisted phytoextraction may involve the risk of groundwater pollution due to metal mobilization and leaching (Wenzel et al. 2003; Robinson et al. 2003). Thus, the efforts for remediation of Se-contaminated soils should remain focused on the use of high biomass plants but without the application of chelators.

16.2.2 Phytovolatilization

Green plants are capable of converting toxic inorganic forms of Se into volatile organic selenocompounds, which are less toxic than the original chemical forms. The process in which plants absorb contaminants from soil and release them as volatile chemical species into the atmosphere is termed as phytovolatilization. The main advantage of phytovolatilization is that it can remove the pollutant without the need for plant disposal. Volatilization of Se by plants was first reported for Se hyperaccumulator *Astragalus bisulcatus* by Beath et al. (1935). Subsequently, Evans et al. (1968) demonstrated that the typical volatile Se compound released by Se-hyperaccumulator plants was dimethyldiselenide (DMDSe), while dimethylselenide (DMSe) was released from leaves of nonaccumulator plants (*B. oleracea*) (Lewis et al. 1971). Evaluation of different plant species tested for Se volatilization in the greenhouse revealed that *A. bisulcatus* and broccoli (*Brassica oleracea*) showed the highest rates of Se volatilization, followed by tomato (*Solanum lycopersicum*), tall fescue and alfalfa (Duckart et al. 1992). Plant species belonging to Brassicaceae family like broccoli and cabbage (*Brassica oleracea* var. *capitata*) have the potential to volatilize Se >10 g/ha/d (Terry and Zayed 1994). Thus, when phytovolatilization is coupled with phytoextraction potential, the efficiency of phytoremediation in reclaiming Se-contaminated soil can be increased by 2–3 times. The process of volatilization depends upon a number of factors like plant species, Se species, temperature, presence of other ions in the growth medium and microorganisms in the rhizosphere (Terry and Zayed 1994). There is still a need to better evaluate the influence of different factors on phytovolatilization under field conditions. Of course, one must always consider that Se only shifts from one inorganic phase (from the soil) to another inorganic phase (atmosphere) from where it can be re-deposited. Also, there is no control on the final destination of the volatilized forms of Se.

16.2.3 Rhizofiltration

Rhizofiltration is a form of phytoremediation that implements the use of plant root systems to intercept and remove contaminants from flowing water. The process is very similar to phytoextraction in that both pertain to removal of contaminants by trapping them into harvestable plant biomass. The major differences between the two processes are that rhizofiltration is used for the treatment of aquatic environments and allows harvesting of both roots and shoots; while phytoextraction deals with soil remediation and only allows harvesting of the shoot. Suitability of several aquatic species as potential rhizofiltration candidates for Se have been evaluated by conducting short duration experiments in aqueous solutions (Pilon-Smits et al. 1999a; Carvalho and Martin 2001; Miranda et al. 2014) and long-term studies in constructed wetlands (Lin and Terry 2003). Aquatic plants can remove Se from agricultural or industrial wastewater through Se accumulation and volatilization, irrespective of the type of Se species present. Many plant species identified in these studies like parrot's feather (*Myriophyllum brasiliense*), iris-leaved rush (*Juncus xiphioides*), cattail (*Typha latifolia*), saltmarsh bulrush (*Scirpus robustus*), hydrilla (*Hydrilla verticillata* Royle) and dotted duckweed (*Landoltia punctata*) in short duration studies and common reed (*Phragmites australis*) showed a great potential for Se phytoremediation in wetlands. Miranda et al. (2014) reported that biomass of several aquatic plants constitute an attractive feedstock for biofuel production. Thus, the dual utility of using aquatic plants for wastewater treatment and for the production of value-added petrochemicals provides an ecologically friendly and cost-effective solution for remediation of Se-contaminated drainage and industrial wastewater.

16.3 Selecting Plants Suitable for Phytoremediation

Selection of plants suitable for phytoremediation is an important factor for successful field management of Se contamination. Plants selected for phytoextraction of Se should possess large capacity for biomass production, accumulation and volatilization of Se, have a deep root system, and be easy to cultivate and harvest under different growing conditions. Plants that accumulate high concentrations of Se may be usefully employed in Se-deficient areas to provide supplementary fodder for livestock either by direct feeding (Bañuelos and Dhillon 2011) or by incorporating as green manure in deficient soil for raising forage crops (Dhillon et al. 2007, Fässler et al. 2010). Depending upon the extent of risk involved to the food chain, the cultivation of selected plants must be practically feasible and economically attractive under the given site and land use conditions (Robinson et al. 2009). For selecting plants with suitable phytoremediation characteristics, researchers have screened a large number of plant species, including cultivated agricultural crops and trees.

16.3.1 Agricultural Crops

Several agricultural crops have been tested for Se absorption capacity. Plant uptake of Se in inorganic forms (SeO_4^{2-} and SeO_3^{2-}) has been investigated extensively (Mayland et al. 1989; Bañuelos et al. 1991, 1993; Sharma et al. 2010). In general, the average Se content of plants belonging to different families grown in normal alkaline soils varied in the following order: Brassicaceae > Chenopodiaceae > Fabaceae > Poaceae (Dhillon et al. 1977). Among vegetable crops, Se accumulation was the greatest in edible portion of radish (*Raphanus sativus*) and the lowest in onion (*Allium cepa*) in the presence of 1.25 mg/kg selenate-Se in soil (Dhillon and Dhillon 2009b). In soil treated either with Se^{6+} or Se^{4+}, Se accumulation in plant tissues was the highest in Indian mustard followed by Old Man saltbush (*Atriplex nummularia*), creeping saltbush (*Atriplex semibaccata*), tall fescue grass (*Festuca arundinacea*) and *Astragalus incanus* (Bañuelos and Meek 1990). In soil amended with 1.5 mg/kg selenate-Se, Se accumulation by canola (*Brassica napus* cv. Westar) plant tissues was 3–4 times more than tall fescue (*Festuca arundinacea* cv. Fawn) tissues (Ajwa et al. 1998). Among three plant species grown on wetland sediments containing 40 mg Se/kg collected from Kesterson Reservoir, Se accumulation was as high as 470 mg/kg DM in canola plants (*Brassica napus* cv. Westar), 45 mg/kg DM in kenaf (*Hibiscus cannabinus* cv. Indian) and 50 mg/kg DM in tall fescue (*Festuca arundinacea* cv. Alta). Compared to pre-plant Se levels, significant reduction of total soil Se at the final harvest indicated that successively planting of canola and to a lesser extent kenaf and tall fescue, in Se-laden soil had the potential to reduce total soil Se (Bañuelos et al. 1997).

Crop plant species tolerant to Se may be useful for bioextraction of Se from deteriorated agricultural soils (Bañuelos 2000). Among different crops grown on selenite-treated alkaline silty loam soil, Indian mustard proved highly tolerant to Se in the plant tissues followed by corn (*Zea mays*), rice (*Oryza sativa*) and wheat (*Triticum aestivum*) (Rani et al. 2005). The critical levels of Se in plants above which significant decrease in yield would occur were recorded as 104.8 mg/kg in Indian mustard, 76.9 mg/kg in corn, 41.5 mg/kg in rice and 18.9 mg/kg in wheat shoots. Among several agricultural crops and weed plants grown on a naturally contaminated soil, Se accumulation was the highest in weeds (34–365 mg/kg) followed by oilseed crops (19–29 mg/kg), legumes (6–13 mg/kg) and cereals (2–18 mg/kg). The highest accumulation of Se for sunflower (*Helianthus annuus*) and *Brassica* species among the agricultural crops and *Mentha longifolia* among weeds, make these species potentially attractive for phytoextraction of Se from seleniferous soils (Dhillon and Dhillon 2009a). Some agricultural crops like wheat (*Triticum* spp.), corn (*Zea mays*), grain sorghum (*Sorghum bicolor*), and sugarcane (*Saccharum officcinarum*) are considered suitable for bioethanol production. Oilseed crops like sunflower (*Helianthus annuus*), rapeseed (*Brassica napus* var. oleifera D.C.), soybean (*Glycine max*), and cottonseed (*Gossypium arboretum*) are suitable for biofuel production (Bañuelos et al. 2013). Besides plant parts rich in starch, sugar and oils; stover and straw can also be used to produce bioenergy.

In view of the available information as summarized above, Brassica oilseed species (e.g. *B. juncea* and *B. napus*) may prove to be the most suitable candidate for managing Se in seleniferous soils. In order to further increase the environmental and economic sustainability of plant based Se-management system, post-harvest strategies now include creating new value-added biofortified agricultural products (Bañuelos 2009). For example, canola and mustard have been adopted as sources of biodiesel fuel crops (Bañuelos et al. 2013). Oil from both the sources have high energy content per unit weight and also act as the most efficient sources of bioenergy. One ton of seed produced from canola grown on Se-contaminated soil was successfully processed to produce 380 liters of 100% biodiesel (BD100) or 1900 liters of BD20 biodiesel – a mixture of 20% vegetable oil and 80% petrodiesel (Stapleton and Bañuelos 2009).

Concerning potential risks associated with biomass produced on Se-contaminated soils, growing tobacco (*Nicotiana tabacum*) for phytoremediation of Se-contaminated soils may prove doubly beneficial. Selenium may be involved in anticarcinogenic activities in humans (Schrauzer 2000; Combs 2005), and thus it may be beneficial when contained within tobacco products. Significant increase in Se content of tobacco plants has been recorded with increasing levels of Se in soil (Chortyk et al. 1984). In addition to removal of Se, significant quantities of Se-rich tobacco leaves will be available for producing cigarettes. Pyrolysis of the cured tobacco showed that about 45% of the Se could be transferred to tobacco smoke (Chortyk et al. 1984). Others have demonstrated that dietary Se inhibits pulmonary cell proliferation in cigarette smoke-exposed mice, indicating that Se is inhibiting cell proliferation independently of smoke exposure (Li et al. 2005). There are reports that the concentration of Se in tobaccos from low lung cancer-incidence countries is three time higher than that in tobaccos from high lung cancer-incidence countries (Bogden et al. 1981). Thus smoking cigarettes produced from Se-rich tobacco leaves may prove ironically beneficial for human health, but is still not recommended.

16.3.2 Tree Crops

Using trees as a vegetation cover has potential for the phytoremediation of Se-contaminated lands (Pulford and Watson 2003). Due to easy propagation, fast growth habits and accumulation of large biomass, willow (*Salix* spp.) and poplar (*Populus* spp.) are grown worldwide for pulp and bioenergy production in cold and temperate regions of the Northern Hemisphere. Hardwood species such as eucalyptus (*Eucalyptus* spp.), Arjuna (*Terminalia arjuna*), shisham (*Dalbergia sissoo*) and moderately hardwood species like acacia tree (*Acacia tortillas*, jambolin (*Syzygium cumini*) and dek (*Melia azedarach*) are used for the production of pulp, timber, furniture, charcoal and fuelwood. Such tree species were able to accumulate significant amounts of Se when grown on Se-contaminated soils (Pilon-Smits et al. 1998; Dhillon et al. 2008a; Bañuelos and Dhillon 2011). Total Se uptake in 1-year-old trees (including stem, leaves and roots) was greatest in Arjuna, followed by eucalyptus, mulberry, dek, jambolin, shisham and acacia (Dhillon et al. 2008a). Among these trees, shisham proved highly sensitive to the presence of selenate in the soil. Excessive accumulation of Se takes place in tree leaves and the abscission of leaves could deposit organic Se into the surrounding ecosystems. Leaves accumulated the highest Se concentrations followed by bark and wood in 5–10-year-old *Populus deltoides*, *Morus alba*, *Eucalyptus* spp., *Acacia tortillas*, *Syzygium cumini* and *Melia azedarach* (Dhillon et al. 2008a). Hybrid poplar (*Populus tremula* × *alba*) has been described as a highly promising tree species for phytoremediation through Se accumulation, as well as volatilization (Pilon-Smits et al. 1998). Growing trees for phytoremediation of Se-contaminated soils may help in reducing dependence on natural forests for fulfilling day to day requirements for wood and bioenergy.

16.4 Assessing Potential of Phytoremediation Technology Under Field Conditions

Although research on different aspects of Se phytoremediation continues to grow, intensive and long-term field investigations are the key to develop ecofriendly and economically viable strategies for decontamination of seleniferous soils and sediments. Success in the application of technology developed under laboratory and greenhouse conditions to actual field situations was evaluated in long-term field experiments and the results of practical utility are discussed below.

16.4.1 Remediation of Se-Contaminated Soils and Waters in Central California, USA

Natural-occurring Se-contaminated soils and ground waters on the west side of central California are also characterized by high salinity, boron, and sulfate levels. Growing crops on a sustainable basis to manage soluble Se requires a comprehensive knowledge of a wide range of site-specific factors like soil salinity, presence of toxic and competitive ions, agronomic practices, source of irrigation water, etc. Multi-year field phytoremediation studies were undertaken from 1993–2009 at different locations in the Westside of central California, where pockets of high Se soils and irrigation waters exist (Zayed et al. 2000; Bañuelos 2002; Bañuelos and Dhillon 2011). Typical biomass yields are shown in Table 16.1 for the experimental sites where soils and ground waters of central California have high sulfate/chloride salinity (>8 dS/m) and soluble B levels (>10 mg/L). Excessive accumulation of both chloride and B ions by the plants may lead to decrease in biomass production, but clipping plant tissue helped in preventing B and Cl concentrations from reaching toxic levels in the plants. Therefore, clipping practice is highly recommended on a regular basis for perennial crops grown under high salinity and B contents (Bañuelos and Dhillon 2011).

Under the field experimental conditions in California, Se concentrations ranged from 2 to 6 mg/kg in different plant species (Table 16.1). High sulfate concentrations in the soils and waters used for irrigation resulted in reduced Se accumulation by plants due to the competitive effect of sulfate on Se accumulation. Thus, under high sulfate soil field conditions, phytoremediation via Se accumulation will require a longer time to remove soluble Se compared to lower sulfate soils present in Northwestern India. Alternatively, volatilization of Se may be a more effective vehicle for removing Se from a high sulfate soil system (Table 16.1). Under the prevailing field conditions, the average rate of Se volatilization ranged from 48 to 108 g/ha; rates were greatest with canola/mustard and lowest for poplar trees.

For long-term management of Se-contaminated soils in the west side of central California, phytoremediation crops were grown in suitable rotations from 1992 to 1995 (Bañuelos et al. 1997). The potential of four different crop rotations (Table 16.2) was evaluated for Se removal from contaminated soils. Kenaf produced the greatest amount of biomass followed by Indian mustard, tall fescue and birds-foot trefoil. Tissue Se concentrations for all the crops remained under 1 mg/kg, except for Indian mustard, which exceeded 2 mg/kg. Compared to other crop rotations, the greatest reduction in total Se content of the soil was observed with the Indian mustard based cropping system after 4 years (Table 16.2). In another long-term field study (Bañuelos et al. 1995), tall fescue was grown for 4 years in a Se-contaminated soil

Table 16.1 Estimated amount of soluble Se biologically removed from soil (0–30 cm) with different crops under field conditions in the Westside of Central Valley, California[a]

Crop	Dry matter yield (Mg/ha)	Average plant Se (mg/kg)	Se removed via Harvest (g/ha)	Se removed via Volatilization[b] (g/ha)	[c]Estimated percentage of soluble Se removed (%)	Growing season length (months)
Canola/ Mustard leaves	13±0.6[d]	6±0.3	78±6	108±10	47	6
Broccoli	12±0.5	5±0.3	60±4	72±6	33	4
Saltbush	14±0.3	3±0.2	42±3	75±4	30	10
Saltgrass	9±0.2	3±0.2	27±2	83±3	28	10
Tall Fescue	11±0.2	2±0.1	23±2	60±2	21	10
Poplar tree leaves	10±0.4	3±0.1	27±2	48±2	19	8
Paulownia tree leaves	5±0.3	3±1	15±1	NA	4	8

[a]Estimated 400 g of soluble Se/ha initially available from a depth of 0–30 cm
[b]Volatilization amounts were calculated by multiplying mean daily rate ($\mu g/m^2/d$) and number of days in growing season of each crop
[c]Percentage of original soluble Se available removed by plant uptake and biological volatilization of Se per ha
[d]Values represent the means from different sites ± the standard error

near Los Banos, CA. Being a perennial grass with extensive deep root system, cultivation of tall fescue helped in reducing Se content of soil by 25% in the surface layer (0–45 cm depth) and also by 25% in the subsurface layer (45–90 cm depth).

In another field study, five clones of prickly pear cactus (*Opuntia ficus-indica*) were evaluated for their growth behavior and Se accumulation and volatilization in highly saline and Se-contaminated soils (Bañuelos and Lin 2010). After 3 years of growth, all clones exhibited significant decreases up to 20% in height and fruit production compared with control-grown clones. Mean Se concentrations ranged, however, from 4.9 to 9.8 mg/kg in cladodes, 1.5–2.5 mg/kg in fruit flesh, and 4.5–10.1 mg/kg in seeds. Rates of Se volatilization varied from 20 to 80 $\mu g/m^2/d$ among the tested clones. The successful growth of prickly pear cactus and its accumulation and volatilization of Se under adverse soil conditions could serve as an ideal alternative drought-tolerant crop for a gentle phytoremediation of Se in the west side of the San Joaquin Valley in central California.

16.4.2 Remediation of Se-Contaminated Soils in Northwestern India

Selenium toxicity problems were recognized in the soil-plant-animal-human continuum in northwestern India in the 1980s. The alkaline and calcareous seleniferous soils are located between 30.9634° to 31.2175° N and 76.1163° to 76.3516° E

Table 16.2 Changing in naturally occurring Se concentrations from 0 to 60 cm depth under crops grown in different rotations for phytoremediation of Se-contaminated soil in California during 1992–1995[a]

Sr No.	Crop rotation	Dry matter yield (kg/m^2)	Shoot Se conc. (mg/kg DM)	Total soil Se conc. (mg/kg)		Percent change in total Se conc. in soil
				Preplant	Postharvest	
1	Bare plots	–	–	1.32±0.08	1.01±0.08	17
2	Indian mustard – Indian mustard – Tall fescue – Tall fescue	Indian mustard – 1.21–1.33	Indian mustard – 1.70–2.15	1.20±0.06	0.51±0.08	60
		Tall fescue – 0.40–0.70	Tall fescue – 0.39–0.41			
3	Birdsfoot trefoil – Birdsfoot trefoil/ Tall fescue mixture – Birdsfoot trefoil/Tall fescue mix – Tall fescue	Birdsfoot trefoil – 0.44	Different plants – 0.36–0.61	1.18±0.12	0.78±0.07	34
		Plant mix – 0.72–0.90				
		Tall fescue – 1.12				
4	Kenaf – Kenaf – Tall fescue – Tall fescue	Kenaf – 3.12–3.45	Kenaf – 0.59–0.70	1.41±0.09	0.83±0.06	41
		Tall fescue – 0.35–0.80	Tall fescue – 0.42–0.52			

Source: Bañuelos (2000)

[a]Indian mustard (*Brassica juncea*) and kenaf (*Hibiscus cannabinus*) were planted and harvested and then replanted the following year. Tall fescue (*Festusa arundinacae*) and birdsfoot trefoil (*Lotus corniculatus*) were only planted once in their respective plots and then clipped as per requirements. Values shown are the mean of six replicates ± standard error

(Dhillon and Dhillon 1991, 2014). A group of researchers led by Karaj S. Dhillon at Punjab Agricultural University successfully demonstrated the suitability of the phytoremediation technology for managing Se-contaminated soils located in northwestern India (Dhillon and Dhillon 1997, 2009c; Dhillon et al. 2008a). Field experiments were conducted during 1998–2008 on-farm locations to assess the Se-removal potential of plant-based strategies, i.e. cropping patterns with Brassica species, agroforestry and flowering plants, for removal of total Se from affected soils (Dhillon et al. 2008a; Dhillon and Dhillon 2009a, b, c, d; Bañuelos and Dhillon 2011). The results are summarized in Table 16.3 and described in the following sections.

16.4.2.1 Brassica Based Cropping Systems

Biomass yields of both rapeseed (*Brassica napus*) and pigeon pea (*Cajanus cajan*) crops (Table 16.3) grown on Se-contaminated soils were comparable to the typical yields of these crops in low Se soils of this region. By the time the crops reached maturity stage, leaf litter was deposited onto the soil due to senescence. The average amount of leaf biomass re-deposited on the soil surface ranged between 1.1 to 1.5 Mg/ha. The high concentrations of Se present in the seleniferous soils did not affect the overall biomass productivity of crops, except that plant products were rendered unsuitable for animal and human consumption due to the potentially toxic levels of Se in the plant material (Dhillon and Dhillon 2009a, c). Selenium removal from seleniferous soil through harvested biomass of rapeseed was 6–7 times more than that of pigeon pea crop. In a field experiment with different crop rotations, Se removal ranged between 4 and 13 g/ha/year in sulfate-rich environment of California soils (Bañuelos et al. 1997) compared to 737 and 949 g/ha/year from Se-contaminated calcareous and low sulfate soils of northwestern India (Table 16.3). The amount of Se recycled through leaf fall was greater in case of rapeseed (193.3 ± 7.6 g/ha) as compared to pigeon pea (23.7 ± 2.5 g/ha). Compared to the Se volatilization rate observed in California soils, *in situ* Se volatilization rates by different crops under Indian field conditions were low and ranged from 1.5 to 14.1 mg/d/ha (Dhillon and Dhillon 2009c).

16.4.2.2 Agroforestry Farming Systems

In the case of agroforestry farming systems, Bañuelos and Dhillon (2011) observed that Se removed by a combined poplar *(Populus deltoides)*-sugarcane *(Saccharum officcinarum)*-wheat *(Triticum aestivum)* system was 1.5 times greater than that of a poplar-spearmint *(Mentha viridis)*-wheat system (Table 16.3). Including sugarcane as intercultural crop improved the efficiency of Se removal within the agroforestry farming systems. Comparing the Se removal efficiency of the two farming systems in 1 year, the Brassica-based cropping systems was 1.1 to 1.4 times more efficient than the poplar-sugarcane-wheat system; 1.5 to 2.1 times more efficient than the

Table 16.3 Changes in Se concentrations in the soil from 0 to 60 cm depth and mean Se concentrations in crops grown in different cropping systems for phytoremediation of Se-contaminated soils in northwestern India

Cropping system	Crops	Total biomass (Mg/ha/year)	Se content (mg/kg)	Se removed by biomass (g/ha)	Initial Se in soil (g/ha)	Se remaining in soil at the end of expt (g/ha)	Se removed by biomass as % of initial Se	Unaccounted Se (g/ha)	Unaccounted Se as % of initial Se
Rapeseed –Pigeon pea (2 years)	Rapeseed	9.89 ± 0.73	Straw – 55.5±17.4	1268 + 205 = 1473	28,467	21,189	5.2	5805	20.4
			Grains – 93.6±28.4						
	Pigeon pea	7.89 ± 0.93	Straw – 9.1±1.1						
			Grains – 29.3±9.6						
Rapeseed – Pigeon pea (3 years)	Rapeseed	8.53 ± 0.44	Straw – 83.3±20.4	2528 + 318 = 2846	15,734	11,324	18.1	1564	9.9
			Grains – 140.7±43.1						
			Oil – 2.5±0.9						
	Pigeon pea	12.4 ± 2.56	Straw – 16.3±8.0						
			Grains – 68.4±10.9						
Poplar-Spearmint -Wheat (7 years)	Poplar	272	Leaves – 32.6 ± 3.8	1959 + 226 + 1335 = 3520	17,350	–	20.3	–	–
			Stem – 5.2 ± 1.25						
			Branches – 7.3 ± 2.13						
	Spearmint	3.58 ± 1.43	Shoots – 32.1 ± 2.45						
			Oil – 2.3 ± 0.44						
	Wheat	6.48 ± 0.52	Straw – 38.4 ± 7.24						
			Grain – 44.3 ± 9.87						

(continued)

Table 16.3 (continued)

Cropping system	Crops	Total biomass (Mg/ha/year)	Se content (mg/kg)	Se removed by biomass (g/ha)	Initial Se in soil (g/ha)	Se remaining in soil at the end of expt (g/ha)	Se removed by biomass as % of initial Se	Unaccounted Se (g/ha)	Unaccounted Se as % of initial Se
Poplar-Sugarcane –Wheat (7 years)	Sugarcane	98 ± 6.2	Cane – 7.8±0.9	1959 + 2006 + 1335 = 5300	17350	–	30.5	–	–
			Leaves – 21.8±6.2						
Flowers – Low biomass plants (4–6 months)	Coreopsis, Dimorpothica Helichrysum	1.0–2.5	Flowers – 13.8–57.6	22–27	–	–	–	–	–
			Stem – 10.3–30.1						
Flowers – High biomass plants (4–6 months)	Calendula, Gaillardia, Marigold	3.6–9.7	Flowers – 20.5–30.6	100 – 238	–	–	–	–	–
			Stem – 21.6–59.6						

Source: Dhillon and Dhillon (2009c), Bañuelos and Dhillon (2011)

Note: Only in case of poplar, total biomass reported is for 7 years. Spearmint/sugarcane was grown for the initial 2 years followed by wheat for the next 5 years as intercultural crops in poplar. The scientific names of crops are: Rapeseed (*Brassica napus*), arhar (*Cajanus cajan*), poplar (*Populus deltoides*), Spearmint (*Mentha viridis*), sugarcane (*Saccharum officcinarum*), wheat (*Triticum aestivum*), Coreopsis (*Coreopsis gladiate*), Dimorpotheca (*Dimorphotheca pluvialis*), Helichrysum (*Helichrysum orientale*), Calendula (*Calendula officinalis*), Gaillardia (*Gaillardia aristata*), African marigold (*Tagets erecta*) poplar-spearmint-wheat system, and 2.6 to 3.7 times more efficient than the cultivation of only poplar trees. Poplar trees are completely denuded each year during winter season and consequently leaf biomass between 15 to 20 Mg/ha was re-deposited in the soil during the 7-year growth period of poplar. Selenium removal efficiency of poplar based farming systems can be increased by 500–660 g/ha, if leaf biomass re-deposited in the soil every year is removed away from the field before sowing of intercultural crops.

16.4.2.3 Flower Farming

With flower cultivation in a seleniferous soil containing 4.2 mg/kg Se, total Se removal was the greatest with Gaillardia (*Gaillardia aristata*) followed by Calendula (*Calendula officinalis*), African marigold (*Tagets erecta*), French marigold (*Tagetes patula*), Coreopsis (*Coreopsis gladiate*), Dimorpothica (*Dimorphotheca pluvialis*) and Helichrysum (*Helichrysum orientale*) (Bañuelos and Dhillon 2011). Thus, cultivation of Gaillardia, Calendula and African marigold should also be encouraged in nonsaline seleniferous regions. Although Se removal by flowers was quite low compared to the cereal and oilseed crops (Table 16.3), their production is considered highly remunerative. The major advantage of cultivation of flowers is that these crops do not form a part of the food chain for animals and humans. Thus, adoption of floriculture will help in achieving the ultimate objective of the Se phytoremediation technology while producing an economically valuable product, and completely reduce the entry of Se into the food chain and thereby avoid any potential toxic effects of Se on animal and human health. Thus, we propose that farmers in seleniferous region of northwestern India be encouraged to initiate the cultivation of flowers. An added value of the cultivation of flowers in this seleniferous region is that it provides a pleasing visual stimulation and helps to increase the perceived happiness of the local population.

Overall, the amount of Se recovered in plant biomass constituted only 5.2–30.5% of the initial total Se in soil under different cropping strategies. Ten to twenty percent of total Se lost from the soil cannot be explained only by plant uptake (Table 16.3). Other unaccounted factors, like Se volatilization by plants, soil micro-organisms, leaching of Se beyond the root zone, Se entrapped in the plant roots, and spatial variability are also responsible for Se losses from soil. In the present situation, volatilization of Se by crops may not be playing a significant role in Se losses, since the rate of *in situ* Se volatilization by different crops was found to be extremely low ranging from 1.5 to 14.1 mg/d/ha. Critical examination of Se removal data (Table 16.3) indicates that that it may require regular cultivation of about 20–35 cycles of rapeseed-pigeon pea sequence (1 year/cycle) or 4–8 cycles of poplar-based farming systems (7 year/cycle) to lower the level of Se in contaminated soils to <0.5 mg/kg, which is the level considered safe for producing forages and grains without any potential danger to animal and human health. To accurately predict the long-term removal of Se over time, more data needs to be generated.

16.4.3 Phytoremediation of Se-Contaminated Waters in Constructed Wetlands

Constructed wetlands (CWs) are designed to take advantage of many of the same processes that occur in natural wetlands, but only within a more controlled environment. Phytoremediation through CWs has been used to improve the quality of contaminated waters by acting as a sink for various contaminants discharged from sewage, industrial and agricultural wastewaters etc. (Rai 2008; Vymazal 2010). However, the technology of CWs for wastewater treatment is still in the nascent stage In spite of the shortcomings, CWs are now being recognized as an economically viable treatment option for heavy metals and metalloids present in wastewaters. Management of Se-contaminated agricultural drainage water is one of the most important environmental issues in central California. From a remediation perspective, volatilization of Se is an attractive method for removing Se from wastewater because it minimizes the entry of Se into the food chain. Moreover, most of the volatile forms of Se (e.g. dimethylselenide, DMSe) are relatively nontoxic (Terry and Zayed 1994). For this purpose, when 20 aquatic plant species were screened for their ability to accumulate and volatilize Se, several plant species showed Se volatilization and accumulation rates (per unit surface area) equivalent to Indian mustard (*Brassica juncea* L., the most suitable terrestrial plant species for Se phytoremediation). Among the wetland species selected from this study were parrot's feather (*Myriophyllum brasiliense*), iris-leaved rush (*Juncus xiphioides*), cattail (*Typha latifolia*) and saltmarsh bulrush (*Scirpus robustus*), which show a great potential for Se phytoremediation in wetlands (Pilon-Smits et al. 1999a). The suitability of constructed wetlands for remediation of Se-laden drainage water and the role of biological volatilization in Se removal was studied in detail by Lin and Terry (2003). The monthly monitoring study revealed that vegetated wetlands (10 flow-through cells) constructed in 1996 in Corcoran, California were capable of reducing Se by an average of 69.2% of the total Se mass present in the inflow drainage water. Most of the Se was retained in sediment, and <5% of the Se was accumulated in plant tissues. Selenium volatilization was highest in the rabbit-foot grass (*Polypogon monspeliensis*) wetland cell, where 9.4% of the Se input was volatilized over a 2-year period. Volatilization was greater in spring (35%) and summer (48%) than in fall and winter months (<5%). The role of biological Se volatilization in reducing Se load of the wastewater was also investigated in a 36-ha constructed wetland vegetated with different aquatic plant species (Hensen et al. 1998). The highest mean rates of Se volatilization for the sites vegetated with rabbit-foot grass, cattail (*Typha angustifolia*), and saltmarsh bulrush (*Scirpus acutus* subsp.), were 190, 180, and 150 $\mu g/m^2/d$, respectively. During a study period of 16 weeks, selenite-contaminated wastewater inflow containing 20–30 μg/L decreased to <5 μg/L in the outflow; 89% of the Se was removed. Most of the Se was removed by immobilization into sediments and plant tissues where Se concentrations reached ~5 and ~15 mg/kg, respectively. Biological volatilization may have accounted for as much as 10–30% of the Se removed. Hence, biological Se volatilization is a significant pathway of Se removal in wetlands.

The suitability of macrophytes, e.g. *Typha latifolia* and *Phragmites australis* (Cav.) for Se phytoremediation in CWs, was investigated by Shardendu-Salhani et al. (2003). Selenium was supplied continuously to the subsurface-flow CWs vegetated with *Typha and Phragmites*. The horizontal subsurface-flow constructed wetland (HF CW) was a large gravel- and sand-filled channel that was planted with aquatic vegetation. The wastewater flowed horizontally through the channel. In the *Typha* bed, Se migrated faster than in the *Phragmites* bed. In the *Typha* bed, about 54% of the Se inlet concentration remained in the outlet water after Se supplementation for 25 days. In the *Phragmites* bed, Se was removed completely from the water after passing through 3/4 of the bed length. After 65 d of Se supplementation, the highest amount of Se (2.8 μg/g DM) was measured in the organic material of the *Typha* bed. Roots and rhizomes accumulated 2.2 and 1.8 μg/g DM, respectively. *Phragmites* accumulated Se in the leaves (1.8 μg/g DM) and stems (0.6 μg/g DM), but not in the rhizomes.

Constructed and naturally created wetlands in the Las Vegas Valley watershed in the USA were studied to characterize and understand their potential role for improving ecosystem services (Adhikari et al. 2011). Nutrient and metal removal was assessed at four sites dominated by wetland vegetation comprising of cattail (*Typha domingensis*), common reed (*P. australis*) or bulrush (*Schoenoplectus acutus*). Irrespective of the type of plant species present, Se uptake by plants was dependent on the ambient Se concentrations in water and sediments of specific wetlands. Bulrushes were more efficient than cattails in taking up Se. Averaging all the wetland sites and plant species, Se removal was found to be 0.38 kg/ha/year.

The fundamental research based on wetland phytoremediation may not be enough for removing toxic metals and metalloids. The successful transfer of the CW technologies from the laboratory to the field is a crucial step for the future development of wetland phytoremediation. Field trials are necessary to forecast and certify that the wetland plants have detoxified contaminants from outflow water posing minimal residual risks to humans and the environment. While using CWs technology, it becomes necessary to document the Se content of the outflowing wastewater, to understand its cycle and extent of accumulation in the environment, and to evaluate the threat it may pose to fish and wildlife on the long-term basis (Lemly and Ohlendorf 2002).

16.5 Enhancing Phytoremediation Potential Through Transgenic Plants

In the Se phytoremediation technology developed by Bañuelos et al. (2002a), the initial step relates to screening different plant species for their Se removal efficiency from contaminated soil and water, followed by manipulation of agronomic practices for increasing their phytoremediation efficiency under a given set of environmental conditions. The suitability of different plant species for phytoremediation is judged on the basis of important characteristics like fast growth, high biomass accumulation, deep root system, high tolerance to Se and high economic value. All the characteristics important for phytoremediation may potentially be ameliorated further either through conventional breeding techniques or via genetic engineering. The advantage of genetic engineering is that its effects can be achieved much faster than conventional breeding, and it is possible to introduce relevant genes from other plant species. Thus, it is possible to introduce those properties into plants, which otherwise cannot be introduced via conventional plant breeding techniques.

Biotechnology has proven useful in gaining better insight into the genetic and biochemical mechanisms that control Se tolerance, accumulation, and volatilization in plants, and the resulting transgenics with enhanced levels of these processes show great promise for use in phytoremediation. When the transformed plant is propagated, the foreign gene is inherited by its offspring. In addition to, or prior to, transforming a large biomass phytoremediation species, the same gene construct may be transformed to a model plant species like *Arabidopsis thaliana*. This small plant with its short generation time and high seed production is very suitable to test in a short time whether a biotechnological approach really works in enhancing Se phytoremediation.

16.5.1 Selenium Metabolism in Plants

Among different Se species present in the environment, selenate is the predominant form of bioavailable Se in oxic soils and selenite is more abundant in anoxic wetland conditions. Plants readily take up selenate (SeVI) or selenite (SeIV) from their environment and incorporate it into organic compounds using sulfate assimilation enzymes (Pilon-Smits and Quinn 2010). The reduction of selenate to selenite appears to be a rate-limiting step in the Se assimilation pathway, since selenite-supplied plants accumulated organic Se and selenate-supplied plants accumulated predominately selenate (de Souza et al. 1998). Nonspecific incorporation of Se-amino acids into proteins may lead to Se toxicity in plants. Plants are also capable of decomposing SeCys either into non-toxic elemental Se or methylate into methyl-SeCys and volatile dimethyldiselenide (DMDSe). Methyl-SeCys does not enter proteins, but its safe accumulation in Se-hyperaccumulators is considered to be a key mechanism for their Se tolerance (Neuhierl et al. 1999).

16.5.2 Potential of Various Transgenic Approaches

The transgenic plants exhibiting new or improved phenotypes are engineered by the over expression and/or introduction of genes from other organisms. In a study by Misra and Gedamu (1989), they observed that plants such as *Brassica napus* can be genetically engineered for heavy metal tolerance/sequestration and eventually for partitioning of heavy metals in non-consumed plant tissues. Some transgenic approaches employed to improve the performance of rate-limiting steps for enhancing Se accumulation, tolerance, and volatilization by plants are discussed in the following sections.

16.5.2.1 Enhancing Assimilation of SeO_4^{2-} to SeCys

To manipulate the initial rate-limiting step, i.e. reduction of selenate to selenite in the assimilation of selenate to organic Se, the mediating enzyme ATP sulfurylase (APS) from *Arabidopsis thaliana* was overexpressed in Indian mustard (Pilon-Smits et al. 1999b). As a consequence of overexpression of APS in Indian mustard, the transgenic plants accumulated two to three-fold more Se than the wild type. The rate of Se volatilization, however, remained unaltered in the APS transgenics.

16.5.2.2 Enhancing Conversion of SeCys to DMeSe

Another enzyme, namely cystathionine-γ-synthase (CgS), active for converting SeCys to SeMet in the Se assimilation pathway, could be the rate-limiter for DMSe volatilization. Indeed, the overexpression of the *A. thaliana* CgS enzyme in *B. juncea* resulted in 2–3 times higher volatilization rates compared to untransformed plants (Van Huysen et al. 2003).

16.5.2.3 Methylation of SeCys

Changes in the protein structure and functions due to nonspecific incorporation of SeCys lead to Se toxicity in plants (Brown and Shrift 1981). In order to prevent SeCys incorporation into proteins, SeCys methyltransferase (SMT) from the Se hyperaccumulator *A. bisulcatus* was overexpressed in two different host plants, *A. thaliana* and *B. juncea* (Ellis et al. 2004; LeDuc et al. 2004). In both species, the SMT transgenics showed enhanced Se accumulation in the form of methyl-SeCys, as well as enhanced Se tolerance. The overexpression of SMT also resulted in increased rates of Se volatilization, with more volatile Se produced in the form of DMDSe.

16.5.2.4 Conversion of SeCys to Se(0)

Another genetic engineering approach to manipulate plant Se metabolism targeted SeCys, and particularly the prevention of the toxicity process of its nonspecific incorporation into proteins. A mouse SL (SeCys lyases) enzyme was expressed in *A. thaliana* and *B. juncea* (Pilon et al. 2003; Garifullina et al. 2003). This enzyme specifically breaks down SeCys into alanine and elemental Se and thus directs Se away from incorporation into proteins. All the transgenic SL plants showed enhanced Se accumulation, up to twofold compared to wild type plants.

16.5.2.5 Double Transgenics for Enhancing Se Assimilation

Although the expression of SMT in Indian mustard enhanced Se tolerance, accumulation, and volatilization (LeDuc et al. 2004), these effects were more substantial when the SMT-expressing plants were exposed to selenite relative to selenate. Possibly, the reduction of selenate to selenite provided a rate-limiting barrier to the production of SeCys, greatly diminishing any potential benefit of overexpressing SMT. To overcome this rate-limitation, APS and SMT transgenics were crossed to create double-transgenic plants that overexpress both APS and SMT (APS × SMT plants). The APS × SMT double transgenics accumulated up to ninefold higher Se levels than the wild type (LeDuc et al. 2006). Most of the Se in the double transgenics was in the form of methyl-SeCys: the APS × SMT plants accumulated up to eightfold more methyl-SeCys than wild type and nearly twice as much as the SMT transgenics. Selenium tolerance, however, remained similar in the single as well as double transgenics.

16.5.2.6 Upregulation of Se Volatilization

A potential strategy for enhancing plant Se tolerance rather than accumulation is the upregulation of Se volatilization. Most plants can produce volatile dimethylselenide (DMeSe) from methionine, whereas hyperaccumulators produce dimethyldiselenide (DMeDSe) from SeMeSeCys. S-adenosyl-L-Met:L-Met S-methyltransferase (MMT) is the key enzyme responsible for the methylation of Se-Met to Se-methyl Se-Metionine (SeMeMet), which is the precursor of volatile DMeSe (Tagmount et al. 2002). However, whether manipulation of SeMeMet can be effective in increasing Se volatilization has yet to be tested. Further identification of different volatile Se compounds could aid in the dissection of Se metabolic pathways in plants of different species, as well as in transgenics (Kubachka et al. 2007).

16.5.2.7 Searching the Genetic Basis for Selenium Tolerance

There are three highly conserved homologues of the mammalian 56-kD selenium-binding protein (SBP) in the *Arabidopsis* genome. To study the function of SBP in this model plant, Agalou et al. (2005) used a transgenic approach by constitutively overexpressing and down-regulating the endogenous *Atsbp1* gene. In the latter case, both a conventional antisense method and gene silencing by intron-containing hairpin RNAs were employed. Under standard growth conditions, *Atsbp1*-overexpressing and silenced plants were found phenotypically normal when compared with wild type plants. Transgenic plants exhibited different growth responses to exogenously supplied selenite, which correlated with the expression levels of *Atsbp1*. Plants with increased *Atsbp1* transcript levels showed enhanced tolerance to selenite, while plants with reduced levels were more sensitive. These results suggest that although *Atsbp1* does not play a detectable role in the regulation of developmental processes under normal growth conditions, it appears to be involved in processes controlling tolerance of *Arabidopsis* to Se toxicity (Agalou et al. 2005).

In a study carried out by Tamaoki et al. (2008), genomic and biochemical comparison of two *A. thaliana* accessions that differ in selenite tolerance showed that the tolerant accession had higher expression levels of genes involved in jasmonic acid (JA) and ethylene production and response, including sulfur assimilation-related genes. Mutants in the tolerant accession that are impaired in JA/ethylene production showed reduced selenite tolerance, while external supply of JA or ethylene to the sensitive accession enhanced tolerance. Thus, JA and ethylene appear to be important hormones for plant Se tolerance; they may confer tolerance by upregulation of S assimilation.

To understand the genetic basis of selenate tolerance, Zhang et al. (2006) mapped the quantitative trait loci (QTL) associated with selenate tolerance in *A. thaliana* accessions Landsberg *erecta* and Columbia using recombinant inbred lines (RILs). The selenate tolerance index was found to be fourfold higher for parental line Col-4 (59%) than for parent Ler-0 (15%). Among the 96 F8 RILs, selenate tolerance index ranged from 11% to 75% (mean 37%). Using composite interval mapping, three QTL were found on chromosomes 1, 3 and 5, which together explained 24% of variation in selenate tolerance index and 32% of the phenotypic variation.

The results from these studies offer insight into the genetic basis of selenate tolerance, and may be useful for identification of selenate-tolerance genes. These results may ultimately be applicable for breeding plants with higher levels of Se tolerance and plants that accumulate higher levels of selenocompounds.

16.5.2.8 Potential of Transgenics for Phytoremediation Under Field Conditions

The transgenic approaches discussed above have shown increased Se tolerance, accumulation, and assimilation from inorganic to organic Se, and volatilization under laboratory conditions. To evaluate the potential of trangenics for phytoremediation, transgenic plants were tested for their capacity to accumulate Se from naturally seleniferous soil and from Se-contaminated sediment. The first report appeared in 2005, showing that genetically engineered plants for phytoremediation can perform successfully under field conditions (Bañuelos et al. 2005). In the field experiment, three transgenic Indian mustard (*Brassica juncea*) lines overexpressed genes encoding the enzymes either adenosine triphosphate sulfurylase (APS), γ-glutamyl-cysteine synthetase (ECS), or glutathione synthetase (GS), respectively. In another experiment, two new transgenic Indian mustard lines overexpressing genes encoding the enzymes either selenocysteine lyase (cpSL) or selenocysteine methyltransferase (SMT) were tested for their phytoremediation capacity (Bañuelos et al. 2007). The results clearly demonstrated that cpSL and SMT transgenic lines have significantly greater Se phytoremediation potential than wild type Indian mustard. These studies offer hope that through genetic manipulation of high biomass, fast-growing plants, Se phytoremediation can be developed into a viable option, while producing crops with possibly better nutritional quality.

16.6 Phytoremediation and Ecological Implications

Plants that can accumulate and tolerate moderately elevated Se levels (<1000 mg/kg) are known as Se-accumulators and plant species that can accumulate from 0.1% to 1.5% Se, are called Se-hyperaccumulators (Rosenfeld and Beath 1964). Hyperaccumulation is the phenomenon through which some plant species accumulate toxic elements to very high concentrations and it is most prevalent in the Brassicaceae family (Baker et al. 2000). Researchers have proposed that Se-accumulating plants belonging to Brassicaceae family could be grown in Se-rich soils or Se-contaminated water, harvested, and removed as Se-enriched plant material (Bañuelos et al. 2002a). Growing plants for phytoremediation could attract and interact with a large number of pathogens, insects and other animals. It may be interesting to note that the level of Se in plants used for phytoremediation may prove lethal to other interacting organisms, yet cause no toxicity in hyperaccumulators.

Some ecological implications of Se-hyperaccumulation in plants were recorded as early as the 1930s (Rosenfeld and Beath 1964). Substantial livestock losses have occurred in the western USA due to consumption of plants containing toxic levels of Se. Some grazing livestock have learned to avoid feeding on Se-rich forages that emit garlicky offensive odour, while some insects and mites (e.g. beetles, seed-chalcids, grasshoppers etc.) were able to develop resistance to Se and complete their life cycle while feeding on Se-rich seeds and plant tissues. Recently, while tracing the evolution of this trait known as *Se hyperaccumulation* in some specific plants, El-Mehdawi and Pilon-Smits (2012) have described the ecological implications due to Se-hyperaccumulation in plants more precisely as:

1. Selenium hyperaccumulator plants may affect the local soil Se distribution and chemical speciation *via* litter deposition, root turnover and root exudation;
2. Redistribution and chemical speciation may also affect soil microbial composition and species abundance, soil fauna populations and neighboring vegetation;
3. Selenium accumulation in plant tissues and Se volatilization may strongly influence the interactions between plants and pathogens, herbivores as well as the pollinating insects.

The transfer of Se from soil to crop, from crop to insect, from insect to insect (predator), and also from insect to mammals (including humans) are the pathways for Se that should be addressed when monitoring the biological fate of removed Se. An illustrative case study highlighted that Se biotransfer from soil through drainage water to aquatic vegetation was responsible for the deleterious health effects on waterfowl and other birds and animals feeding in a Se-rich area at Kesterson Reservoir (Ohlendorf and Hothem 1995). In order to understand the ecological reverberations of Se hyperaccumulation in more details, systematic studies have been undertaken by Elizabeth A. H. Pilon-Smits and her associates during the last decade in the Biology Department of Colorado State University at Fort Collins. The knowledge gained will certainly be useful in the management of seleniferous areas and the agricultural production of Se-rich crops for phytoremediation. Experimental evidence and discussion supporting different ecological implications due to phytoremediation (hyperaccumulation) of Se-contaminated soils are presented in the following sections.

16.6.1 Phytoremediation and Herbivore Interactions

Selenium is a well-known toxic element for animals at high levels, and hence Se accumulation may act as a defense system in plants against herbivore attacks. As a response to this defense, herbivores may learn to avoid high-Se plant material. This type of elemental defense mechanism in plants has been studied in great detail either by offering plants pre-treated with Se as feed to herbivores to test for deterrence/toxicity under controlled conditions. Bañuelos et al. (2002b) evaluated biotransfer of Se accumulated by different plants to several insects and animals by monitoring mortality, deterrence, and biomagnification of Se (Table 16.4). The growth and survival rate of the tested insects and animals was inversely related to Se concentration in leaves of host plants, but increased levels of Se were observed in tissues of vital organs like kidney, liver, heart and spleen (Table 16.4). In a similar study by Hanson et al. (2003), Se-rich mustard leaves proved lethal to the caterpillars of the cabbage white butterfly (*Pieris rapae*), but the snails (*Mesodon ferrissi*) continued to feed on leaves without showing any toxicity symptoms. Selenium-rich *B. juncea* plants also proved resistant to fungal infection by *Alternaria brassicicola*, as well as *Fusarium* species. The observed protection from caterpillar herbivory and fungal infection due to Se accumulation in *Brassica juncea* plants must also prove protective under field conditions.

Similarly, Freeman et al. (2007) reported that Se could protect plants against other leaf chewing herbivores like crickets (*Acheta domestica*) and grasshoppers via deterrence and toxicity. In addition to leaf chewers, phloem-feeding aphids were also deterred by high-Se *B. juncea* plants, and suffered toxicity at plant Se levels as low as 10 mg/kg (Hanson et al. 2004). In a field survey, Se was also shown to protect *B. juncea* and *S. pinnata* plants from a vertebrate herbivore—the black-tailed prairie dog (Quinn et al. 2008; Freeman et al. 2009). While measuring arthropod loads in Se hyperaccumulator habitats in Colorado, Galeas et al. (2008) observed that Se hyperaccumulating plant species containing 1000–14,000 mg/kg DM, harbored significantly fewer arthropods and fewer arthropod species compared with nonaccumulator plant species that contained <30 mg Se/kg DM.

Table 16.4 Herbivory response of different animals, insects and pathogens feeding on Se-rich plants for phytoremediation

Sr. No	Experimental insect/ animal/pathogen	Experimental crop	Se treatment	Mean Se conc in leaves (mg Se kg^{-1})	Se conc. in experimental insect/animal (mg kg^{-1} DM)	Body part	Survival rate	Source
1.	Cabbage looper (*Trichoplusia ni*)	Indian mustard (*Brassica juncea*)	Control[a]	<1	<0.00 4	Body tissue	34–45 moths	Bañuelos et al. (2002b)
			1 mg Se L^{-1}	401–465	2.68–3.17	---do---	12–16 moths	
2.	Beet armyworm (*Spodoptera exigua*)	Saltbush plant lines (*Atriplex spp.*)	Control	0.08–0.40	–	Not analyzed	16.4–30 days	
			1 mg Se L^{-1}	33.2–61.5	–	---do---	9.2–27 days	
3.	Lambs (*Ovis aries*)	Canola (*Brassica napus*)	Control	ND	0.22–1.51	Different vital organs	NA	
			Irrigated with 75–100 µg Se L^{-1}	2.0–3.63	0.53–2.10	---do---	NA	
4.	Cows (*Bos taurus*)	Alfalfa/grains	Control	0.11–0.24	0.063–0.067	Blood	NA	
		Canola (*Brassica napus*)	Irrigated with 75–100 µg Se L^{-1}	3.75	0.062–0.090	---do---	NA	
5.	Dutch Rabbits (*Oryctolagus cuniculus*)	Alfalfa (*Medicago sativa* var. Salado)	Control	0.05	0.84–2.61	Different vital organs	NA	
			Se-rich soil	6.62	4.95–57.14	---do---	NA	

6.	Caterpillar (*Pieris rapae*)	Indian mustard (*Brassica juncea*)	Control	10	3	Body tissue	100%	Hanson et al. (2003) Results of nonchoice feeding expts.
			+Se [b]	1600	90	---do---	0%	
7.	Snails (*Mesodon ferrissi*)	Indian mustard (*Brassica juncea*)	Control	<1	2	Body tissue	100%	
			+Se	750	50	---do---	100%	
8.	Fungal pathogen (*A. brassicicola*)	Indian mustard (*Brassica juncea*)	Control	<4	5	Infection lesions/leaf	NA	
			+Se	300	1	---do---	NA	
9.	Cricket (*Acheta domestica*)	Indian mustard (*Brassica juncea*)	Control	ND	ND	Body tissue	100%	Freeman et al. (2007) nonchoice feeding expts
			+Se	447	20	---do---	0%	
10.	Grasshopper (mix of *Orthoptera* spp.)	Se-hyperaccumulator (*S. pinnata*)	Control	3	0.5	Body tissue	100%	
			+Se	230	12	---do---	0%	

[a] Control means that experimental plants were grown without Se addition

[b]+Se means that plant seedlings were grown in half strength Hoagland solution containing 20 μM sodium selenate for 6–8 rows and 40 μM sodium selenate solution for 9–10 rows. *ND* Not detected, *NA* Not applicable

Table 16.5 Selenium concentrations (mg/kg) in different biological systems, soil and crop found at different field sites for phytoremediation

Biological system found at experimental sites			Se conc. in biological system collected at:		Mean Se conc. at experimental site in:	
Group	Biological species	Body part analyzed	Se-site	Control-site	Soil[a]	Crop
Rodents	Gopher (*Thomomys bottae*)	Liver	3.75 ± 0.4[b]	0.30 ± 0.01	8.21± 0.6	6.25 ± 0.7
Birds	Dove (*Zenaida macraeva*)	Egg shells	0.64 ± 0.03	0.19 ± 0.01	5.65± 0.5	4.35 ± 0.5
Insects		Embryos	0.32 ± 0.01	0.12 ± 0.01	5.65±0.6	4.35 ± 0.5
	Loopers (*Trichoplysia ni*)	Body tissue	1.10 ± 0.05	0.42 ± 0.01	4.92± 0.5	4.65 ± 1.4
	Aphids (*Aphidae*)	Body tissue	0.24 ± 0.01	0.10 ± 0.01	6.51 ± 0.6	15.21 ± 1.5
	Borer (*Paranthrene dollii*)	Body tissue	0.75 ± 0.03	0.14 ± 0.01	2.72 ± 0.3	2.75 ± 0.3

Source: Bañuelos (2006a)
[a]The soils from control sites recorded total Se conc. of < 0.27 mg/kg
[b]Values represent mean Se concentration ± standard error

Studies conducted so far provide ample support for the Se element defense hypothesis, which states that Se hyperaccumulation has evolved as a defense mechanism. Even Se levels as low as 10 mg/kg DW can protect plants from herbivores, due to deterrence and toxicity. Therefore, Se absorbed by plants used for phytoremediation can be transferred biologically in an intentional or unintentional manner to insects and animals. Plants enriched with Se may discourage the infestation by most insect species, but may lead to bioaccumulation of Se by Se-tolerant insects, whose feeding on the plants may exert deleterious effects on birds and mammals that eat the insects.

16.6.2 Selenium Biotransfer Potential to Biological Systems

While monitoring the fates of Se in biological organisms associated with different phytoremediation experimental sites in California, Bañuelos (2006a) recorded significant accumulation of Se in the livers of gophers (*Thomomys bottae*) (1.2–6.6 mg/kg DM) and jack rabbits (*Lepus californicus*), which may lead to serious health problems (Table 16.5). Among birds, Se concentrations did not exceed 0.25 mg/kg in the yolk, or 0.60 mg/kg in the egg shell of killdeer (*Charadrius vociferous*) and mourning dove (*Zenaida macraeva*) found nesting in both saladograss and cordgrass fields. Selenium concentrations in biological systems depend on the plant species used for phytoremediation, concentration and form of available Se in soil, feeding behavior of the insect, and amount of feed consumed by the bird/animal/insect selected for the study. Consumption of selenium-enriched plant tissue consumed by any biological species may not only have negative consequences for the insect (Trumble et al. 1998), but also for the invertebrates and vertebrates that feed on them (Wu et al. 1995). Biotransformation of Se through three trophic levels – the host crop plant alfalfa (*Medicago sativa)* fed upon by the beet armyworm (*Spodoptera exigua*) larva, which in turn is parasitized by a wasp (*Cotesia marginiventris)* was examined by Vickerman et al. (2004). Selenium concentrations of alfalfa were 327 ± 0.2 mg/kg DW for the plants irrigated with water containing selenate-Se. Selenium speciation in each trophic level revealed that alfalfa partially transformed selenite to organoselenium. Beet armyworm contained only organoselenium, both directly absorbed from alfalfa and transformed from selenite. Wasp cocoons analyzed after larval emergence contained only organic-Se derived from the host. Presence of trimethylselenonium like species in adult parasitoids and the cocoons suggests that adults and pharates can detoxify excess Se through methylation and volatilization. The results indicate that *C. marginiventris* can be used to control *S. exigua* damage to *M. sativa* being used to remove Se from soils. Moreover, the presence of such insects may improve phytoremediation by increasing biotransformation of inorganic Se and release of volatile Se species. This study also demonstrates that Se can move through a terrestrial food chain from a Se-irrigated plant, through an herbivore, and into a developing insect parasitoid. The presence of insects may improve phytoremediation by increasing biotransformation of inorganic Se and release of volatile Se species.

16.6.3 Phytoremediation and Pollinators Interaction

Selenium is an essential trace element for insects (Zhang and Gladyshev 2009), and therefore a Se-enriched diet may promote bee health. On the other hand, if ingested in excess, Se may have a toxic effect on bees. The ecological impacts of Se accumulation in flowers were investigated by monitoring honey bees visiting *B. juncea* and *S. pinnata* plants containing low and high Se levels respectively (Quinn et al. 2011a). No differences in floral visitation were observed for either plant species, in spite of large variations in Se content (Pilon-Smits et al. 1999b). Analysis of different bee species foraging on *S. pinnata* revealed that European honey bees (*Apis mellifera*) contained 14 times less Se than the native bumble bees (*Bombus* sp.). While the bumble bees contained predominantly C-Se-C (presumably the non-toxic MeSeCys), the honey bees contained more toxic forms of Se. In other biological systems, tissue Se levels of 10–90 mg/kg DM were found to be lethal to Se-sensitive Lepidoptera larvae of *Pieris rapae* and *Plutella xylostella* (Hanson et al. 2003; Freeman et al. 2006). It is possible that the European honey bee, which is native to a seleniferous area, has evolved Se tolerance; and can forage on Se hyperaccumulators. The lower Se levels and presence of different forms of Se in the European honey bees may reflect different foraging behavior (visiting both hyper- and non-hyperaccumulators), or may be an indication of reduced Se tolerance.

Large variations in reproductive functions of different plants have also been observed due to Se accumulation in flowers. Pollen germination was significantly affected in *B. juncea* but not in *S. pinnata* (Prins et al. 2011). On the other hand, higher rate of pollen viability and germination was observed due to Se application to olive (*Olea europaea*) cultivars (Tedeschini et al. 2015).

16.6.4 Phytovolatilization and Risk of Se Inhalation

Volatilization of Se is a biological process carried out by all types of microorganisms (Doran 1982) and efficiency of this process is enhanced in the presence of plants. In a detailed discussion on the role of microbes in Se volatilization by plants, Terry and Zayed (1994) observed that bacteria could volatilize Se independently and may also assist plants in volatilizing Se either through promotion of Se uptake or reduction of selenate to forms more easily processed by the S-assimilation pathway. It is now evident that the majority of the agricultural crops are capable of volatilizing Se with rice, broccoli and cabbage as the superior volatilizers and sugar beet, lettuce, onion at the lowest end.

As soon as Se is methylated into volatile species such as DMSe, it is released into the ambient environment, diluted and dispersed further by air currents far away from the contaminated site. DMSe reacts with OH and NO_3 radicals and ozone (O_3) within a few hours to yield products that are unknown at present (Atkinson et al. 1990). The calculated life time of DMSe due to reactions with these species ranged from 5 min to 6 h. However, it is likely that these oxidized products may be scavenged into aerosols or sorbed onto particulate matter that have a relatively long residence time (7–9 days) in the atmosphere and can travel considerable distances away from the source of origin.

Compared to aqueous SeO_3^{2-} and SeO_4^{2-} ions, DMSe is 500–700 times less toxic to rats (Frankenberger and Karlson 1994). In an acute toxicity study on the inhalation of DMSe by rats, 85 adult rats were exposed to the concentrations varying from 0 to 34,000 mg/m^3 (or 8034 ppm) for 1 h. After the treatment, animals were kept under watch for one week for recording clinical abnormalities if any, but all the animals were normal. The half-life of DMSe in animals appears to be short and the compound is eliminated mainly via lungs. Thus, it may be concluded that the inhaled DMSe vapour is nontoxic to the rats at concentration up to 34,000 mg/m^3.

The amount of volatile Se (DMSe) being released into the atmosphere from the Se-contaminated sites through plants, as well as soil, together appears to be quite low, and it will never be reaching concentration more than 34000 mg/m^3 in the environment, which is safe for animals. Concentrations of DMSe will be relatively higher in the ambient environment of the crop than a few meters away from the point of release and it may or may not be able to affect the life of insects, animals living in burrows, birds nesting in the immediate environment of plants or humans passing through the cropped area for various field operations related to the crop management. Thus, there is need to study the behavior of insects, animals, birds and their predators living in the cropped area on long-term basis and undertake periodic assessment of Se accumulation in their body tissues.

16.6.5 Phytoremediation and Se-Rich Organic Residue

Phytoremediation of Se-contaminated soils involves cultivation of selected plants for removal of Se from the affected soil. Effective removal of Se is a function of harvested biomass, which is actually removed away from the contaminated site. As soon as the crop reaches maturity, leaves fall on the ground due to senescence. Physical removal of fallen leaves and root biomass from the land is not considered economical and thus are left on-site and are mixed into the soil during land preparation for the second crop. It has been estimated that when *Brassica*-based cropping sequences are practiced for Se phytoremediation, mean leaf and root biomass redeposited to the soil could range from 0.8 to 2.3 and 0.9 to 2.5 Mg (or metric ton)/ha/year, respectively (Dhillon and Dhillon 2009c). Similarly, addition of about 17.2 and 19.2 Mg/ha of leaf and root biomass, respectively, may take place under agroforestry (poplar) based cropping systems (Bañuelos and Dhillon 2011). Mean Se concentration in the leaves of different crops ranged between 40 and 220 mg/kg and that in roots ranged between 20 and 30 mg/kg. Thus, inadvertent addition of a large quantity of Se-rich, easily decomposable biomass may gradually lead to serious ecological implications.

As in case of herbivory of Se-rich plants, Se toxicity to microbial population may lead to slow rate of decomposition of Se-rich leaf biomass deposited in the soil. However, Quinn et al. (2011b) observed that, when plant material with varying Se concentrations was left to decompose in a seleniferous area, high-Se *Astragalus bisulcatus* tissues containing 350–550 mg/kg (DM) Se decomposed faster than low-Se *Medicago sativa* tissues containing 1–2 mg/kg (DM) of Se. The fast decomposition of the high-Se litter in seleniferous habitat suggests that the local microbial population has developed substantial Se tolerance. Presence of Se-tolerant microorganisms and fast rate of decomposition may have significant effect on the balance among different Se forms and species in soil. Shift in balance among Se species in favour of organic Se may affect decomposing microorganisms differently as compared to inorganic Se. Consequently this process in the long-run may result in qualitative as well as quantitative impact on Se fluxes through the local ecosystem. Thus a regular monitoring of changes in Se transformation in the contaminated soil must be undertaken to answer several questions related to the success of the phytoremediation process and to assess associated risks.

In a number of studies, significant decrease in Se uptake by plants has also been observed with the addition of some organic amendments like crop residue and animal manure (Ajwa et al. 1998); high molecular weight organic compounds

(Pezzarossa et al. 2007), and press mud and poultry manure (Dhillon et al. 2010). Reduction in Se uptake by different crops with the presence of various organic amendments implies appreciable reduction of Se load in the food chain, as well as the efficiency of phytoremediation in the long run. The positive impact is that the rate of Se volatilization increased significantly with the addition of organic resources (Dhillon et al. 2010). Volatilization of Se in soil-plant systems can further be promoted by stimulating indigenous microorganisms with specific organic amendments (Doran 1982).

16.7 Strategies for Increasing Acceptability of Phytoremediation Technology

For reducing Se in soils to safer levels, the phytoremediation strategy developed by Bañuelos et al. (2002a) has emerged as an efficient, environment friendly and low cost technology; however, the process still has certain limitations (Dhillon and Dhillon 2016). Some of the limitations and the possible solutions are discussed in the following sections.

16.7.1 Safe Disposal/Utilization of Se-Rich Biomass

Among different sources of Se, organic Se is considered as more bioavailable to animals and humans (Muñiz-Naveiro et al. 2006). It can be easily inferred from the published data that the cost of Se phytoextraction can be offset from the sale of organic Se-enriched biomass, grains and fruit as Se feed and food supplements for meeting daily Se intake of animals and humans in Se-deficient areas. *Brassica* species play an important role in remediation of Se-contaminated soils. To encourage its adoption, some viable economic measures are discussed in the following sections.

16.7.1.1 Selenium-Enriched Brassica and Use of Its Oil as Biofuel

In central California's San Joaquin Valley (SJV), growing canola and mustard plants for Se phytomanagement produces an array of Se-biofortified products, including the byproduct biofuel. Vegetable oils and their mono-fatty acid low alkyl esters are used for biodiesel or as a diesel fuel lubricity additive (Kulkarni et al. 2007). Biodiesel derived from vegetable oils, e.g. *Brassica*, is rapidly gaining market share as a diesel fuel extender worldwide for environmental benefits and for a new renewable resource of biodiesel (Dalai et al. 2006; Stapleton and Bañuelos 2009). It is possible to extract oil at a conservative rate of 2.7–4.5 Mg from canola and mustard seeds (Bañuelos 2006a) adopted as a source of biodiesel fuel (McDonnell et al. 2000). Key to the production of biofuel from plants used in the phytoremediation of Se is that otherwise unproductive Se-laden sites are used. There should be no competition for food crop production. Even a partial shift from conventional diesel fuels to blended biodiesels within the agro-industry will certainly encourage growers using *Brassica* species and particularly canola as a Se phytoremediation crop.

16.7.1.2 Selenium-Enriched Brassica Seed Meal as Animal Feed

Canola is grown for long-term phytomanagement of Se in high saline/B soils of the SJV, with total biomass and seed yields over 12–13 Mg/ha and ~ 2 Mg/ha, respectively. After pressing and extracting oil from canola seed, the residual seed meal is usable as a Se-enriched feed meal. Importantly, the Se concentration in the canola meal produced from the high sulfate and Se-laden soils of the SJV was generally less than 2 mg/kg DM (Bañuelos 2006b), which allows its safe use as part of the daily feed ration (Bañuelos et al. 2010). Moreover, this Se-enriched meal can be of special economic importance for successfully providing additional and bioavailable Se to central California's livestock and dairy industries (Bañuelos et al. 2010).

16.7.1.3 Selenium-Enriched Brassica Biomass as Forage for Animals

Careful mixing of Se-rich vegetative biomass generated at the phytoremediation sites with other feed stuffs may lead to improved Se status of animals (Bañuelos and Mayland 2000). Similarly, Echevarria et al. (1988) have suggested Se as an additional organic supplement should be incorporated into animal feed rations to ensure adequate Se levels in animals. The bioavailability of organic Se (generally as SeMet) supplied to animals as Se-enriched feed and its subsequent absorption by animal tissues may be greater for some animals than inorganic forms of Se (Agbossamey et al. 1998; Bañuelos and Mayland 2000; Hall et al. 2013). When Se-enriched canola plants harvested from a phytoremediation site were fed to marginally Se-deficient lambs and cows as a part of their daily ration (Bañuelos 2006a), an appreciable increase in Se concentration of different body tissues and excreta was recorded (Table 16.6). Keeping these results in view, it can be safely predicted that the biomass generated during phytoremediation of Se-contaminated soils in central California ranging from 9 to 14 Mg/ha/y (Bañuelos and Dhillon 2011) is sufficient for feeding 100–170 Se-deficient animals for improving their Se status to normal levels. In Enshi, China, the so-called "World Capital of Se," Yuan et al. (2012) have suggested growing clover (*Trifolium repens*) and alfalfa (*Medicago sativa*) for producing Se-biofortified fodder.

An alternative utilization of Se-rich plant biomass relates to its incorporation in nonseleniferous soils for raising forage crops (Bañuelos et al. 1991, 1992; Dhillon et al. 2007). Incorporation of straw up to 1.0% (20 Mg/ha) did not show any detri-

Table 16.6 Selenium concentration in feces, urine and some body tissues of lambs and cows fed Se-enriched canola plants grown with irrigation water containing 75–100 μg/L Se

	Lambs		Cows	
Parameters	Se-enriched	Control[a]	Se-enriched	Control
Feces (μg Se/kg DM)	10 ± 2 – 704 ± 120	94 ± 2 – 71 ± 10	38 ± 1 – 73 ±2	35 ± 1 – 52 ± 1
Urine (μg Se/L)	9 ± 2 – 301 ± 50	Not detected	21 ± 1 – 59 ± 2	23 ± 1 – 33 ± 1
Blood (μg Se/L)	229 – 301	47 – 102	62 ± 2 – 90 ± 3	63 ± 2 – 67 ± 2
Milk (μg Se/L)	Not applicable	Not applicable	32 ± 1 – 71 ± 2	31 ± 1 – 65 ± 3
Heart (μg Se/kg DM)	682 ± 14	360±11	Not analyzed	Not analyzed
Liver (μg Se/kg DM)	809 ± 42	429 ± 25	Not analyzed	Not analyzed
Kidney (μg Se/kg DM)	2100 ± 99	1507 ± 59	Not analyzed	Not analyzed
Spleen (μg Se/kg DM)	525 ± 66	216 ± 10	Not analyzed	Not analyzed

Source: Bañuelos et al. (2002b)
[a]Canola plants used for control treatment contained <0.1 mg Se/kg DM

mental effect on dry matter yield and Se concentration of the forage crops was also within safe levels for animal consumption. Providing animals with Se-enriched plant or seed material, or growing known forage crops (like alfalfa, corn, sorghum) in soils amended with Se-laden plant materials are both practical and economically viable strategies for disposing of plants used for sustainably phytomanaging Se on Se-laden soil sites. Other uses also include applying the Se-rich plant material as an organic source of Se fertilizer to crops grown for animal feed.

16.7.1.4 Developing Se-Enriched Food Products

In the phytoremediation strategy developed by Bañuelos et al. (2002a), they discussed that for a phyto-based system to be practical and sustainable for managing Se in Se-laden soils, cultivation of selected plant candidates should not result in economic losses for growers implementing such a system. Chances for greater widespread acceptance and usage of a Se phytoremediation strategy could increase if there are potential marketable products from the harvested plant biomass. For example, *Brassica* plants like broccoli (known to have an affinity for S) will still absorb Se and result in Se-enriched broccoli. This new Se-biofortified product may be a potential source of supplemental dietary and bioavailable Se for humans in Se-deficient regions, as studied extensively by J.W. Finley and G. Combs at the Human Research Center in Grand Forks, ND (Combs 2000; Finley 2006, 2007). In contrast to the USA, a low Se intake by human inhabitants in China, the UK, and other countries in Europe and Africa is of great concern (Gao et al. 2011; Lyons 2010). Consequently, to ensure an adequate level of Se in human nutrition, Se biofortification strategies have been practiced in different regions of the world for the purpose of producing Se-enriched crops for Se-deficient regions (White and Broadley 2009; Rayman et al. 2008).

16.7.1.5 Organic Se Fertilizer

In contrast to adding Se to the soil as sodium selenite incorporated into commercial fertilizers (Alfthan et al. 2010), incorporating Se-enriched seed meal (after oil extraction) or Se-rich plant material into soil as organic sources of Se fertilizer is another option similar to that reported for boron (Robinson et al. 2007). The addition of seed meals or plant materials harvested from plants used in Se phytoremediation to soils supporting food crops can lead to increased concentrations of Se in edible crops. The gradual breakdown of the seed meal or plant material and the slow release of Se make these organic sources of Se ideal sources of biofertilizer. Important examples on the production of Se-enriched products are as follows: Se-enriched strawberries with incorporation of Se-rich *Brassica* seed meal into soil (Bañuelos and Hanson 2010), Se-enriched carrots and broccoli with incorporation of Se-enriched *S. pinnata* into soil (Bañuelos et al. 2015), and Se-enriched forage crops like alfalfa, corn, and sorghum with incorporation of Se-laden plant tissues into soil (Bañuelos et al. 1991, 1992; Dhillon et al. 2007). In countries such as China, India, or Africa where Se deficiencies are observed in populations consuming rice grains or corn low in Se, the application of organic Se and inorganic Se-enriched fertilizers (White and Broadley 2009) could be very useful. Incorporating Se-enriched meals into Se-deficient soils not only has the potential to increase Se levels in food crops and in animal forages, but also develop a beneficial and practical way of disposing of Se-enriched plants or residual seed meal material after oil extraction.

16.7.2 Producing Nutraceutical Foods

In a field study, Banuelos and Lin (2010) have observed that a drought-tolerant prickly pear cactus (*Opuntia ficus-indica*) could serve as an ideal alternative crop for a gentle phytoremediation of Se-contaminated soils in central California. The cactus plants are capable of accumulating significant amounts of Se and the rate of Se volatilization varied from 20 to 80 $\mu g/m^2/d$. Analysis of different tissues of Se-biofortified crop revealed that cactus pear fruit and cladodes could serve as an excellent source of Se and other important mineral nutrients (Bañuelos et al. 2012). When grown on agricultural drainage sediments, fruit harvested from cactus pear plants exhibited nutraceutical qualities and may represent a useful Se-enriched chemotherapeutic food crop suitable for promoting human health. Methylselenocysteine

Table 16.7 Impact of leaf fall and root biomass on Se removal efficiency of different phytoremediation strategies

Phytoremediation strategies	Se removed by harvested biomass (g/ha/year)	Se re-deposited in soil *via*		Projected Se removal by total plant biomass generated (col 2+3+4) (g/ha/year)	Expected increase in Se removal efficiency as compared to col 2 (%)	Computed from published data sources
		Leaf fall (g/ha/year)	Root biomass (g/ha/ year)			
1	2	3	4	5	6	7
Brassica-based cropping sequences	917–1152	220–640	19–107	1156–1899	26–65	Dhillon and Dhillon (2009c)
Agroforestry-based cropping systems	280–760	71–86	69–88	420–934	23–50	Bañuelos and Dhillon (2011)

(MeSeCys) is considered to be the least toxic and the most anticarcinogenic Se compound (Ip and Ganther 1992), and it constituted 16% to 24% of total soluble Se in cactus fruit. The organic Se chemical forms, overall mineral nutrition characteristics and antioxidants contained within cactus pear demonstrated the potential high market value of growing such a unique food crop on this type of poor quality soil. The development of unique value-added Se-biofortified nutraceutical food products from cactus pear could help improve profitability in these hypersalinized areas and increase the realistic potential of producing fruit from a truly specialized arid land food crop.

16.7.3 Physically Improving Se Removal Efficiency of Phytoremediation Strategies

As discussed earlier, among different phytoremediation strategies, *Brassica*-based cropping sequences are able to remove 917–1152 g/ha/year through the harvested biomass in India (Dhillon and Dhillon 2009c). Although leaf and root biomass removal involves economic liability, but if removed along with the above-ground harvested biomass, Se removal capacity of Brassica based cropping sequences can be increased by 26% to 65% (Table 16.7). Similarly, removal of Se-rich fallen leaves every year and the root biomass at the harvesting stage along with branches and stem portion, Se removal capacity of agroforestry based systems can be increased by 23–50% (Table 16.7).

16.7.4 Exploring Non-food Plants with Economic Returns for Phytoremediation

Phytoremediation technology for remediation of Se-contaminated soils is eco-friendly and involves very low operational costs. However, on many sites, this process will have to be repeated for many years to bring the level of Se within safe limits in the soil and, consequently, the operational costs may increase considerably. Moreover, consumption of Se-enriched food or fodder is not without risks, as Se readily becomes toxic at elevated concentrations. To keep a check, one has to follow a rigorous schedule to determine and control Se concentrations in food and feed products, and also monitor the rate of Se accumulation in different body tissues at regular intervals. The narrow range between dietary deficiency (< 40 μg/d) and toxic levels (> 400 μg/d) of Se (WHO 1996) makes it extremely necessary to carefully regulate Se intake by humans and other animals. Therefore, a relatively safer option is to utilize non-food plants with assured economic returns for phytoremediation. It may provide favorable perspectives for the affected farmers to generate non-food crops profitably, but the effective soil decontamination process may progress at relatively slow pace. On the other hand, by growing non-food plants in Se-contaminated soils, it will be possible to minimize or completely ban the entry of Se into the food chain.

Some of the non-food plants potentially suitable for cultivation on Se-contaminated soils are listed in Table 16.8. The non-food plants selected for phytoremediation should be fast growing, easy to propagate and cultivate, deep rooted, relatively high Se uptake capacity and have high biomass production. The cultivation of non-food plants must be practically feasible and economically attractive under the given site and land use conditions. Some of the tree species like eucalyptus (*Eucalyptus* spp.), poplar (*Populus* spp.), and acacia (*Acacia nelotica)* are grown for pulp, timber and fuel wood. Thus utilization of Se-contaminated soils for wood production will certainly reduce the pressure for wood on natural forests. Some plant species like giant reed and palm leaves provide raw material for preparing high-end products like baskets, hats etc. Fresh flowers raised on seleniferous soils can fetch handsome price depending upon the demand in a particular region and season. In addition to this, flower cultivation could lead to significant improvement in the environmental outlook of the seleniferous site.

16.8 Conclusions

Phytoremediation has been recognized as an economically viable and ecofriendly technology for remediation of Se-contaminated soils. In recent years, *Brassica juncea*, *B. carinata* and *B. napus* have emerged as the most suitable crops for Se phytoextraction and phytovolatilization processes. Proposed commercialization of *Brassica* biomass for utilization as animal feed in Se-deficient areas, along with use

Table 16.8 Potential non-food plant species with economic returns suggested for remediation of Se-contaminated soils

Potential non-food plant species for phytoremediation	Revenue earning products	Se removed by some non-food plants[a]	Source
Giant reed (*Arundo donax*); Cattail (*Typha latifolia*); Palm leaves	Baskets, hats, mats etc.	Cattail – 380 g/ha/y	Adhikari et al. (2011)
Jute (*Corchorus capsularis*); Dhaincha (*Sesbania aculeate*), *Cannabis sativa;* Sunn hemp (*Crotalaria juncea*), Cotton (*Gossypium spp*)	Rope, rough/fine cloth etc.	Cotton – 160–530 g/ha Sunn hemp – 110–240 g/ha	Dhillon and Dhillon (2009c)
Morning glory (*Ipomea carnea*), *Ipomea tricolor, Datura stramonium, Salvia divinorum*	Sacred plants used for religious ceremonies	*I. carnea* known for Se-rich seeds and leaves	Sabogal and Borkowski (2007)
Prosopis juliflora and *Acacia nelotica*	Fuel wood	*Acacia Nelotica*, Stem – 5 mg/kg	Dhillon et al. (2008a)
Poplar (*Populus deltoides*), Eucalyptus (*Eucalyptus hybrid*), Shisham (*Dalbergia sissoo*)	Pulp, furniture, construction work etc.	1.4–2.4 kg/ha in a life cycle of different trees varying from 10 to 24 years	Dhillon et al. (2008a); Bañuelos and Dhillon (2011)
Calendula (*C. officinalis*), African marigold (*T. erecta*) French marigold (*T. patula*), Gaillardia (*G. aristata*)	Fresh flowers	100–240 g/ha by different flowering plants	Bañuelos and Dhillon (2011)
Tobacco (*Nicotiana tabacum*)	Cigarettes	Presence of high Se level in tobacco leaves used for preparation of cigarettes is desirable due to its chemopreventive role in lung carcinogenesis	Chortyk et al. (1984)

[a]Se removed by cotton, sunn hemp and flowering plants pertains to one growing season of the crop varying from 5 to 7 months

as biofuel for diesel engines and the seed by-products after oil extraction as animal feed meal may result in financial gains to the grower in a Se-contaminated region. Adopting agroforestry copping system for wood, pulp and some valuable chemicals have the capability to turn phytoremediation technology into a financial asset for the grower. Developing genetically improved transgenic Brassica crops with several times higher Se-removal capacity than non-transgenic crops will certainly lead to reduction in length of time needed for Se reduction to safe level in the affected soil. If insects are present, their presence may contribute to Se phytoremediation by increasing biotransformation of inorganic Se to volatile Se species. The biotransformation and apparent methylation of Se by phytophagous insects, and their parasitoid adds a new dimension to Se remediation.

16.9 Future Research Prospects

Success of phytoremediation strategies is extremely dependent upon the type of plant species or crop rotations selected for Se removal from contaminated sediments and water. Harvesting of Se-rich biomass of oilseed crops like canola and Indian mustards may yield products of potential economic importance for the grower. These may include Se-rich plant biomass as supplemental animal feed, *Brassica* oil mixed with diesel fuel for the production of biofuel for diesel engines, and seed by-products after oil extraction as animal feed meal. In addition to canola, oil from other oleaginous crops like sunflower, safflower (*Carthamus tinctorius*), soybean, cotton, and peanut (*Arachis hypogea*) must be considered as potential alternative fuel stocks for diesel engines.

The usefulness of feeding Se-rich canola harvested from Se-phytoremediation sites to animals has been practically demonstrated. Before considering this method for routine disposal of Se-rich plants, strict monitoring of Se concentrations in various plants used for phytoremediation, pattern of Se accumulation in animal tissues, Se excretion in milk and follow-up feeding schedule is highly recommended. Economics of transporting and processing Se-rich plants from phytoremediation sites to Se-deficient regions for sustainable animal production must also be examined carefully.

During the process of phytoremediation, significant changes in nutritional parameters are expected in the plants growing on the poor quality agricultural sediment soils laden with salt, B and Se. Before using plants products as Se-fortified products, quality must be evaluated in terms of organic Se forms, changes in contents of other minerals, total phenolics and antioxidant status or free radical scavenging capacity etc. This type of certification will ensure potentially high market value of plant products obtained from Se-enriched food crops grown on poor quality soil. Chances for widespread acceptance of phytoremediation can be dramatically increased with the production of Se-rich phyto-products of good quality for use in Se-deficient regions of the world.

There is a tremendous increase in our knowledge of plant processes involved in the rate-limiting steps for phytoremediation of Se-contaminated soils. Significant increase in phytoremediation potential of plants has been achieved due to positive manipulation of natural plant processes involved in uptake, assimilation, or detoxification mechanisms of Se. Results from field studies tend to confirm results from initial lab and greenhouse trials. Clearly, plant biotechnological approaches could play an important role in moving the field of phytoremediation forward for better acceptance by the actual user. It is important that new transgenic plants continue to be carefully tested in the field. New strategies are required to improve the acceptability of using genetically engineered plants for remediation projects. It will be helpful if regulatory restrictions can be regularly re-evaluated to make the use of transgenics for phytoremediation less cumbersome.

Also, to provide more mechanistic insight into selenate tolerance, results from QTL mapping of selenate tolerance may be compared with results from biochemical studies, mutant studies, and studies on the effects of overexpression of key enzymes from the S/Se assimilation pathway. Genomic and biochemical studies may help in locating genes that encode specific transporters of selenocompounds into and within hyperaccumulators. Such key genes will be the ultimate candidates for over-expression studies, with the potential of transferring the complete Se hyperaccumulator profile into high-biomass plant species. Impact of plant Se on the local ecosystem in terms of risk to pollinators, herbivores or microbial activity must be evaluated for developing a large-scale sustainable phytoremediation strategy. Understanding the contributions of ecological partners and the effects of Se on ecological partners will certainly be helpful in minimizing potential harmful effects of accumulated Se.

Acknowledgements G.S. Bañuelos acknowledges financial support from California State University, Fresno Agricultural Research Initiative, California Department of Water Resources Proposition 204, and OECD Cooperative Research Programs, Paris, France.

References

Adhikari AR, Acharya K, Shanahan SA, Zhou X (2011) Removal of nutrients and metals by constructed and naturally created wetlands in the Las Vegas Valley, Nevada. Environ Monit Assess 180:97–113

Agalou A, Roussis A, Spaink HP (2005) The Arabidopsis selenium-binding protein confers tolerance to toxic levels of selenium. Funct Plant Biol 32(10):881–890

Agbossamey Y, Petit H, Seoane J, St-Laurent G (1998) Performance of lambs fed either hay or silage supplemented with canola or fish meals. Can J Anim Sci 78:135–141

Ajwa HA, Bañuelos GS, Mayland HF (1998) Selenium uptake by plants from soils amended with inorganic and organic materials. J Environ Qual 27:1218–1227

Alfthan G, Aspila P, Ekholm P, Eurola M, Hartikainen H, Hero H, Aro A (2010) Nationwide supplementation of sodium selenate to commercial fertilizers: History and 25-year results from the finnish selenium monitoring programme. In: Thompson B, Amoroso L (eds) Combating micronutrient deficiencies: food-based approaches. FAO/CAB International, Rome, pp 312–337

Atkinson R, Aschmann SM, Hasegawa D, Thompson-Eagle ET, Frankenberger WT Jr (1990) Kinetics of the atmospherically important reactions of dimethyl selenide. Environ Sci Technol 24:1326–1332

Baker AJM, McGrath SP, Reeves RD, Smith JAC (2000) Metal hyperaccumulator plants: a review of the ecology and physiology of a biological resource for phytoremediation of metal polluted soils. In: Terry N, Bañuelos GS (eds) Phytoremediation of contaminated soil and water. CRC Press, Boca Raton, pp 85–107

Bañuelos GS (2000) Factors influencing field phytoremediation of selenium-laden soils. In: Terry N, Bañuelos GS (eds) Phytoremediation of contaminated soil and water. CRC Press, Boca Raton, pp 41–59

Bañuelos GS (2002) Irrigation of broccoli and canola with boron and selenium-laden effluent. J Environ Qual 31:1802–1808

Bañuelos GS (2006a) Multi-faceted considerations for sustainable phytoremediation under field conditions. For Snow Landsc Res 80:235–245

Bañuelos GS (2006b) Phyto-products may be essential for sustainability and implementation of phytoremediation. Environ Pollut 144:19–23

Bañuelos GS (2009) Phytoremediation of selenium contaminated soil and waters produces biofortified products and new agricultural by-products. In: Bañuelos GS, Lin ZQ (eds) Biofortification and development of new agricultural products. CRC Press, Boca Raton, pp 57–70

Bañuelos GS, Dhillon KS (2011) Developing a sustainable phytomanagement strategy for excessive selenium in Western United States and India. Int J Phytoremediation 13(Suppl. 1):208–228

Bañuelos GS, Hanson BD (2010) Use of selenium-enriched mustard and canola seed meals as potential bioherbicides and green fertilizer in strawberry production. Hort Science 45:1567–1572

Bañuelos GS, Lin ZQ (2010) Cultivation of the Indian fig Opuntia in selenium-rich drainage sediments under field conditions. Soil Use Manag 26:167–175

Bañuelos GS, Mayland HF (2000) Absorption and distribution of selenium in animals consuming canola grown for selenium phytoremediation. Ecotoxicol Environ Saf 46:322–328

Bañuelos GS, Meek D (1990) Accumulation of selenium in plants grown on selenium-treated soil. J Environ Qual 19:772–777

Bañuelos GS, Mead R, Akohoue S (1991) Adding selenium-enriched plant tissue to soil causes the accumulation of selenium in alfalfa. J Plant Nutr 17:701–713

Bañuelos GS, Mead R, Wu L, Beuselinck P, Akohoue S (1992) Differential selenium accumulation among forage plant species grown in soils amended with selenium-enriched plant tissue. J Soil Water Conserv 47:338–342

Bañuelos GS, Cardon GE, Phene CJ, Wu L, Akohoue S, Zambrzuski S (1993) Soil boron and selenium removal by three plant species. Plant Soil 148:253–263

Bañuelos GS, Mackey B, Wu L, Zambrzuski S, Akohoue S (1995) Bioextraction of soil boron by tall fescue. Ecotoxicol Environ Saf 31:110–116

Bañuelos GS, Ajwa HA, Mackey B, Wu L, Cook C, Akohoue S, Zambrzuski S (1997) Evaluation of different plant species used for phytoremediation of high soil selenium. J Environ Qual 26:639–646

Bañuelos GS, Lin ZQ, Wu L, Terry N (2002a) Phytoremediation of selenium-contaminated soils and waters: fundamentals and future prospects. Rev Environ Health 17:291–306

Bañuelos GS, Vickerman DB, Trumble JT, Shannon MC, Davis CD, Finley JW, Mayland HF (2002b) Biotransfer possibilities of selenium from plants used in phytoremediation. Int J Phytoremediation 4:315–329

Bañuelos GS, Terry N, LeDuc DL, Pilon-Smits EAH, Mackey B (2005) Field trial of transgenic Indian mustard plants shows enhanced phytoremediation of selenium contaminated sediment. Environ Sci Technol 39:1771–1777

Bañuelos GS, LeDuc DL, Pilon-Smits EAH, Tagmount A, Terry N (2007) Transgenic Indian mustard overexpressing selenocysteine lyase, selenocysteine methyltransferase, or methionine methyltransferase exhibit enhanced potential for selenium phytoremediation under field conditions. Environ Sci Technol 41:599–605

Bañuelos GS, da Roche J, Robinson J (2010) Developing Se-enriched animal feed and biofuel from canola planted for managing Se-laden drainage waters in the Westside of central California. Int J Phytoremediation 12:243–253

Bañuelos GS, Stushnoff C, Walse SS, Zuber T, Yang SI, Pickering IJ, Freeman JL (2012) Biofortified, selenium enriched, fruit and cladode from three *Opuntia Cactus* pear cultivars grown on agricultural drainage sediment for use in nutraceutical foods. Food Chem 135:9–16

Bañuelos GS, Dhillon KS, Banga SS (2013) Oilseed Brassicas. In: Singh BP (ed) Biofuel crops: production, physiology and genetics. CAB International, Wallingford, pp 339–368

Bañuelos GS, Arroyo I, Pickering IJ, Yang SI, Freeman JL (2015) Selenium biofortification of broccoli and carrots grown in soil amended with Se-enriched hyperaccumulator *Stanleya pinnata*. Food Chem 166:603–608

Beath OA, Gilbert CS, Eppson HF (1935) The use of indicator plants in locating seleniferous soils in the Western United States. I General. Am J Bot 26:257–269

Bogden JD, Kemp FW, Buse M, Thind IS, Louria DB, Forgacs J, Llanos G, Moncoya-Terrones I (1981) Composition of tobaccos from countries with high and low incidences of lung cancer. I. Selenium, polonium-210, Alternaria, tar, and nicotine. J Natl Cancer Inst 66:27–31

Bolan N, Kunhikrishnan A, Thangarajana R, Kumpiened J, Parke J, Makinof T, Kirkhamg MB, Scheckelh K (2014) Remediation of heavy metal(loid)s contaminated soils – to mobilize or to immobilize? J Hazard Mater 266:141–166

Brown T, Shrift A (1981) Exclusion of selenium from proteins in selenium-tolerant Astragalus species. Plant Physiol 67:1951–1953

Carvalho KM, Martin DM (2001) Removal of aqueous selenium by four aquatic plants. J Aquat Plant Manag 39:33–36

Chaney RL (1983) Plant uptake of inorganic waste constituents. In: Parr JF, Marsh PD, Kla JM (eds) Land treatment of hazardous wastes. Noyes Data Corporation, Park Ridge, pp 50–76

Chaney RL, Brown SL, Li YM, Angle JS, Homer FA, Green CE (1995) Potential use of metal hyperaccumulators. Minist Environ Mag 3:9–11

Chaney RL, Malik M, Li YM, Brown SL, Brewer EP, Angle JS, Baker AJM (1997) Phytoremediation of soil metals. Curr Opin Biotechnol 8:279–284

Chortyk OT, Chaplin JF, Schlotzhauer WS (1984) Growing selenium-enriched tobacco. J Agric Food Chem 32:64–68

Combs GF (2000) Food system-based approaches to improving micronutrient nutrition: the case for selenium. BioFactors 12:39–43

Combs GF Jr (2005) Current evidence and research needs to support a health claim for selenium and cancer prevention. J Nutr 135:343–347

Cunningham SD, Ow DW (1996) Promises and prospects of phytoremediation. Plant Physiol 110:715–719

Cunningham SD, Berti WR, Huang JW (1995) Phytoremediation of contaminated soils. Trends Biotechnol 13:393–397

Dalai A, Kulkarni M, Meher L (2006) Biodiesel productions from vegetable oils using heterogeneous catalysts and their applications as lubricity additives. Paper presented at the EIC Climate Change Technology, 2006 IEEE

de Souza MP, Pilon-Smits EAH, Lytle CM, Hwang S, Tai J, Honma TSU, Yeh L, Terry N (1998) Rate-limiting steps in selenium assimilation and volatilization by Indian mustard. Plant Physiol 117:1487–1494

Dhillon KS, Dhillon SK (1991) Selenium toxicity in soils, plants and animals in some parts of Punjab, India. Int J Environ Stud 37:15–24

Dhillon KS, Dhillon SK (1997) Distribution of seleniferous soils in northwest India and associated toxicity problems in the soil-plant-animal-human system. Land Contam Reclam 5:313–323

Dhillon KS, Dhillon SK (2003a) Quality of underground water and its contribution towards selenium enrichment of the soil-plant system for a seleniferous region of northwest India. J Hydrol 272:120–130

Dhillon KS, Dhillon SK (2003b) Distribution and management of seleniferous soils. Adv Agron 79:119–184

Dhillon KS, Dhillon SK (2009a) Selenium concentrations of common weeds and agricultural crops grown in the seleniferous soils of northwestern India. Sci Total Environ 407:6150–6156

Dhillon KS, Dhillon SK (2009b) Accumulation and distribution of selenium in some vegetable crops grown in selenate-Se treated clay loam soil. Front Agric China 3:366–373

Dhillon SK, Dhillon KS (2009c) Phytoremediation of selenium-contaminated soils: the efficiency of different cropping systems. Soil Use Manag 25:441–453

Dhillon SK, Dhillon KS (2009d) Characterization and management of seleniferous soils of Punjab. Research Bulletin No. 1/2009. Department of Soil Science, Punjab Agricultural University, Ludhiana, pp 1–87

Dhillon KS, Dhillon SK (2014) Development and mapping of seleniferous soils in northwestern India. Chemosphere 99:56–63

Dhillon KS, Dhillon SK (2016) Phytoremediation of selenium contaminated soils: strategies and limitations. In: Bañuelos GS, Lin ZQ, Moraes MF, Guilherme LRG, Reis ARD (eds) Global advances in selenium research from theory to application. CRC Press/Balkemma, Leiden, pp 201–202

Dhillon KS, Randhawa NS, Sinha MK (1977) Selenium status of some common fodders and natural grasses of Punjab. Indian J Dairy Sci 30:218–224

Dhillon KS, Bawa SS, Dhillon SK (1992) Selenium toxicity in some plants and soils of Punjab. J Indian Soc Soil Sci 40:132–136

Dhillon SK, Hundal BK, Dhillon KS (2007) Bioavailability of selenium to forage crops in a sandy loam soil amended with Se-rich plant materials. Chemosphere 66:1734–1743

Dhillon KS, Dhillon SK, Thind HS (2008a) Evaluation of different agroforestry tree species for their suitability in the phytoremediation of seleniferous soils. Soil Use Manag 24:208–216

Dhillon SK, Dhillon KS, Kohli A, Khera KL (2008b) Evaluation of leaching and runoff losses of selenium from seleniferous soils through simulated rainfall. J Plant Nutr Soil Sci 171:187–192

Dhillon KS, Dhillon SK, Dogra R (2010) Selenium accumulation by forage and grain crops and volatilization from seleniferous soils amended with different organic materials. Chemosphere 78:548–556

Doran JW (1982) Microorganisms and the biological cycling of selenium. Adv Microb Ecol 6:1–32

Duckart EC, Waldron LJ, Doner HE (1992) Selenium uptake and volatilization from plants growing in soil. Soil Sci 153:94–99

Echevarria M, Henry P, Ammerman C, Rao P, Miles R (1988) Estimation of the relative bioavailability of inorganic selenium sources for poultry. 2. Tissue uptake of selenium from high dietary selenium concentrations. Poult Sci 67:1585–1592

Ellis DR, Sors TG, Brunk DG, Albrecht C, Orser C, Lahner B, Wood KV, Harris HH, Pickering IJ, Salt DE (2004) Production of Se-methylselenocysteine in transgenic plants expressing selenocysteine methyltransferase. BMC Plant Biol 4:1–11

El-Mehdawi AF, Pilon-Smits EAH (2012) Ecological aspects of plant selenium hyperaccumulation. Plant Biol 14:1–10

Elrashidi MA, Adriano DC, Workman SM, Lindsay WL (1987) Chemical equilibria of selenium in soils: a theoretical development. Soil Sci 144:141–152

Elrashidi MA, Adriano DC, Lindsay WL (1989) Solubility, speciation and transformations of selenium in soils. In: Jacobs LWG (ed) Selenium in agriculture and the environment, SSSA Spec. Publ. 23, Madison, WI, pp 51–63

Esringü A, Turan M (2012) The roles of diethylenetriamine pentaacetate (DTPA) and ethylenediamine disuccinate (EDDS) in remediation of selenium from contaminated soil by Brussels sprouts (*Brassica oleracea* var. gemmifera). Water Air Soil Pollut 223:351–362

Evangelou MWH, Ebel M, Schaeffer A (2007) Chelate assisted phytoextraction of heavy metals from soil: effect, mechanism, toxicity, and fate of chelating agents. Chemosphere 68:989–1003

Evans CS, Asher CJ, Johnson CM (1968) Isolation of dimethyl diselenide and other volatile selenium compounds from *Astragalus Racemosus* (Pursh.) Aust J Biol Sci 21:13–20

Fässler E, Robinson BH, Stauffer W, Gupta SK, Papritz A, Schulin R (2010) Phytomanagement of metal-contaminated agricultural land using sunflower, maize and tobacco. Agric Ecosyst Environ 136:49–58

Finley JW (2006) Bioavailability of selenium from foods. Nutr Rev 64:146–151

Finley JW (2007) Increased intakes of selenium-enriched foods may benefit human health. J Sci Food Agric 87:1620–1629

Fleming GA (1980) Essential micronutrients II: iodine and selenium. In: Davis BE (ed) Applied soil trace elements. Wiley, New York, pp 199–234

Fordyce F (2013) Selenium deficiency and toxicity in the environment. In: Selinus O (ed) Essentials of medical geology. Springer, Dordrecht, pp 375–416

Frankenberger WT Jr, Karlson U (1994) Microbial volatilization of selenium from soils and sediments. In: Frankenberger WT Jr, Benson S (eds) Selenium in the environment. Marcel Dekker, New York, pp 369–388

Freeman JL, Quinn CF, Marcus MA, Fakra S, Pilon-Smits EAH (2006) Selenium-tolerant diamondback moth disarms hyperaccumulator plant defense. Curr Biol 16:2181–2192

Freeman JL, Lindblom SD, Quinn CF, Fakra S, Marcus MA, Pilon-Smits EAH (2007) Selenium accumulation protects plants from herbivory by Orthoptera via toxicity and deterrence. New Phytol 175:490–500

Freeman JL, Quinn CF, Lindblom SD, Klamper EM, Pilon-Smits EAH (2009) Selenium protects the hyperaccumulator *Stanleya pinnata* against black-tailed prairie dog herbivory in native seleniferous habitats. Am J Bot 96:1075–1085

Galeas ML, Klamper EM, Bennett LE, Freeman JL, Kondratieff BC, Pilon-Smits EAH (2008) Selenium hyperaccumulation affects plant arthropod load in the field. New Phytol 177:715–724

Gao J, Liu Y, Huang Y, Lin ZQ, Bañuelos GS, Lam MHW, Yin X (2011) Daily selenium intake in a moderate selenium deficiency area of Suzhou, China. Food Chem 126:1088–1093

Garifullina GF, Owen JD, Lindblom SD, Tufan H, Pilon M, Pilon-Smits EAH (2003) Expression of a mouse selenocysteine lyase in *Brassica juncea* chloroplasts affects selenium tolerance and accumulation. Physiol Plant 118:538–544

Gondi F, Pantó G, Feher J, Bogye G, Alfthan G (1992) Selenium in Hungary. Biol Trace Elem Res 35:299–306

Hall JA, Bobe G, Hunter JK, Vorachek WR, Stewart WC, Vanegas JA, Estill CT, Mosher WD, Pirelli GJ (2013) Effect of feeding selenium-fertilized alfalfa hay on performance of weaned beef calves. PLoS One 8:e58188

Hanson B, Garifullina GF, Lindblom SD, Wangeline A, Ackley A, Kramer K, Norton AP, Lawrence CB, Pilon-Smits EAH (2003) Selenium accumulation protects *Brassica juncea* from invertebrate herbivory and fungal infection. New Phytol 159:461–469

Hanson B, Lindblom SD, Loeffler ML, Pilon-Smits EAH (2004) Selenium protects plants from phloem feeding aphids due to both deterrence and toxicity. New Phytol 162:655–662

Hensen D, Duda P, Zayed AM, Terry N (1998) Selenium removal by constructed wetlands: role of biological volatilization. Environ Sci Technol 32:591–597

Hira CK, Partal K, Dhillon KS (2004) Dietary selenium intake by men and women in high and low selenium areas of Punjab. Public Health Nutr 7:39–43

Ip C, Ganther HE (1992) Relationship between the chemical form of selenium and anticarcinogenic activity. In: Wattenberg I, Lipkin M, Boon CW, Kellott GJ (eds) Cancer chemoprevention. CRC Press, Boca Raton, pp 479–488

Johnsson L (1991) Selenium uptake by plants as a function of soil type, organic matter content and pH. Plant Soil 133:57–64

Kubachka KM, Meija J, LeDuc DL, Terry N, Caruso JA (2007) Selenium volatiles as proxy to the metabolic pathways of selenium in genetically modified *Brassica juncea*. Environ Sci Technol 41:1863–1869

Kulkarni MG, Dalai A, Bakhshi N (2007) Transesterification of canola oil in mixed methanol/ethanol system and use of esters as lubricity additive. Bioresour Technol 98:2027–2033

Kumar NPBA, Dushenkov V, Motto H, Raskin I (1995) Phytoextraction: the use of plants to remove heavy metals from soils. Environ Sci Technol 92:1232–1238

Lasat MM (2000) Phytoextraction of metals from contaminated sites – a critical review of plant/soil/metal interaction and assessment of pertinent agronomic issues. J Hazard Subst Res 2:1–25

LeDuc DL, Tarun AS, Montes-Bayon M, Meija J, Malit MF, Wu CP, Abdel-Samie M, Chiang CY, Tagmount A, de Souza MP, Neuhierl B, Bock A, Caruso JA, Terry N (2004) Overexpression of selenocysteine methyltransferase in Arabidopsis and Indian mustard increases selenium tolerance and accumulation. Plant Physiol 135:377–383

LeDuc DL, AbdelSamie M, Móntes-Bayon M, Wu CP, Reisinger SJ, Terry N (2006) Overexpressing both ATP sulfurylase and selenocysteine methyltransferase enhances selenium phytoremediation traits in Indian mustard. Environ Pollut 144:70–76
Lemly DA, Ohlendorf HM (2002) Regulatory implications of using constructed wetlands to treat selenium-laden wastewater. Ecotoxicol Environ Saf 52:46–56
Lewis BG, Johnson CM, Broyer TC (1971) Cleavage of *Se*-ethylselenomethionine selenonium salt by a cabbage leaf enzyme fraction. Biochim Biophys Acta 237:603–605
Li L, Xie Y, El-Sayed WM, Szakacs JG, Franklin MR, Roberts JC (2005) Chemopreventive activity of selenocysteine prodrugs against tobacco-derived nitrosamine (NNK) induced lung tumors in the A/J mouse. J Biochem Mol Toxicol 19:396–405
Lin ZQ, Terry N (2003) Selenium removal by constructed wetlands: quantitative importance of biological volatilization in the treatment of selenium-laden agricultural drainage water. Environ Sci Technol 37:606–615
Lyons G (2010) Selenium in cereals: improving the efficiency of agronomic biofortification in the UK. Plant Soil 332:1–4
Mayland HF, James LJ, Panter KE, Sonderegger JL (1989) Selenium in seleniferous environments. In: Jacobs LWG (ed) Selenium in agriculture and the environment, SSSA Spec. Publ. 23, Madison, WI, pp 15–50
McDonnell KP, Ward SM, McNulty PB, Howard-Hildige R (2000) Results of engine and vehicle testing of semirefined rapeseed oil. Trans Am Soc Agric Eng 43:1309–1316
Miranda AF, Muradov N, Gujar A, Stevenson T, Nugegoda D, Ball AS, Mouradov A (2014) Application of aquatic plants for the treatment of selenium-rich mining wastewater and production of renewable fuels and petrochemicals. J Sustain Bioenergy Syst 4:97–112
Misra S, Gedamu L (1989) Heavy-metal tolerant transgenic *Brassica napus* and *Nicotiana tabacum* plants. Theor Appl Genet 78:161–168
Muñiz-Naveiro Ó, Domínguez-González R, Bermejo-Barrera A, Bermejo-Barrera P, Cocho JA, Fraga JM (2006) Study of the bioavailability of selenium in cows' milk after a supplementation of cow feed with different forms of selenium. Anal Bioanal Chem 385:189–196
Neal RH (1995) Selenium. In: Alloway BJ (ed) Heavy metals in soils. Blackie Academic and Professional, London, pp 260–283
Neuhierl B, Thanbichler M, Lottspeich F, Bock A (1999) A family of S-methyl- methionine dependent thiol/selenol methyltransferases: role in selenium tolerance and evolutionary relation. J Biol Chem 274:5407–5414
Ohlendorf HM, Hothem RL (1995) Agricultural drainwater effects on wildlife in Central California. In: Hoffman DF, Rattner BA, Burton GA Jr, Cairns J (eds) Handbook of ecotoxicology. Lewis Press, Boca Raton, pp 577–595
Pezzarossa B, Petruzelli G, Petacco F, Malorgio F, Ferri T (2007) Absorption of selenium by *Lactuca sativa* as affected by carboxymethylcellulose. Chemosphere 67:322–329
Pilon M, Owen JD, Garifullina GF, Kurihara T, Mihara H, Esaki N, Pilon-Smits EAH (2003) Enhanced selenium tolerance and accumulation in transgenic *Arabidopsis thaliana* expressing a mouse selenocysteine lyase. Plant Physiol 131:1250–1257
Pilon-Smits EAH, Quinn CF (2010) Selenium metabolism in plants. In: Hell R, Mendel R (eds) Cell biology of metals and nutrients. Springer, Berlin/Heidelberg, pp 225–241
Pilon-Smits EAH, de Souza MP, Lytle CM, Shang C, Lugo T, Terry N (1998) Selenium volatilization and assimilation by hybrid poplar (*Populus tremula x alba*). J Exp Bot 49:1889–1892
Pilon-Smits EAH, de Souza MP, Hong G, Amini A, Bravo RC, Payabyab ST, Terry N (1999a) Selenium volatilization and accumulation by twenty aquatic plant species. J Environ Qual 28:1011–1018
Pilon-Smits EAH, Hwang S, Lytle CM, Zhu Y, Tai JC, Bravo RC, Chen Y, Leustek T, Terry N (1999b) Overexpression of ATP sulfurylase in Indian mustard leads to increased selenite uptake, reduction, and tolerance. Plant Physiol 119:123–132
Prins CN, Hantzis LJ, Quinn CF, Pilon-Smits EAH (2011) Effects of selenium accumulation on reproductive functions in *Brassica juncea* and *Stanleya pinnata*. J Exp Bot 62:5633–5640

Pulford ID, Watson C (2003) Phytoremediation of heavy metal-contaminated land by trees – a review. Environ Int 29:529–540
Quinn CF, Freeman JL, Galeas ML, Klamper EM, Pilon-Smits EAH (2008) The role of selenium in protecting plants against prairie dog herbivory: implications for the evolution of selenium hyperaccumulation. Oecologia 155:267–275
Quinn CF, Prins CN, Freeman JL, Gross AM, Hantzis LJ, Reynolds RJB, Yang S, Covey PA, Bañuelos GS, Pickering IJ, Pilon-Smits EAH (2011a) Selenium accumulation in flowers and its effects on pollination. New Phytol 192:727–373
Quinn CF, Wyant KA, Wangeline AL, Shulman J, Galeas ML, Valdez JR, Self JR, Paschke MW, Pilon-Smits EAH (2011b) Enhanced decomposition of selenium hyperaccumulator litter in a seleniferous habitat – evidence for specialist decomposers? Plant Soil 341:51–61
Rai PK (2008) Heavy metal pollution in aquatic ecosystems and its phytoremediation using wetland plants: an ecosustainable approach. Int J Phytoremediation 10:133–160
Rani N, Dhillon KS, Dhillon SK (2005) Critical levels of selenium in different crops grown in an alkaline silty loam soil treated with selenite-Se. Plant Soil 277:367–374
Ravikovitch S, Margolin M (1957) Selenium in soils and plants. Ktavim 7:41–52
Rayman MP, Infante HG, Sargent M (2008) Food-chain selenium and human health: spotlight on speciation. Br J Nutr 100:238–253
Robinson B, Fernandez JE, Madejon P, Maranon T, Murillo JM, Green S, Clothier B (2003) Phytoextraction: an assessment of biogeochemical and economic viability. Plant Soil 249:117–125
Robinson B, Green S, Chancerel B, Mills T, Clothier B (2007) Poplar for the phytomanagement of boron contaminated sites. Environ Pollut 150:225–233
Robinson BH, Bañuelos G, Conesa HM, Evangelou MWH, Schulin R (2009) The phytomanagement of trace elements in soil. Crit Rev Plant Sci 28:240–266
Rosenfeld I, Beath OA (1964) Selenium: geobotany, biochemistry, toxicity, and nutrition. Academic Press, New York
Sabogal A, Borkowski D (2007) Estado actual de la investigación sobre *Ipomoea carnea*: toxicidad en ganado caprino. Revista de Química (Lima, Perú: Pontificia Universidad Católica del Perú), 29–35 Jan–Dec, ISSN 1012-3946
Salt DE, Blaylock M, Kumar NPBA, Dushenkov V, Ensley BD, Raskin I (1995) Phytoremediation: a novel strategy for the removal of toxic metals from the environment using plants. Biotechnology 13:468–474
Schrauzer GN (2000) Anticarcinogenic effects of selenium. Cell Mol Life Sci 57:1864–1873
Schwarz K, Foltz CM (1957) Selenium as an integral part of factor 3 against dietary necrotic liver degeneration. J Am Chem Soc 79:3292–3293
Shardendu-Salhani N, Boulyga SF, Stengel E (2003) Phytoremediation of selenium by two halophyte species in subsurface flow constructed wetland. Chemosphere 50:967–973
Sharma S, Bansal A, Dhillon SK, Dhillon KS (2010) Comparative effects of selenate and selenite on growth and biochemical composition of rapeseed (*Brassica napus*). Plant Soil 329:339–348
Sims JT, Johnson GV (1991) Micronutrient soil test. In: Mortvedt JJ (ed) Micronutrients in agriculture. SSSA, Madison, pp 427–476
Stapleton JJ, Bañuelos GS (2009) Biomass crops can be used for biological disinfestation and remediation of soils and water. Calif Agric 63:41–46
Tagmount A, Berken A, Terry N (2002) An essential role of S-adenosyl-L-methionine:L-methionine-S-methyltransferase in selenium volatilization by plants: methylation of selenomethionine to selenium-methyl-L-selenium-methionine, the precursor of volatile selenium. Plant Physiol 130:847–856
Tamaoki M, Freeman JL, Pilon-Smits EAH (2008) Cooperative ethylene and jasmonic acid signaling regulates selenite resistance in *Arabidopsis thaliana*. Plant Physiol 146:1219–1230
Tedeschini E, Proietti P, Timorato V, D'Amato R, Nasini L, Buono DD, Businelli D, Frenguelli G (2015) Selenium as stressor and antioxidant affects pollen performance in *Olea europaea*. Flora 215:16–22

Terry N, Zayed AM (1994) Selenium volatilization by plants. In: Frankenberger WT Jr, Benson S (eds) Selenium in the environment. Marcel Dekker, New York, pp 343–369

Trumble GT, Kund GS, White KK (1998) Influence of form and quantity of selenium on the development and survival of an insect herbivore. Environ Pollut 101:175–182

Van Huysen T, Abdel-Ghany S, Hale KL, LeDuc D, Terry N, Pilon-Smits EAH (2003) Overexpression of cystathionine-γ-synthase in Indian mustard enhances selenium volatilization. Planta 218:71–78

Vickerman DB, Trumble JT, George GN, Pickering IJ, Nichol H (2004) Selenium biotransformations in an insect ecosystem: effects of insects on phytoremediation. Environ Sci Technol 38:3581–3586

Vymazal J (2010) Constructed wetlands for wastewater treatment. Water 2:530–549

Wang Z, Gao Y (2001) Biogeochemical cycling of selenium in Chinese environments. Appl Geochem 16:1345–1351

Wenzel WW, Unterbrunner R, Sommer P, Pasqualina S (2003) Chelate-assisted phytoextraction using canola (*Brassica napus*) in outdoors pot and lysimeter experiments. Plant Soil 249:83–96

White PJ, Broadley MR (2009) Biofortification of crops with seven mineral elements often lacking in human diets – iron, zinc, copper, calcium, magnesium, selenium and iodine. New Phytol 182:49–84

World Health Organization, WHO (1996) Trace elements in human nutrition and health. World Health Organization, Geneva

Wu L, Chen JC, Tanji KK, Bañuelos GS (1995) Distribution and bioaccumulation of selenium in a restored upland grassland contaminated by selenium from agriculture drain water. Environ Toxicol Chem 14:733–742

Yuan L, Yin X, Zhu Y, Li F, Huang Y, Liu Y, Lin Z (2012) Selenium in plants and soils, and selenosis in Enshi, China: implications for selenium biofortification. In: Yin X, Yuan L (eds) Phytoremediation and biofortification. Springer, Dordrecht, pp 7–31

Zayed A, Pilon-Smits EAH, de Souza M, Lin ZQ, Terry N (2000) Remediation of selenium polluted soils and waters by phytovolatilization. In: Terry N, Bañuelos GS (eds) Phytoremediation of contaminated soil and water. CRC Press, Boca Raton, pp 61–83

Zhang Y, Gladyshev VN (2009) Comparative genomics of trace elements: emerging dynamic view of trace element utilization and function. Chem Rev 109:4828–4861

Zhang L, Byrne PF, Pilon-Smits EAH (2006) Mapping quantitative trait loci associated with selenate tolerance in *Arabidopsis thaliana*. New Phytol 170:33–42

Index

E.A.H. Pilon-Smits et al. (eds.), *Selenium in plants*, Plant Ecophysiology 11,
DOI 10.1007/978-3-319-56249-0

MIX
Papier aus verantwortungsvollen Quellen
Paper from responsible sources
FSC® C105338

If you have any concerns about our products,
you can contact us on
ProductSafety@springernature.com

In case Publisher is established outside the EU,
the EU authorized representative is:
Springer Nature Customer Service Center GmbH
Europaplatz 3, 69115 Heidelberg, Germany

Printed by Libri Plureos GmbH
in Hamburg, Germany